BRAZIL

Titles in ABC-CLIO's *Latin America in Focus Series*

Venezuela   Elizabeth Gackstetter Nichols and Kimberly J. Morse

Titles in ABC-CLIO's *Africa in Focus Series*

Eritrea   Mussie Tesfagiorgis G.
Ethiopia   Paulos Milkias

Titles in ABC-CLIO's *Asia in Focus Series*

China   Robert André LaFleur, Editor
Japan   Lucien Ellington, Editor
The Koreas   Mary E. Connor, Editor

BRAZIL

*Antonio Luciano de Andrade Tosta
and Eduardo F. Coutinho,
Editors*

ABC-CLIO™

An Imprint of ABC-CLIO, LLC
Santa Barbara, California • Denver, Colorado

Library of Congress Cataloging-in-Publication Data

Brazil / Antonio Luciano de Andrade Tosta and Eduardo F. Coutinho, editors.
    pages cm. — (Latin America in focus)
  Includes bibliographical references and index.
  ISBN 978-1-61069-257-1 (alk. paper) — ISBN 978-1-61069-258-8   1. Brazil—Description and travel.   I. Tosta, Antonio Luciano de Andrade, editor.   II. Coutinho, Eduardo de Faria, editor.
  F2517.B694 2016
  981—dc23   2015025746

ISBN: 978-1-61069-257-1
EISBN: 978-1-61069-258-8

20 19 18 17 16   1 2 3 4 5

This book is also available on the World Wide Web as an eBook.
Visit www.abc-clio.com for details.

ABC-CLIO
An Imprint of ABC-CLIO, LLC

ABC-CLIO, LLC
130 Cremona Drive, P.O. Box 1911
Santa Barbara, California 93116-1911

This book is printed on acid-free paper ∞
Manufactured in the United States of America

Contents

Preface

The interest in Brazilian studies on the part of North American scholars can be clearly measured when we think about the number of courses on Portuguese language and Brazilian literature and culture that are frequently offered by American universities. The majority of these courses on literature and culture are taught, however, at the graduate level, for they require a reasonable domain of the language and are usually meant for those who want to specialize in the area. This book on Brazil, published by ABC-CLIO, is not only intended to fill out this gap but also to provide a broad and at the same time solid view of the country in as many of its aspects as possible. It is designed for high school and undergraduate students, but also for the general public who may be interested in visiting the country or in spending some time there, either for professional reasons or for simply having some kind of intercultural experience.

The book is divided into seven chapters or parts. It begins with a presentation of the country's geography, both physical and human, and proceeds to a condensed report of its history from the earliest human settlements to the present. It then focuses on the government systems the country has gone through and offers a view of its political life since colonial times. The following chapter is devoted to the country's economy and level of development. These four chapters are followed by two entitled "Society" and "Culture." The first includes "Religion and Thought," "Social Classes and Ethnicity," "Gender, Marriage, and Sexuality," and "Education"; the second is subdivided into "Language," "Family and Etiquette," "Literature and Drama," "Art," "Architecture," "Music," "Food," "Leisure and Sports," and "Cinema." Chapter 7, finally, is centered on a selection of issues that have been playing a relevant role in contemporary times. This chapter is followed by a glossary of terms that have a special significance for Brazilian life and information on important facts and figures, the major Brazilian holidays and festivals, and a list of the country-related organizations. The reader willl also find an annotated bibliography and a thematic index, extremely useful for further research.

All the chapters were written by scholars who specialize in their specific topic. They are either Brazilians or North Americans who have lived in Brazil or who are very familiar with the country's culture and way of life. In any case, however, extensive research was done for the writing of each chapter, and the result meets our expectations. Aware of the fact that this book will bring a relevant contribution to the understanding of Brazil both as a nation and on the international scene, we invite readers to initiate a pleasurable voyage by delving into its pages.

BRAZIL

Introduction

The world turned its attention to Brazil when, by a stroke of luck, or perhaps due to vision, politics, and strategic planning, the country was selected to be the site for the 2014 World Soccer Cup and the 2016 Summer Olympics. Brazil is, of course, no stranger to the international community. It has the world's eighth economy, and it is one of the world's largest coffee producers and exporters. Home to over more than half of the world's largest tropical rainforest, Brazil often hits the small and big screens when the topic is nature and, in particular, ecology. The Amazon River, the world's second-longest river and the largest in terms of water flowing, the Atlantic forest, the Iguaçu Falls, the Pantanal Wetlands, and the sand dunes of the Lençóis Maranhenses are some of the natural sites that have helped to raise the interest in the largest South American nation. Few other countries are as biodiverse as Brazil. Every year 700 new animal species are found there. A new plant species is discovered practically every two days. It is not surprising, therefore, that animals such as the *sabiá-laranjeira* (rufous-bellied thrush) and the toucan, native to the country, regularly appear as national symbols.

Sports, of course, have helped Brazil to gain international popularity. Widely known as the "Country of Soccer," a reputation that it has worked hard to keep, Brazil has won the World Soccer Cup five times so far, more than any other country ever did. Pelé, known as the "King of Soccer," Ronaldo, "the Phenomenon," and the talented Neymar, currently playing for the FC Barcelona in Spain, are names revered by most soccer fans. Besides soccer, Brazil has been very competitive in volleyball; the male and female national teams have won Olympic medals, as well as distinguished themselves in competitions such as the World Championship, the FIVB World League, and the Volleyball World Grand Champions Cup. Although basketball is only the third most played sport in the country, Brazilian players such as Oscar Schmidt and Hortência Marcari have won international recognition. There are currently 14 Brazilian players in the NBA. Tiago Splitter (San Antonio Spurs), Leandro Barbosa (Golden State Warriors), Nenê Hilário (Washington Wizards), and Anderson Varejão (Cleveland Cavaliers) are some of the best-known ones. Brazilian jiu-jitsu and capoeira, the Afro-Brazilian artistic expression that mixes martial arts, dance, game, and sports, are now popular all over the world. Every so often, a Brazilian rises in the international sports scene; among them are famed racecar drivers Ayrton Senna, Nelson Piquet, and Emerson Fittipaldi; celebrity tennis players Gustavo Kuerten and Maria Esther Bueno; the gymnast Daiane dos Santos; and Olympic swimmers Fernando Scherer, a.k.a. "Xuxa," and César Cielo.

Brazilian musical genres such as samba, bossa nova, and even the more regional forró and axé music have also reached the global musical scene. The rise in Brazilian immigration to countries in North America and Europe as well as to Paraguay and Japan as of the 1980s has helped to increase the presence of Brazilian music in these regions as well. The works of Brazilian authors such as Jorge Amado, Clarice Lispector, Machado de Assis, Moacyr Scliar, and Rubem Fonseca have been translated into several languages. Brazilian cinema is also booming. Films such as *City of God* and *Elite Squad* have received international critical acclaim. The success of the Brazilian productions has granted Brazilian directors access into the U.S. film industry. Fernando Meirelles (*The Constant Gardener*, *Blindness*), Walter Salles (*Dark Water*, *The Motorcycle Diaries*), José Padilha (*Robocop*), and Carlos Saldanha (*Ice Age*) are some examples. Cities like Brasília, the country's capital, Rio de Janeiro, São Paulo, and Salvador have attracted tourists and scholars from around the world due to their history, vibrant culture, and idiosyncrasies. Inaugurated in 1960, Brasília is notorious for the modernist architecture of Oscar Niemeyer and Lúcio Costa. Rio de Janeiro's magnificent landscape and popular carnival are only two of the many reasons why the country's former capital is the number one destination in South America. UNESCO conferred it the title of World Heritage Site in 2012. The statue of Christ Redeemer has been recently named one of the New Seven Wonders of the World. The cosmopolitan São Paulo, the world's third-largest city and the largest one in South America, is the business heart of the nation. The city also has a noteworthy ethnic composition, which is reflected in its people, cuisine, and diverse communities. They include Italian, Arab, and Lebanese, and especially Japanese. Brazil has the largest community of people of Japanese descent outside Japan. A substantial number of them live in the state of São Paulo. Salvador, Brazil's first capital, is known for its rich African heritage. The local food, music, and religious characteristics are a sad reminder that the country had the longest slavery in the Americas.

Brazil has also become known for some of its problems. Brazilian *favelas* (slum areas), their poverty, and drug-related role are infamous. Recently, news about rampant corruption in politics and scandals involving major Brazilian political leading figures and national organizations has reached international media. Broadcasts about unemployment rates and income inequality in the country, both of which have actually improved in the past years, are still common. The now-frequent street protests against President Dilma Rousseff, who is in her second term, have also become public knowledge. The world's fifth-largest country in population, a large producer of automobiles and aerospace products, which has the world's third-largest roadways and the second-highest number of airports, still has many challenges in its future. Even though it plays such a major role in the preservation of the world's ecosystem, deforestation remains a great concern in the country. Violence against women and homosexuals is common. Racism, police brutality, and instances of religious intolerance and class prejudice mark Brazilian society. Safety is a huge problem in many urban areas. Brazilians complain about public transportation in most cities, suffer with a poor public health system, and continually affirm that it is only by seriously investing in education that the country can progress. Whereas Brazilian public universities generally do not fall short in quality of their peers all over the world, most elementary and secondary public schools are below satisfactory standards.

There is a lot about Brazil that is still unknown, and this is what justifies this collection. The country's present global leading role has increased the study of Portuguese as a foreign language and boosted study-abroad programs from the United States and European countries. "Brazilian Studies" has become an established field of intellectual inquiry, especially, but not limited to, the humanities, social sciences, and environmental studies. The chapters in this volume cover Brazilian history from the 1500s to the present. Together they provide the reader with a general but critical introduction to a complex country through an interdisciplinary perspective. Brazil shares a great deal historically, politically, socially, and culturally with other Latin American and American nations. Its peculiarities, however, are nonetheless numerous, due to its singular geographical position and size, its linguistic panorama, ethnic composition, political history, and, above all, Brazilians' multifaceted human nature. Brazilians are often said to be the best and most complex aspect of Brazil.

The opening chapter, "Geography," brings information about Brazil's geography and demography. Marcos Cerdeira discusses Brazil's five regions, characterizing each state and its major cities. There are comments about the country's racial composition, and, when relevant, about significant immigrant flows and internal migration. Kara D. Schultz provides an overview of Brazilian history from the Portuguese colonization to contemporary times in the second chapter. Schultz covers Brazil's early history as a colony, the empire, the first republic, the Vargas era, the second republic, the military dictatorship, and the post-dictatorial democratic Brazilian society. In "Government and Politics," Renato Lima de Oliveira departs from the end of the military dictatorship, and Brazil's challenges as a young democracy, to outline the main events in the Brazilian political system since independence from Portugal in 1822. Oliveira also explains Brazil's complex party system and its electoral structure. The third chapter ends with a critical assessment of more recent governments. Rodrigo R. Coutinho shows how international media has covered Brazilian economy's latest successes and difficulties in Chapter 4. He provides a general picture of Brazilian economy and then traces its history from the colonial times to the present.

Chapter 5 has four different subchapters: "Religion and Thought," "Social Classes and Ethnicity," "Gender, Marriage, and Sexuality," and "Education." Joseph Abraham Levi describes the evolution of religious life, moral values, and belief systems in Brazil. Levi explores the importance of the colonial past and the intercultural contact that took place in Brazil for the current religious amalgamation in the country. He explains the Amerindian and African-based religions, Catholicism, Judaism, and Islam, as well as the religious syncretism that has been a part of a segment of Brazilian society for a long time. Simone Bohn examines Brazil's uneven social structure, analyzing the intersection between social class and ethnicity. Her essay underscores the ways in which Brazil's historical inequalities challenge its image as a "racial democracy." Bohn tracks Brazil's economic history to expose the persistence of social disparities among the racial and ethnic groups in the country. In "Gender, Marriage, and Sexuality," Maria Lúcia Rocha-Coutinho provides an historical examination of the status of Brazilian women as compared to men's from the colonial times up to the 21st century, especially regarding gender and parental roles, household duties, schooling, and participation in the workforce and politics. Antonio Lima da Silva and Volnei M. Carvalho delineate the progress of formal education in Brazil from the Jesuitical teachings in the colonial period to the educational structure currently in place. They explain the format and operation of the current educational system, highlighting the differences between public and private education and describing the federal, state, and municipal educational systems. The chapter also differentiates elementary, secondary, and higher education in Brazil, and refers to current state-sponsored educational policies.

There are nine subchapters in Chapter 6: "Language," "Family and Etiquette," "Literature and Drama," "Art," "Architecture," "Music," "Food," "Sports and Leisure," and "Cinema." Helade Scutti Santos destroys the myth of monolingualism that is attributed to Brazil in "Language." She reveals that, besides Portuguese, there are 274 indigenous languages spoken in the Brazilian territory, as well as a number of immigrant languages, some of which are dialects of their European languages such as German and Italian. Santos gives a panorama of Brazil's indigenous and immigrant languages, explains the history of Brazilian Portuguese and its current status around the world, and analyzes the different varieties of Portuguese in Brazil. Her chapter also explores the current role that Brazil plays in the promotion of Portuguese and provides a picture of the teaching of Portuguese as a nonnative language, of foreign language education in Brazil, and of Portuguese presence on the Internet and social media. The chapter ends with a selection of cognates and false-cognates in English and Brazilian Portuguese, and a few "useful expressions" for those of you who shall decide to learn Brazil's most spoken language. Domingos Sávio Pimentel Siqueira defines and subsequently deconstructs the notion of a "typical" Brazilian family. Siqueira shows how economic fluctuations have led to changes in the family structure in Brazil. He points out that Brazil's varied immigration flows have contributed to the country's diverse family routines and traditions. Siqueira's chapter also discusses typical family values such as solidarity, and the participation of extended family members and maids in the household. Moreover, it analyzes the role of religion, machismo, and taboos within the family. Last, the chapter analyzes how Brazil's aging population has generated changes in the domestic space, and the emergence of nontraditional family structures. Eduardo F. Coutinho's chapter studies the history of Brazilian literary production, calling attention to Brazilian literature's struggle to differentiate itself from the European literary tradition from which it arises. The chapter analyzes the influence of the Baroque in literature and theater, and shows how movements such as Romanticism, Realism, and Modernism take up a character of their own in Brazil. The chapter closes with a reference to some of Brazil's most distinguished writers and relevant artistic events. Alice Heeren starts her chapter with a critique of the Portuguese colonization in Brazil, leading to an analysis of the preconquest indigenous art and of Afro-Brazilian art. She discusses the Dutch influence during the colonial period, the association between art and architecture in the 17th and 18th centuries, the influence of the French Artistic Mission in the 19th century, and the importance of the 1922 Week of Modern Art as a founding moment in Brazilian art history. Her chapter follows artistic developments in Brazil chronologically and reveals the ways in which historical events (e.g., the 20-year-long military dictatorship) have often determined the fate of art in Brazil. Doriane Andrade Meyer explains that Brazilian architecture was influenced by the complex interactions between the Amerindians, the Portuguese, and the African slaves, as well as by the immigrant groups that went to Brazil. She also demonstrates that Brazilian architecture was affected by political, social, and economic events in Brazil and the world. She analyzes the civil, religious, and military architecture of the colonial period, the neoclassical style in imperial Brazil and the early republic, the eclectic late 19th century, the influence of Art Nouveau, and the rise of Modernism and Postmodernism in Brazilian architecture. The chapter underscores the construction of Brasília to become the country's capital and brings information about Brazil's major architects. It ends with an examination of Brazil's contemporaneous architecture. Angela Lühning, whose essay was translated by Michael Iyanaga, calls attention to the importance of music to Brazilian people. Lühning surveys Brazil's rich and diverse musical production, explaining how its history of colonization and immigration helps to account for the country's cultural, and particularly, musical plurality. She writes about better-known musical genres such as samba, bossa nova, and Brazilian Popular Music, but also includes discussions about Brazilian indigenous and Afro-Brazilian musical styles, about the standing of orchestras, bands, and ensembles, and about the interconnection between music and poetry in the Northeast region. Moreover, Lühning notes the intrinsic connection between music and dance in Brazilian cultural traditions, particularly in Afro-religious ones. The chapter also pays attention to certain musical genres and their particular audiences, such as the romantic *sertanejo* (country), *vanerão*, and *arrocha*, and styles such as funk and *tecnobrega*, which are more popular among youth audiences. Elisa Duarte Teixeira explores the origins of Brazilian cuisine, highlighting the importance of the indigenous, African, and Portuguese elements. Her chapter also stresses the many contributions that immigrant groups such as the Japanese, Italians, and Germans gave to Brazilian gastronomy. It discusses Brazilian staple food and everyday eating habits, as well as explores the country's major regional cuisines and their main foods and dishes. Teixeira provides a concise food glossary at the end of her chapter. In "Leisure and Sports," Erin Flynn McKenna describes sports and leisure activities practiced in Brazil from an historical and contemporary perspective. She discusses how soccer and capoeira have become national symbols. The chapter examines indigenous sporting traditions and points out how a mixture of African recreational and athletic traditions were transformed in Brazil, leading to artistic expressions such as the now global capoeira. Another Brazilian martial art it discusses in detail is the national version of jiu-jitsu, which is also recognized and acclaimed worldwide. McKenna also calls attention to the European influence in modern sports played in Brazil, especially the British impact, who brought horse racing and soccer to the country. She comments on Brazil's unmatched success in the World Soccer Cups, and the magnitude of the sport in the country. Her chapter describes Brazilian participation and performance in internationally competitive sports such as beach volleyball, basketball, auto racing, gymnastics, swimming, and tennis, and women's participation is sports. There are also sections with the portrayal of some of Brazil's most famous athletes, and about recreational sports and games such as sand soccer and frescoball. The last section is about leisure activities in the country. In their chapter, Carla Alves da Silva and Simone Cavalcante da Silva discuss the history of Brazilian cinema from its origins in 1898 to the contemporary national productions that have gained international critical acclaim. They suggest that Brazilian films have followed trends in the world cinema industry but also developed their own tendencies and movements. The chapter highlights the government influence on the production and distribution of national films, and pays special attention to the influential socially driven *Cinema Novo* (New Cinema) movement of the 1960s, and the *Retomada* (Renewal) and the subsequent *Post-Retomada*, the two phases of Brazilian cinema in the 1990s and 2000s.

Rodrigo R. Coutinho pens the book's last chapter on "Contemporary Issues." In his opinionated piece, Coutinho writes about economy and politics, as well as present-day issues such as abortion and family planning, affirmative action, racism, the role of the Catholic Church, and nationalism in Brazil. This collection also contains a glossary of important Portuguese terms and acronyms, a timeline of events both historical and cultural, tables of relevant facts and figures, a list of major Brazilian holidays and organizations, and an annotated bibliography to assist readers with further research. As you finish this book, we encourage you to search for more opportunities to learn about Brazil. Visit the country, if you can. We truly hope that the volume will inspire you to further your studies about this important Latin American nation, its history, society, and people.

We, the editors, would like to thank all the contributors who have participated in this project. We managed to gather a fine group of scholars from the United States, Brazil, and Canada to write about Brazil from the perspective of their specific areas of expertise. We would also like particularly to thank Carla Alves da Silva and Renato Lima de Oliveira for their help with the "Country-Related Organizations" section, Doriane Andrade Meyer for her assistance with the "Facts and Figures," and all of the contributors who helped with the Glossary. We cannot fail to recognize the invaluable assistance of Kaitlin Ciarmiello, Senior Acquisitions Editor at ABC-CLIO, who has helped to keep the project on track from its inception.

Luciano Tosta would like to dedicate this book to his daughters Ana and Lillian Tosta, who are always the reason for, and the meaning of, everything. Some of his former students are contributors to this volume. He would also like to dedicate the book to them.

Eduardo F. Coutinho would like to dedicate this book to all those interested in Brazilian culture, and particularly to his students of Brazilian Literature and Culture of the academic year 2011–2012, when he was visiting professor at the University of Illinois (Urbana-Champaign).

Geography

Marcos Cerdeira

OVERVIEW

Brazil is a Portuguese-speaking country located at the northeastern section of South America. Although it is impossible to reduce a country to a single word, if one were asked to do this for Brazil that word would be "superlative." It is the largest nation in South America and the third largest in the Western Hemisphere. It is over 8.5 million square kilometers in size, making it the fifth-largest country in the world behind only Russia, Canada, the United States, and China. For a better understanding of its size, we can say that it is slightly bigger than the continental United States (i.e., without Hawaii and Alaska). Brazil has three time zones; however, most of the population lives in only one, the Brasília time zone, which includes Brazil's three major cities (Brasília, São Paulo, and Rio de Janeiro). It shares borders with 10 of the 12 other countries that make up South America, the exceptions being Ecuador and Chile. On the north Brazil is bordered by Colombia, Venezuela, Guyana, Suriname, and French Guiana; on the west it is bordered by Peru and Bolivia; on the south it is bordered by Paraguay, Argentina, and Uruguay; and on the east it is bordered by only the Atlantic Ocean. Its sheer size means that there are a variety of different geographic and climactic regions found within the country, which will be discussed later in this chapter.

Apart from being the largest country in size in South America, Brazil is also a large and complex country demographically. Demographers list Brazil as having a little over 202 million people. In English the adjective used to describe these people is "Brazilian" in the singular or "Brazilians" in the plural. Thus, over 202 million Brazilians together make it the fifth-largest country by population in the world, behind China, India, the United States, and Indonesia. However, despite this large population, Brazil is still a sparsely populated country with a population density of only 22.4 inhabitants per square kilometer. Brazil has always been sparsely populated. At the turn of the 19th century there were only 17.4 million people in Brazil. By comparison the U.S. Census of 1900 recorded a population of 76.2 million people covering a similar amount of territory. However, it is important to remember that this "scarcity" is just a national average, and so one may "feel" this number differently depending on where one is situated. The northern state of Roraima, for example, has a population density of 2.01 inhabitants per square kilometer, whereas the southeastern state of Rio de Janeiro has a population density of 365.23 people per square kilometer. The comparison between these two states also signals regional differences that are present in other social areas in Brazil. Half of Brazil's population or slightly over 80 million people live in the southeastern region of the country. The next most populous region is the Northeast, with slightly over 53 million people, and then the Southern region with 27 million, followed by the Northern and Midwest region with just under 16 and 14 million, respectively. The North and Midwest remain less populated than the other three regions. However, their respective share of national population has increased, while that of the other regions has decreased (IBGE 2010).

Another characteristic of contemporary Brazilian society and life is urbanization. While the countryside remains important for a variety of cultural and economic reasons, Brazilian society is overwhelmingly urban, with 87 percent of the population living in urban areas (Baer 2). Furthermore, 21.02 percent of the population lives in the 15 largest municipalities (IBGE 2012). According to the latest estimates, by far the largest of these municipalities is São Paulo with an estimated population of 11.8 million people. The next largest city is Rio de Janeiro, the former capital of the country, with an estimated 6.5 million inhabitants in 2014. After that, the next four largest municipalities all have between 2.5 million and 3 million people: Salvador (2.9 million), Brasília (2.9 million), Fortaleza (2.6 million) and Belo Horizonte (2.5 million).

Unlike most other countries in Latin America, Brazilians speak mostly Portuguese. Indeed Brazil, along with Portugal, Angola, Mozambique, Cape Verde, Guinea-Bissau, Macau, East Timor and São Tomé and Principe, recognizes Portuguese as an official language. Although Portuguese is certainly the dominant language spoken in the country, Brazil is a polylinguistic nation, meaning that the many different people who claim Brazil as home often speak other languages apart from Portuguese. Some of these alternative languages include Italian, Japanese, and German, which arrived with immigrant waves from these countries in the 20th century. In addition, Native Brazilians make up a remarkable amount of the linguistic diversity in Brazil, accounting for 274 different languages spoken in the country (IBGE 2012).

Like many other countries in the Americas, Brazil's racial composition is one of its most complex and difficult facets to explain. Traditionally, the history of Brazil has been told as what has been now described as the "myth of the three races." In it, authors argued that "the Brazilian" derived from a combination of the "African," the "Indian," and the "Portuguese." While this may still be a popular belief, scholars recognize that this is not true today nor has it ever been true in the past. The myth of the three races simplifies a culturally rich and ethnically diverse colonial past. For example, a prominent social division between early colonial Portuguese was that of the "new" versus "old" Christian, a distinction that arose out of the end of the religious conflict in the Iberian Peninsula between Christians and Muslims in which non-Christians (Muslims and Jews) were forced to convert in order to remain in the kingdom. However, after their conversion they were soon seen as untrustworthy and their sincerity began to be questioned. As a result, there were several restrictions placed on new Christians, who were barred from certain social and government positions. Many new Christians migrated to Brazil. Thus, we can find many Sephardic influences in Brazilian society. Non-Europeans also migrated to Brazil, most notably Japanese and Arabic people.

It is impossible to discuss race in Brazil without discussing the role Afro-Brazilians have played in the development of the country. As in the rest of the Americas, in Brazil the first Africans to arrive were slaves. Indeed, slavery as an institution was central to the Brazilian economy for many centuries. As a result, one finds still to this day higher distribution of Afro-Brazilians in the areas that housed higher quantities of slaves. The Brazilian Institute of Statistics and Geography lists 10.47 percent of the population as black (*preta*) and 59.78 percent as being mixed (*parda*) in the Northeast region of Brazil. In the Southeast 8.56 percent of the population is black and 34.84 percent of the population is listed as mixed. This is higher than the national average for both populations, which lists 8.21 percent of the population as black and 43.07 percent of the population as mixed. This is, in part, the result of the economic history of the areas, which saw many African slaves brought to both regions due to economic booms (first, sugar in the Northeast and then gold and coffee in the Southeast).

Race has a geographic aspect as well. Unlike in the United States where slavery was geographically contained to the states of the Southeast, in Brazil the institution of slavery was widespread throughout the national territory. However, slavery was not solely a plight of Africans and their descendants. Many natives were also held in bondage or in effective bondage (Metcalf 1992). The preference for slaves (Africans vs. Natives) depended on the amount of money the slave owners had. African slaves were more expensive. Therefore, African slaves were more concentrated in areas with greater economic dynamism, which during the early colonial times was the Northeastern section of Brazil.

In pure numbers, Brazil's racial composition is easily described by the census, which recognizes five different racial categories: "white," "black," Asian, "mixed," and "indigenous." According to the latest census, which occurred in 2010, in total, a plurality of Brazilians identify themselves as white, accounting for 47.7 percent of the population. The second highest is "mixed," with 43.1 percent and then black, with 7.6 percent. Then come Asians with 1.09 percent and indigenous with 0.4 percent. However, there are important regional differences that reflect the socioeconomic history of each region. In the Northeast, for example, which had the longest continuous history of African slavery, 9.5 percent of the population consider themselves "black." That is almost two percentage points higher than the rest of Brazil. The Southeast and South have higher proportions of the population that considers themselves "white," 55 percent and 78.5 percent, respectively (IBGE 2010). This reflects the influx of European immigrants to both regions at the turn of the 20th century.

Despite the natural borders and ecosystems, Brazilians—as all humans do—divide their national territory in several ways, which facilitates our comprehension of the country. In the broadest categorical sense, Brazil is divided into five regions

(Northeast, Southeast, South, Midwest, and North). These divisions, although commonly used, serve no direct administrative purpose. These regions have no governors or congressmen or any such officials. Such government officials represent citizens only at the state level (e.g., the citizens of the southeastern state of São Paulo elect governors, not the Southeast region). The regions were defined and created officially in 1942 for the purposes of organizing information (e.g., census data) about each geographic sector. In other words, this division is a way to make information about the country visible and manageable to people interested in studying its composition.

THE NORTHEAST

One could say that the region of the Northeast is the "oldest" region in Brazil, because it is certainly famous for the colonial architecture found in many of its cities. However, that is always a dubious statement because of two reasons: First, there were natives already living in all of Brazil at the time of the arrival of the Portuguese, thus making the arguments of age somewhat complicated since one would have to mark down fixed positions for questions that are ultimately unanswerable (e.g., what constitutes a Brazilian? When does this Brazilianess arise?). The second point is that even if one were to assume the traditional position that Brazilian history begins when the Portuguese captain Pedro Álvares Cabral landed in what is Porto Seguro (a town in the southern part of what is today Bahia) in AD 1500, then one would be in the precarious position of ignoring the fact that the first Portuguese settlement, São Vicente, was not in the Northeast, but in the Southeast, in what is today the state of São Paulo. Yet there is no denying that the Portuguese focused their early colonial enterprise on the northeast of Brazil. In part, this had to do with the ease with which colonists and their slaves could cultivate sugar—the major agricultural export during colonial times—in the region. In 1549 Tomé de Sousa established Salvador as the first capital of colonial Brazil, thus solidifying the prominence of the Northeast.

Today, the Northeast of Brazil is the regional area with most states in the country, numbering nine: Maranhão, Piauí, Ceará, Rio Grande do Norte, Paraíba, Pernambuco, Alagoas, Sergipe, and Bahia. Interestingly, all the states that compose this region have access to the Atlantic Ocean. Most likely because of its coastal accessibility, it was the first region encountered by the Portuguese explorers in April 1500. The Northeast is the third-largest region measuring roughly 1.5 million square kilometers, for approximately 18 percent of the national territory. Demographically, it is the second most populated region in Brazil, with, as of the most recent census in 2010, more than 53 million inhabitants, or 28 percent of the total Brazilian population. Despite its large population, the Northeast is the country's poorest region, accounting for only 13.5 percent of the nation's gross domestic product (GDP). Still according to *The Economist*, in recent years the region has increased its economic performance steadily, raising its GDP by 4.2 percent in the first decade of the 21st century, impressive when compared to the 3.6 percent of the whole nation (2011). The Brazilian Institute of Geography and Statistics (IBGE) notes that the region has the third-largest livestock output in the country, with approximately 16 percent of the total livestock production.

The environment of the Northeast is unique. Areas of the Bahia, Maranhão, and Piauí states are made up of a biome called *Cerrado*. As it is most common in the Midwest region, it will be discussed in that section. There is also the presence of the *Mata Atlantica* (Atlantic Forest), which will also be discussed later. However, what makes the Northeast unique is the ecosystem found in its interior, known as the *Caatinga*. Although less famous internationally than the Amazon biome, the *Caatinga* is just as fascinating. It is almost 850,000 square kilometers in size, which means it covers 11 percent of the national territory. It is an especially precious environment, because unlike the other ecosystems found in Brazil, the *Caatinga* is the only one that is found exclusively within Brazil (Silva 2003, p. 7). Thus, there is literally no other place on earth like the *Caatinga*. It is found in the interior of the states of the Northeast: Maranhão, Piauí, Ceará, Rio Grande do Norte, Paraíba, Pernambuco, Alagoas, Sergipe, and Bahia, as well as the northern part of Minas Gerais. The *Caatinga* has two main climatic divisions: the dry and wet seasons. Although popularly thought of as a dry and somewhat desolate place, it is also a fantastically diverse ecosystem, which, according to the Brazilian Ministry of the Environment, contains 178 species of mammals, 591 species of birds, 177 species of reptiles, 79 species of amphibians, and 241 species of fish and bee as well.

The IBGE divides the Northeast into four environmental regions: Meio-Norte/Mata dos Cocas (Mid-North/Palm Grove Forest), Sertão (Backlands), Agreste Nordestino (Northeastern Wilderness), and Zona da Mata (Forest). The Meio-Norte (Mid-North) is a transition area between the Amazon forest and the dry backlands, located in the state of Maranhão and the west of Piauí. It is also known as Mata dos Cocais because of the Babassu and Carnauba palm trees encountered there. The Sertão encompasses most of the Northeast region, being its biggest geographical zone. This is a dry region with irregular rain patterns. It is a biome unique to Brazil, containing the endemic vegetation, which is what the Brazilian Ministry of the Environment calls the *Caatinga*. The Agreste Nordestino (Northeastern Rural) is a transition area between the Zona da Mata and the Sertão. It is located in the Borborema highlands, the high ground that prevents rain from reaching the Sertão. It encompasses areas of Rio Grande do Norte all the way to the south of Bahia. The Zona da Mata is the wettest zone of the region and gets its name, which translates to Forest Zone, from the abundant forests that used to cover the area. The forests were eventually substituted with agricultural crops, mostly sugar and cacao. It extends from Rio Grande do Norte to the south of Bahia.

Like the biomes, the topography of the region is very diverse. It contains coastal and interior plains, plateaus, highlands, valleys, pre-coastal hills and landings, and the Espinhaço Mountain range and plateau of the Diamantine.

Maranhão

Located to the north center of Brazil, Maranhão is the second-largest state in the region covering an area of almost 332,000 square kilometers. Despite its large size, the state has one of the lowest population densities of the region, with only 19.8 inhabitants per square kilometer according to the latest census. As of 2014 the Brazilian government estimates that 6.9 million inhabitants call Maranhão home.

The capital and largest city of Maranhão is São Luis. São Luis is located to the north of the state in an island bearing the city's name on the São Marcos Bay. The city has two major ports, Madeira and Itaqui, from which the state's biggest product, iron ore, is exported.

Piauí

Located to the north between Maranhão and Ceará, Piauí has the shortest coastline of any of the northeastern states. Additionally, Piauí is one of the least densely populated states in the region with only 12.4 inhabitants per square kilometer. Demographers estimate that as of 2014 3.2 million people live within the state's nearly 252,000 square kilometers. The largest city and capital of the state is Teresina, which holds the distinction of being the only non-coastal capital of the northeastern states.

In Piauí we can find a UNESCO World Heritage Site, the Serra da Capivara National Park. According to UNESCO, "Many of the numerous rock shelters in the Serra da Capivara National Park are decorated with cave paintings, some more than 25,000 years old. They are an outstanding testimony to one of the oldest human communities of South America" (2013).

Ceará

Ceará is located in the north coast of Brazil between Piauí and Rio Grande do Norte. It is the third most populated state of the region, with an estimated 8.8 million inhabitants. Ceará covers an area of nearly 149 square kilometers, giving it a population density of 56.76 inhabitants per square kilometer.

The capital and largest city of Ceará is Fortaleza. Fortaleza has an estimated population of just over 2.5 million inhabitants, making it the second-largest city in the region and the fifth-largest city in the whole country. Fortaleza and Ceará as a whole are known for their long and beautiful coastline, which includes the Jericoacoara National Park.

Rio Grande do Norte

Like its name suggests, Rio Grande do Norte is located in the north coast of Brazil, between Ceará and Paraíba. The most recent estimates conducted in 2014 predict that the state has a population of almost 3.4 million inhabitants and covers an area of nearly 53 million square kilometers, giving it a population density of almost 60 inhabitants per square kilometer.

The capital and largest city of Rio Grande do Norte is Natal. The city gets its name, which means Christmas, because it was founded on December 25, 1599, close to the medieval fortress of the Wise Men. Natal's economy, like that of the state as a whole, relies on its coastline, which brings tourists from all over the world, and provides access to one of the state's biggest export: sea food.

Paraíba

Paraíba is located on the east coast of Brazil between the south of Rio Grande do Norte and the north of Pernambuco. The state measures roughly 56,000 square kilometers, and as of 2014, there are approximately 3.9 million inhabitants there. The most recent census data indicates that Paraíba has a population density of almost 67 inhabitants per square kilometer.

Paraíba's capital, João Pessoa, is the easternmost city in all of the Americas. The city gets its name as homage to one of the state's governors and vice presidential candidate to Getúlio Vargas, João Pessoa Cavalcânti de Albuquerque, who was assassinated months before the Revolution of 1930 (Aguiar 2005).

Pernambuco

Pernambuco is located on the east coast of Brazil to the south of Paraíba and the north of Alagoas. It is the second most populated state of the region, with an estimated 9.3 million inhabitants. Pernambuco covers an area of over 98,000 square kilometers, and has a population density of almost 90 inhabitants per square kilometer.

Its capital and largest city is Recife. Because of its location between the Beberibe River and the Capibaribe River and because of its abundance of bridges, the city has the moniker of the "Brazilian Venice." From 1630 to 1654 Recife was occupied by the Dutch, which made it the capital of its Brazilian colony (Marley 2005, p. 696).

Alagoas

Alagoas is located on the east coast between the states of Pernambuco and Sergipe. It is the most densely populated state in the region and the fourth most densely populated state in all of Brazil, with 112.3 inhabitants per square kilometers. Alagoas is the second-smallest state of the region and the second-smallest state in the country (excluding the Federal District), measuring just less than 28,000 square kilometers. It has an estimated population of about 3.3 million inhabitants.

Alagoas's capital and largest city is Maceió, located between the Mundaú Lake and the coast. Both Maceió and Alagoas derive their name from the lagoons and lakes located in the state. Alagoas is also home to the most famous Quilombo (Maroon society) of Brazil, the *Quilombo dos Palmares* (Maroon of Palmares) and its famous leader Zumbi.

Sergipe

Located on the east coast between Alagoas and Bahia, Sergipe is the smallest state of the Brazilian Federation (excluding the Federal District). At slightly smaller than 22,000 square kilometers, Sergipe is roughly the same size as Israel. As of 2014 the Brazilian Institute of Geography and Statistics estimates that around 2.2 million inhabitants live in Sergipe, making it the second most densely populated state in the region and the fifth most densely populated state in the country, with 94 inhabitants per square kilometer.

The capital and largest city of Sergipe is Aracaju. Like Brasília, Aracaju is also a planned city, except that its planning and development happened in 1855, 100 years prior to that of the Federal District (Crocitti and Vallance 2012, p. 599). Aracaju, like the rest of Sergipe, is known for its vast production of sugarcane. It currently has over half a million inhabitants.

Bahia

Located at the southernmost point of the Northeast, Bahia is the largest geographically—560,000 square kilometers— and the most populated state of the region with over 15 million people estimated to claim it as their home state. Interestingly, Bahia's population density resembles the national average at 24.8 inhabitants per square kilometer. Bahia is the fourth most populated state in Brazil and the largest one not located in the Southeast region.

Bahia's capital and largest city is Salvador (or São Salvador da Bahia de Todos os Santos). The culturally bustling historic center of Salvador, also known as Pelourinho, is a UNESCO World Heritage Site. According to UNESCO ("Historic Centre of Salvador de Bahia"),

As the first capital of Brazil, from 1549 to 1763, Salvador de Bahia witnessed the blending of European, African and Amerindian cultures. It was also, from 1558, the first slave market in the New World, with slaves arriving to work on the sugar

View of the town of Cachoeira in the state of Bahia. Cachoeira is an important historical city in Brazil. It preserves its colonial architecture and is famous for its tobacco and preservation of Afro-Brazilian traditions such as Candomblé *and* Capoeira. *(Jan Butchofsky-Houser/Corbis)*

plantations. The city has managed to preserve many outstanding Renaissance buildings. Special features of the old town are the brightly colored houses, often decorated with fine stuccowork.

Partly due to legacies of slavery, West African culture greatly influences Bahian culture. Well-known Afro-Brazilian cultural productions and customs associated with Bahia include Candomblé, Capoeira, samba, and Axé Music (Pinho 2010). Salvador's Carnaval (Carnival) festivities also attract tourists from all over Brazil and the world, losing only to Rio de Janeiro in numbers.

THE NORTH REGION

The Northern region of Brazil contains the states of Acre, Amazonas, Roraima, Rondônia, Pará, Amapá, and Tocantins. It is bordered by the nations of Bolivia, Peru, Colombia, Venezuela, Guyana, and Suriname along with the French overseas region of French Guiana. It is the largest of the five statistical regions, measuring slightly more than 3.8 million square kilometers in size. In comparative terms, the Amazon is over 550 square kilometers larger than the nation-state of India. Despite this massive size it is relatively underpopulated in comparison to other states. There are, for example, only 450 municipalities. In 2010, there were 15.9 million people living in this region. Just in the state of Minas Gerais, there are more municipalities (853) and inhabitants (over 19 million) than in the entire Northern region of the country. Most of the nonindigenous Brazilian population comes from various other regions.

Undoubtedly, the region remains most famous for the biome found within it, the *Amazônia* (Amazon) and the wildlife. Certainly out of the five biomes found in Brazil, the Amazon is the most well known. Thanks to great local and global interest in the region, in 2012 the BBC reported that the deforestation in the Amazon was down to a record low. In 1994 an American-made movie *The Burning Season* focused on the life of rubber tapper Francisco Alves Mendes Filho, better-known as Chico Mendes. Both productions demonstrate that the fascination with the amazon rainforest extends far beyond Brazil's borders. However, it is important to recognize that although it is the most prominent feature of the "wilderness" in those areas, the Amazon biome is certainly not the only biome present in those states. For example, one finds the *Cerrado* (Savannah) in the states of Roraima, Amapá, and Amazonas. Beyond clearly defined biomes, there are also ecoton regions or transition ecosystems where one biome meets the other (e.g., where the Amazon meets the Savannah).

The interest in the Amazon biome is due to two main elements: The first reason is its sheer size. The Amazon biome measures over 4.1 million square kilometers, which makes it the largest in Brazil (Ministério do Meio Ambiente). In comparative terms, it is almost half the size of Brazil and bigger than the 3.2 million square kilometers that comprises the nation of India. It is found in all of the Brazilian states that make up the North region (i.e., Acre, Amazonas, Roraima, Pará, Amapá, Rondônia, and parts of Maranhão) and also the state of Mato Grosso (which is in the Midwest region of the country). The second reason for heightened interest is its biodiversity. As of January 2013, the Brazilian Ministry of the Environment's website lists a massive

A hut on the Amazon River in Brazil. One of the longest rivers in the world, the Amazon holds the record for having the greatest volume of water. (Jakazvan/Dreamstime.com)

quantity of flora alone, containing 2,500 species of trees and 30,000 species of plants. It is precisely this biodiversity that leads to great concern among environmentalists over environmental degradation in the region, especially deforestation.

In addition, while this biome is often associated with trees and dense "jungle," it is also immensely important in terms of water. Again, this can be expressed in sheer size. The Brazilian Ministry of the Environment lists the Amazon watershed as the world's largest at over 6 million square kilometers. According to the Brazilian government the entire Amazon River Basin (which includes the parts of the Amazon that do not belong to Brazil) accounts for the draining of 20 percent of the freshwater in the world (Federative Republic of Brazil 2013). The river system is so extensive and the forest so dense that many inhabitants of the region travel via boats especially if they do not have access to airplanes.

The geographical relief of the region is diverse as well. Much of the topography is lowland plains, particularly around the rivers. There are also two sets of highland areas found in the Northern region. The northern one occurs along the border with Venezuela and Guyana (particularly with the state of Roraima and to a lesser extent in Amazonas state). It is also present in the states of Pará and Amapá. This is often referred to as the Guiana highland or the Guiana shield. There are also southern Amazonian highlands that run across parts of Rondônia and the southern parts of the states of Amazonas and Pará, as well as Mato Grosso.

Pará

In terms of the population size, the largest state in the region is Pará. In 2014 government demographers estimated that over 8 million people live in Pará. It is also the second-largest state in the region, with slightly over 1.2 million square kilometers in size. Like the rest of the Northern region, it too has few municipalities, only 144. The largest of these municipalities, Belém (which is Portuguese for Bethlehem) is also the state's largest city. Slightly greater than 1.4 million inhabitants are estimated to reside in Belém, making it the second most populous city in the entire Northern region, only after Manaus.

Amazonas

The state of Amazonas is the biggest in the region in terms of geographical size. It is slightly larger than 1.5 million square kilometers in size. However, it has fewer people (fewer than 3.9 million inhabitants) and municipalities (62) than Pará state. The capital city, Manaus, is the largest population center with an estimated population of over 2 million inhabitants as of 2014. The temperature in Manaus is fairly stable, with the daily temperatures ranging from lows in the middle 70 degrees Fahrenheit and the highs in the mid-80s to low 90 degrees Fahrenheit throughout the year. The notable climatic divergence occurs with rainfall. While there is significant rainfall throughout the year, the rain level begins to increase in the month of November and then starts to decline from May to June.

Amapá

The state of Amapá lies just to the north of Pará. Comparatively smaller than the other states in the region, it is 142,000 square kilometers in size, or slightly larger than the nation of Greece. Over 750,000 people are estimated to live in the entire state, continuing the pattern of low-density population found throughout the region. As of 2015 most of the population, or nearly 447,000 people live in the state capital, Macapá. The climate in Macapá is similar to climates in the region, generally with little variation in the overall temperature, with lows in the mid-70s and highs in the middle 80 degrees Fahrenheit. Again we see the main climatic oscillation occurring between the "dry" (more or less June–November) and "rainy" seasons (more or less, December–July).

Roraima

Roraima, along with Amapá, is the northernmost state in Brazil. It is slightly smaller than 225,000 square kilometers in size. Again, if we compare it to a country, it is slightly larger than the nation of Guyana. However, few people live in this state. In fact, with 2014 estimates calculating fewer than 500,000 inhabitants, Roraima is the least populous state in the Northern region. In fact, according to the latest census, it is also the least densely populated state in Brazil, with only 2.01 inhabitants per square kilometer.

Acre

The state of Acre is the westernmost state in Brazil. It borders the state of Amazonas to the north and east and the nations of Bolivia and Peru to the south and west. It is

164,000 square kilometers in size and has a population of 790,000 people spread out over 22 municipalities. Rio Branco, the capital of Acre state, is the largest city with an estimated population of slightly fewer than 364,000 residents.

Rondônia

Like Acre, the state of Rondônia also borders the nation of Bolivia. It also borders mostly the Amazonas and Mato Grosso states. It shares a brief border with the state of Acre to the west. According to the most recent estimates, over 1.7 million people live in Rondônia. Slightly below 500,000 of those residents live in the state capital, Porto Velho. The state was named after Brazil's famed explorer Candido Rondon, who sought to explore the interior of Brazil and connect it with the population centers via telegraph.

Tocantins

Tocantins is the most easterly state in the Northern region. It is one of the newest states, because it used to be a part of Goiás state. Tocantins is located almost in the center of the country. It is bordered by Goiás to the south, Pará to the west, and Maranhão to the northeast as well as, for a brief stretch, the state of Piauí and finally Bahia to the east. Tocantins's capital is the city of Palmas, which has a population of 265,000 people. This represents a little over 17 percent of the total population of the state, which stands at 1.5 million, according to 2014 government estimates. Unlike the rest of the Northern region's biome, which is mostly the Amazon rain forest, the biome most present in Tocantins is the *Cerrado*.

THE SOUTH REGION

The South region of Brazil comprises three states: Paraná, Santa Catarina, and Rio Grande do Sul. The region is bordered to the north by the Southeast region and the Midwest region of Brazil; to the east it is bordered by the Atlantic Ocean; to the west it is bordered by Paraguay and Argentina; finally, to the south it is bordered by Uruguay. Being the smallest region it encompasses only 6.8 percent of the national territory or just over 576,000 square kilometers (Hudson 1997). Because a significant part of its territory is located below the Tropic of Capricorn, the climate of the South region of Brazil shows defined seasons unlike the other regions of the country. Still, as well as a temperate climate the region also contains a tropical climate in its northernmost state of Paraná.

In addition to its diverse climate, the South region contains a diverse topography, with depressions from the Amazon River, levels of the Paraná River Basin—the Paranaense, Sul-Rio-Grandense, and Araucárias plateau, coastal and interior plains, and the mountain ranges on the coast and the east of Santa Catarina. But perhaps its most famous topography is the pampa grassland similar to those found in Uruguay and Argentina, located at its southernmost point. The region also contains the Atlantic Forest and pinewoods, but many of these forested areas were cleared in the decades after World War II (Hudson 1997).

According to the IBGE's report entitled "Região Norte aumenta participação no PIB nacional," (Northern region increases its participation in the national GDP) despite its small size in relation to the other regions of the country, in 2008, the South region contributed 16.6 percent of the country's GDP. Its economy is dominated by agriculture, with soy and corn being its biggest crops. Other important crops include rice, wheat, sugar, cassava, tobacco, and oranges. Interestingly, most of the crops produced in the South come from small farmers who, despite their size, have a high level of productivity (Hudson 1997).

According to the latest census data, the South is also the third most populous region of the country, with a population of just over 27 million people. Despite its great diversity of terrain, the population of the region is concentrated mainly across the coastline (Hudson 1997).

Paraná

Paraná is the northernmost state of the region and the only state in the region that borders other regions of Brazil. The state's capital and largest city is Curitiba, located 105 kilometers west of the seaport of Paranaguá in a plateau known as the Curitiba Plateau.

The state of Paraná measures approximately just less than 200,000 square kilometers and, as of 2014, has an estimated population of just fewer than 11.1 million Paranaenses. This means that the population density of the state, 52.4 inhabitants per square kilometer, is more than double the national average.

IGUAÇU FALLS

In the state of Paraná we find one of the most impressive wonders of nature: the Iguaçu Falls. Located between northeastern Argentina and southwest Paraná, these falls measure 80 meters in height and 3 kilometers in width. The Brazilian part of these falls is located on the Iguaçu National Park, a World Heritage Site, and one of the principal touristic attractions of the state of Paraná. As UNESCO describes it, "The falls are made up of many cascades that generate vast sprays of water and produce one of the most spectacular waterfalls in the world."

The vegetation surrounding the falls is typical of a wet forest giving the area an amazing variety of fauna and flora. In the Iguaçu National Park you can find many endangered species like jaguars, tiger cats, ocelots, giant anteaters, and Brazilian otters. The park also contains an astonishing variety of bird species.

Prior to the conquest of the Americas, the area is known to have been populated by the Caingangues people and later the Tupi-Guarani. The latter gave the falls its present name, "Iguaçu," meaning "big water." In 1541, the Spanish conquistador (conqueror) Alvar Nuñez Cabeza de Vaca encountered the falls. He was the first European to do so. Nuñez Cabeza de Vaca was also one of the first Europeans to reach Florida and other southern areas of the United States.

UNESCO, "Iguazu National Park," World Heritage List. Accessed January 29, 2013. http://whc.unesco.org/en/list/303.

Santa Catarina

Santa Catarina is the middle state of the South region, located below Paraná and above Rio Grande do Sul. Of the three states that compose the South region, Santa Catarina is the smallest both in geographical size and in population. The state measures greater than 95 square kilometers and as of 2014 has a population of around 6.7 million Catarinenses. Despite its small status in the region, its population density of 65.29 inhabitants per square kilometer is almost triple that of the whole country.

The capital city of Santa Catarina is Florianópolis. Interestingly, Florianópolis is composed mostly of an island called the Island of Santa Catarina, located just east of the shore. However, the capital city also has a section located on the mainland, as well as encompassing the surrounding smaller cays.

Rio Grande do Sul

Rio Grande do Sul is the southernmost state of the region as well as of the whole of Brazil. It is also the biggest state of the region measuring a little larger than 281,000 square kilometers. In addition to its big geographical size, it is also the most populous state of the region; as of 2014, over 11.2 million Gaúchos live in Rio Grande do Sul. The term "Gaúcho" is used to refer to the inhabitants of the South American pampas, and therefore, it is also commonly used in other countries that have this particular biome, like Argentina and Uruguay. The capital of Rio Grande do Sul is the city of Porto Alegre, which is located on the northern coast of a large coastal freshwater lagoon called the Lagoa dos Patos (the lagoon of the ducks).

As Beatriz Dornelles narrates, by the end of the 18th century the crown of Portugal was trying to populate the south of Brazil, and to do so it promoted the settlement of people from the Portuguese Azores Islands in the region where Rio Grande do Sul is currently located (2004, p. 29). Therefore, a great number of the population is of Azorean ancestry. Later, in the 19th and early 20th centuries the region received many immigrants from other parts of Europe, particularly from Germany, Italy, and the former Soviet Union (Dornelles 2004, pp. 164 and 216). Because of these migratory waves, many of the inhabitants of this region are of European descent.

THE MIDWEST REGION

The Midwest region of the country has traditionally been a sparsely populated region. It is tied with the South for having the fewest states, comprising itself of only the states of Mato Grosso, Mato Grosso do Sul, and Goiás, as well as the Federal District, which contains the capital of Brazil, Brasília. It is geographically a very large region; at over 1.6 million square kilometers in size, it is larger than the nation of Mongolia. The region was larger but shrank in total size with the creation of the state of Tocantins and that state's subsequent transfer to the Northern region. Originally, Tocantis had been a part of the Goiás state, but politicians partitioned it from the state of Goiás in article 13 of the 1988 Constitution. Further administrative changes occurred in this region. Previously, the states of Mato Grosso and Mato Grosso do

Sul were one administrative unit, but in law number 31 of October 11, 1977, the military regime divided the single state into two: Mato Grosso and Mato Grosso do Sul. The region that previously had only two massive states and the Federal District was subdivided into three states and the Federal District. Still in terms of sheer physical land area, the region accounts for almost 19 percent of the total size of Brazil.

Historically, this region has been lightly populated. To this day the 14 million inhabitants account for slightly more than 7 percent of the nation's inhabitants. Thus, this region is the least populous region in terms of national percentages. This however is, and has been, changing. Along with the Northern region, the Midwest region has experienced the largest relative growth rate of any of the major regions in Brazil according to the latest census, conducted in 2010.

In terms of ecosystems this region is quite diverse. It contains two major biomes: the *Pantanal* and the *Cerrado.* Perhaps the *Pantanal* is the most famous ecosystem among foreign tourists, because it is a popular destination for ecotourists interested in experiencing a different environment and even fishing. The *Pantanal* is an immense wetland region. It stretches from the southern parts of Rondônia state in the North through Mato Grosso state and into Mato Grosso do Sul. Like most other biomes in Brazil, the *Pantanal* does not stop at Brazil's borders; it extends into Bolivia and Paraguay as well. According to the Ministry of the Environment, at slightly over 150,000 square kilometers it is the smallest ecosystem in Brazil accounting for just less than 1.8 percent of the national territory. Furthermore, perhaps due to its sparsely populated status, the *Pantanal* is in some ways "healthier" than other biomes. Almost 2,000 plants have been identified. There are 463 species of birds, 263 species of fish, 132 species of mammals—two of which are exclusively of the area—41 species of amphibians, and 113 species of reptiles (Ministério do Meio Ambiente, "Pantanal," 2013).

The other important biome in the Midwest region is the *Cerrado.* It is a savanna and thus contains prominent dry and rainy seasons. At over 2 million square kilometers, it covers 22 percent of the national territory. It can be found broadly in the states of Goiás, Tocantins, Mato Grosso, Mato Grosso do Sul, Minas Gerais, Bahia, Maranhão, Piauí, Rondônia, Paraná, São Paulo, and the Federal District. It is an incredibly diverse region. The Ministry of the Environment lists 11,627 plant species discovered so far, which, along with the fauna, makes it the most biodiverse savanna in the world (Ministério do Meio Ambiente, "O Bioma Cerrado," 2013). This region is found in the central part of the country. Brasília, Brazil's capital, is located in the *Cerrado.* It is also an incredibly productive agricultural zone, and the region's farmers have been at the forefront of Brazil's increasingly prominent agricultural growth. However, this is not without controversy, as such growth has also led to environmental degradation.

The topography of the region varies as well. There is the *Pantanal* Mato-Grossense and the *Pantanal* of Guaporé. There are also several geographic depressions in this region. There is the southern Amazon depression. Around the border area of Mato Grosso, Goiás, and Tocantins, we find the depressions of the rivers Araguaia, Tocantins, and Xingu, which run from the northern part of the state of Pará down into the Midwest states of Mato Grosso and Goiás. Another major depression is the Paraguay River depression. The São Francisco River mesa is also found along the northeast of the Midwest region where Goiás meets the state of Bahia. A prominent feature of the topography of the region is the planaltos or plateaus. The better known of these plateaus is the Planalto Central Brasileiro (Brazilian Central Plateau). Colloquially, when Brazilians refer to the planalto, they usually mean this one, as it is where the capital, Brasília, is located. The Guimarães and Parecis plateaus are found in this region, as is the Paraná River Basin plateau. At the very north of Mato Grosso state one finds the plateaus of the southern Amazon. There are even brief stretches of mountainous areas found in Mato Grosso and Mato Grosso do Sul known as Serra dos Rios Paraguay and Guaparé. Floodplains are also found in this region.

Although this region is the only one not to border an ocean, water is still a defining feature, as it contains two immensely important rivers not just for Brazil but also for all of South America, the Paraguay and Paraná rivers. Both river systems originate in Brazil and flow southward through Paraguay and Argentina (the Paraguay River also runs through parts of Bolivia). The Paraguay River drainage basin is massive covering an area of over 1 million square kilometers, 33 percent of which lies in Brazil.

Mato Grosso

Mato Grosso is the largest state of the Midwest region. Its surface area is slightly larger than 900,000 square kilometers in size, making it slightly smaller than the nation of Venezuela. As of 2014, demographers estimate slightly more than 3.2 million people claim Mato Grosso as their home. This makes Mato Grosso the least densely populated state in the Midwest region of the country. A little under 18 percent of the population or slightly more than 575,000 people live in the state capital of Cuiabá, located in the southern region of the state. It is bordered by the states of Rondônia to the west, Amazonas and Pará to the north, Tocantins and Goiás to the east, and Mato Grosso do Sul to the south. Furthermore, it shares an international border to its southwest with the nation of Bolivia.

Mato Grosso do Sul

Mato Grosso do Sul is another large state in the Midwest of Brazil. At over 357,000 square kilometers it is slightly larger than the nation of Germany. It has fewer inhabitants (almost 2.6 million) than Mato Grosso state. However, its capital Campo Grande, located in the center of the state, is larger (almost 843,000 people) than Cuiabá. Its image is popularly associated with that of the *Pantanal* region discussed earlier. It is bordered to the north by Mato Grosso and Goiás and to the east by Minas Gerais, São Paulo, and Paraná. Like Mato Grosso, it too has an international border, with Bolivia and Paraguay.

Goiás

Goiás is located almost at the very center of Brazil. It is the most highly populated state in the region. With a population of slightly more than 6.5 million people, it has almost double the population of the second most populous state in the region,

Mato Grosso. It also contains the second most populous city in the region, Goiânia, the state capital, with a little over 1.4 million people. In terms of land area, it is
over 340,000 square kilometers, making it slightly smaller than Germany. Today, it
borders several states: Bahia and Minas Gerais to the east, Mato Grosso do Sul (as
well as parts of Minas Gerais) to the south, Mato Grosso to the west, and Tocantins
to the north. One of its distinctions geographically is that it almost completely sur-
rounds the Federal District (the Federal District shares a brief border with Minas
Gerais as well).

Distrito Federal

The Federal District is one of the newest administrative districts. It is the smallest
of the federal administrative units at less than 5,800 square kilometers. This makes
it slightly larger than the nation of Trinidad and Tobago. However, it contains only
one city, Brasília. Apart from being the capital, Brasília is the largest city in the Midwest region and the country's fourth largest city (after São Paulo, Rio de Janeiro,
and Salvador), with slightly more than 2.9 million people estimated to live there as
of 2014.

THE SOUTHEAST REGION

The Southeast region of Brazil is composed of four states: São Paulo, Minas Gerais,
Espírito Santo, and Rio de Janeiro. In total, it stretches for over 900,000 square kilometers. It is bordered to the north by the state of Bahia, to the west by the states of
Goiás and Mato Grosso do Sul, and to the south by the state of Paraná. To the east
the region ends at the Atlantic coast of Brazil. Perhaps the most significant information about this region is that it is by far the largest region in terms of population,
with slightly more than 80 million people living in the four states mentioned earlier.
The two largest cities in Brazil, São Paulo and Rio de Janeiro, are found in this area.
Furthermore, three of the four states (São Paulo, Minas Gerais, and Rio de Janeiro)
are also the most populous in Brazil.

It is impossible to discuss the Southeast region without mentioning its impact
on the economy of Brazil. According to the 2010 census, the region accounted for
55.4 percent of the total GDP of the country. While São Paulo accounts for the
majority of this, the neighboring states of Rio de Janeiro and Minas Gerais are also
second and third on the list of largest GDPs in Brazil. Much of this economic strength
derives from historic booms in gold and coffee and also from the fact that Brazil's
political center was in the Southeast, which focused a lot of governing attention on
that region.

The major reason there are so many people in this region is migration. Ever since
the discovery of gold in the late 17th century, migrants have been a mainstay in the
region. Many came from other parts of Brazil, but it also took in many immigrants.
Around the last decade of the 19th century and the beginning decades of the 20th
century, immigrants started to arrive in ever larger numbers. The Southeast was immensely important in this process as the largest number of people arrived at the port
city of Santos on the coast of São Paulo state (Skidmore 2010, p. 77).

The Southeast contains two major biomes: the *Cerrado*—discussed earlier—is found in significant parts of the states of São Paulo and Minas Gerais. The other major biome is the *Mata Atlantica*, which unlike the *Cerrado* is found in all of the states of the Southeast. Unfortunately, the Ministry of the Environment declared that the *Mata Atlantica* is perhaps the most decimated of the Brazilian biomes. It originally covered 1.3 million square kilometers in size. Now it has been reduced to 22 percent of its original size. Throughout the *Mata Atlantica* (which extends beyond the Southeast), there is an extensive amount of biodiversity. There is an estimated 20,000 species of flora. It also contains 849 species of birds, 370 species of amphibians, 350 species of fish, 270 species of mammals, and 200 species of reptiles (Ministério do Meio Ambiente. "Mata Atlântica," 2013).

The Southeast region has one of the most diverse topography of any region. There are low-lying areas such as the coastal and floodplains and also areas of geographic depressions. However, the region is well known as a mountainous one. Almost immediately from the Atlantic side one encounters the mountain ranges that form the Brazilian highlands. The most famous of these is known as *Serra do Mar* (Mountain Range of the Sea). Its formations are responsible for some of the most iconic images one finds of the city of Rio de Janeiro, which includes the Sugar Loaf Mountain and the Corcovado Mountain, upon which stands the statue of Christ the Redeemer that overlooks the city. There are also significant highland areas such as the planalto of the Paraná River Basin and the Mesas. The state of Minas Gerais is particularly famous for its mountainous terrain and mesas, especially the mesas of the São Francisco River and Diamantina.

Minas Gerais

Minas Gerais is the only landlocked state of the region, located to the west of Espírito Santo and Rio de Janeiro, and to the north of São Paulo. Minas Gerais gets its name from the mineral-rich highlands, which cover a great part of the state. It is the fourth-largest state of Brazil measuring a little over 586,000 square kilometers, roughly the same size as France. As of 2014 it is estimated that the state has a population of over 20.7 million inhabitants, making it the second most populous state in Brazil. Despite the great number of inhabitants, the latest census, conducted in 2010, found that it has a population density of 33.4 inhabitants per square kilometer, making it the least densely populated state of the region.

The state's largest city and capital is Belo Horizonte, located on hills surrounded by mountains toward the center-east region of the state. In Belo Horizonte you can find the Pampulha Complex, a project of famous architect Oscar Niemeyer (Barnitz 2001, p. 169).

Espírito Santo

Espírito Santo is located on the east coast, just south of the state of Bahia. It is the second-smallest state in the Southeast region measuring 46,095.6 square kilometers. As of 2014 Espírito Santo had a population of close to 3.9 million inhabitants.

During the 2010 census Brazilian demographers calculated that Espírito Santo had a population density of 76.25 inhabitants per square kilometer. Its capital is the city of Vitória, located on an island off the Vitória Bay. Despite being the capital, Vitória is not the largest city of the state, this distinction falling to the coastal city of Vila Velha, which forms part of the greater Vitória metropolitan area.

Like other states in the region, Espírito Santo has a great number of Italian descendants in its population. According to the immigration records of the state, 75 percent of the immigrants who came to Espírito Santo were Italian, particularly from the Veneto region of that country (Immigration Records of Espírito Santo 2013).

Rio de Janeiro

Rio de Janeiro is the only state in the region that borders all other southeastern states. It is the smallest state in the region, measuring only 43,780.17 square kilometers. Despite its relative small size, it is the third most populous state in Brazil, with almost 16.4 million inhabitants. Its massive population and small geographic size makes it the most densely populated state in Brazil (with the exception of the Federal District), with 365.23 inhabitants per square kilometer. The large population is easily explained by noting that it has the second-largest economy in the country, contributing 10.8 percent of the country's GDP, following only neighboring São Paulo.

Rio de Janeiro's capital and biggest city has the same name as the state, Rio de Janeiro. This city was the capital of Brazil from 1763, when the capital was moved from Salvador, until 1960 when Brasília, the current capital, was inaugurated. Rio de Janeiro is the principal touristic attraction of Brazil, containing iconic sites such as the Christ the Redeemer statue, the Copacabana and Ipanema beaches, and Sugar Loaf Mountain, as well as Brazil's premier soccer stadium Maracanã. It is also famous all over the world for its Carnival celebrations.

São Paulo

São Paulo is the southernmost state of the southeastern region. It is the economic center of Brazil contributing an impressive 33.1 percent of the nation's GDP. In terms of population São Paulo is also the largest state in the country, with over 44 million estimated inhabitants. It covers a total area of 248,222.8 square kilometers, making it the second most densely populated state in the country (excluding the Federal State), with 166.25 inhabitants per square kilometer.

The state's capital and the largest city in all of Brazil and South Americas is São Paulo. The city of São Paulo is a great international hub, containing such areas as Liberdade, the largest Japanese community outside of Japan (Campion 2009). Another famous neighborhood in São Paulo, Bexiga, was home to many newly arrived Italian immigrants as well as newly arrived Brazilian migrants of color.

Despite its great industrialization, the state of São Paulo contains impressive national reserves like the UNESCO World Heritage Site of the Atlantic Forest South-East Reserves. According to UNESCO, the reserves "comprises a rich natural environment of great scenic beauty" that contains "mountains covered by dense forests, down to wetlands, coastal islands with isolated mountains and dunes" (UNESCO, "Atlantic Forest South-East Reserves," 2013). Thus, São Paulo state is more than its very famous city.

There is much about Brazil that cannot be conveyed or understood by simply looking at a map. Although any casual observer can see Brazil's sheer massive size, such a view would fail to grasp the true grandeur of Brazil, which comes from its diversity. In ecological terms, the country has a plethora of different biomes and climates. Yet it is also important to remember that Brazil is also diverse in terms of its people. While geography is an essential aspect of any nation, we must remember that geography gains particular significance when it is related to people. Thus, it is not a map that makes Brazil a nation, but rather its inhabitants, the Brazilians.

REFERENCES

Aguiar, Wellington. 2005. *João Pessoa, O Reformador.* João Pessoa: Idéia.

Baer, Werner. 2014. *The Brazilian Economy: Growth and Development.* Boulder: Lynne Rienner Publishers.

Barnitz, Jacqueline. 2001. *Twentieth-Century Art of Latin America.* Austin: University of Texas Press.

Campion, Charles. February 24, 2009. "São Paulo's Top 10 Restaurants," *The Guardian* (London). Accessed January 27, 2013. http://www.theguardian.com/travel/2009/feb/24/sao-paulo-food-drink-restaurants.

Crocitti, John J., and Monique Vallance. 2012. *Brazil Today: An Encyclopedia of Life in the Republic.* Santa Barbara, CA: ABC-CLIO.

Dornelles, Beatriz. 2004. *Porto Alegre em destaque: história e cultura.* Porto Alegre: EDIPUCRS.

Federative Republic of Brazil. "Brazilian Biomes." Geography in Environment. Accessed January 13, 2013. http://www.brasil.gov.br/sobre/environment/geography.

Hudson, Rex A. 1997. *Brazil: A Country Study.* Washington, DC: Federal Research Division, Library of Congress.

Immigration Records of Espírito Santo "Estatística." Accessed January 27, 2013. http://www.ape.es.gov.br/imigrantes/html/estatisticas.html.

Instituto Brasileiro de Geografia e Estatística Brazilian Institute of Gerography and Statistics (IBGE). 2010. "Censo 2010." Brasília. Accessed January 2013. http://www.ibge.gov.br/home/.

Instituto Brasileiro de Geografia e Estatística Brazilian Instituto of Geogaphy and Statistics (IBGE). "IBGE divulga as estimativas populacionais dos municípios em 2012." Brasília. 2012.

Marley, David. 2005. *Historic Cities of the Americas: An Illustrated Encyclopedia, Volume 1.* Santa Barbara, CA: ABC-CLIO.

Metcalf, Alida. 1992. *Family and Frontier in Colonial Brazil: Santa de Parnaíba, 1580–1822.* Austin: University of Texas Press.

Ministério do Meio Ambiente. "Mata Atlântica." Accessed January 13, 2013. http://www.mma.gov.br/biomas/mata-atlantica.

Ministério do Meio Ambiente. "O Bioma Cerrado." Accessed January 21, 2013. http://www.mma.gov.br/biomas/cerrado.

Ministerio do Meio Ambiente. "Pantanal." Accessed January 21, 2013. http://www.mma.gov.br/biomas/pantanal.

Pinho, Patricia De Santana. 2010. *Mama Africa: Reinventing Blackness in Bahia.* Durham, NC: Duke University Press.

Silva, Marina. 2003. "Apresentação." In José Maria Cardoso da Silva, Marcelo Tabarelli, Mônica Tavares da Fonseca, and Lívia Vanucci Lins (eds), *Biodiversidade da Caatinga: áreas e ações prioritárias para a conservação.* pp. 7–8. Brasília DF: Ministério do Meio Ambiente and Universidade Federal de Pernambuco.

Skidmore, Thomas E. 2010. *Brazil Five Centuries of Change.* New York: Oxford University Press.

United Nations Educational, Scientific Cultural Organization (UNESCO). "Atlantic Forest South-East Reserves." World Heritage List. Accessed January 29, 2013. http://whc.unesco.org/en/list/893.

United Nations Educational, Scientific Cultural Organization (UNESCO). "Historic Centre of Salvador de Bahia." World Heritage List. Accessed January 29, 2013. http://whc.unesco.org/en/list/309.

United Nations Educational, Scientific Cultural Organization (UNESCO). "Serra da Capivara." World Heritage List. Accessed January 29, 2013. http://whc.unesco.org/en/list/606.

Suggested Reading

Fausto, Boris. *A Concise History of Brazil.* New York: Cambridge University Press, 1999.

Oliveira, Paulo S., and Robert J. Marquis. *The Cerrados of Brazil: Ecology and Natural History of a Neotropical Savanna.* New York: Columbia University Press, 2002.

Skidmore, Thomas E. *Brazil: Five Centuries of Change.* New York: Oxford University Press, 2010.

Slater, Candace. *Entangled Edens: Visions of the Amazon.* Berkeley: University of California Press, 2003.

History

Kara D. Schultz

TIMELINE OF EVENTS

1494   Treaty of Tordesillas divides the non-European world between the Spanish and Portuguese, conceding all land 370 leagues east of the Cape Verde Islands (including the yet-undiscovered Brazil) to the Portuguese.

1500   Portuguese expedition led by Pedro Alvares Cabral en route to India anchors at Porto Seguro, in the modern-day state of Bahia, and claims the "Land of the Holy Cross" for Portugal. At the time of Cabral's arrival, it is estimated that anywhere between 2 and 7 million indigenous people lived along the Brazilian coast.

1518   First sugar mill constructed in Brazil.

1534   D. João III divides Brazil into 15 hereditary captaincies (*donatários*).

1549   Tomé de Souza founds Salvador da Bahia, the capital of the new governor-generalship of Brazil.

1549   First Jesuits arrive in Brazil.

1550   The French establish a colony, "France Antarctique," as a Protestant refuge on an island in the Bay of Guanabara. The colony lasts 10 years.

1565   Rio de Janeiro is founded near the site of the former French colony.

1580–1640   The Spanish and Portuguese crowns are united under a single monarch, King Philip II of Spain.

1620   The state of Maranhão is created from the captaincies of Ceará, Maranhão, and Pará. Maranhão remains administratively separate from the "state of Brazil" for a century.

1624   The Dutch West India Company sponsors an invasion of the northeastern coast of Brazil, ruling for 30 years from the capital city of Recife.

Under Dutch rule, freedom of worship is permitted, and some 1,000 Dutch Jews of Sephardic descent settle in Pernambuco.

1648 Forces led by the governor of Rio de Janeiro Salvador Correia de Sá e Benavides oust the Dutch from the Angolan port of Luanda, securing a steady supply of West Central African slave labor for Portuguese America.

1654 Dutch expelled from Pernambuco. With the end of Dutch rule, freedom of worship is no longer permitted. Some 23 Dutch Jews depart Recife for New Amsterdam (later New York), the northernmost Dutch colony in the Americas.

1680 Portuguese establish the settlement of Colônia do Sacramento across from Buenos Aires (modern-day Uruguay) to secure access to trade with the Rio de la Plata region.

1693 Discovery of gold in Minas Gerais.

1695 Portuguese forces destroy the *Quilombo* of Palmares. The maroon community (which included runaway slaves, free people of African descent, Indians, and whites) is said to have numbered as many as 100,000 during its peak.

1703 Portugal signs the Metheun Treaty with Great Britain, giving British imports preferential treatment in Portugal and the empire. British imports are paid for with Brazilian gold and diamonds.

1750 Spain and Portugal sign the Treaty of Madrid, revising the boundaries between Spanish and Portuguese America set forth by the Treaty of Tordesillas. *Uti possidetis*, or ownership by occupation, becomes the guiding principle for determining sovereignty. The Portuguese Colônia do Sacramento (modern-day Uruguay) is ceded to Spain, while Spanish Jesuit missions along the eastern part of the Uruguay River are given to Portugal.

1755 An earthquake and subsequent tsunami and fires destroy one-third of the city of Lisbon. Sebastião José de Carvalho, the Marquis of Pombal and de facto head of the Portuguese government since 1750, launches sweeping administrative reforms for Portugal's overseas colonies in an effort to rebuild what was lost.

1759 Pombal expels Jesuits from the Portuguese empire.

1763 Rio de Janeiro replaces Salvador as the capital of the Brazilian colony. The move heralds the increased economic and political importance of Brazil's southern provinces.

1789 A mostly elite group of residents of Ouro Preto, in Minas Gerais, develop a plan to assassinate the governor and declare independence. The "Inconfidência Mineira" ("Minas Gerais Conspiracy") is discovered and its chief conspirators are placed on trial.

1792 Joaquim José da Silva Xavier, also known as "Tiradentes," the scapegoated leader of the Inconfidência Mineira, is hanged, drawn, and quartered.

1798 In Salvador, a diverse group of artisans, soldiers, and other nonelites influenced by the French Revolution conspire, unsuccessfully, for independence and a free, egalitarian society. Their plot is discovered before their plan can come to fruition.

1808   Fleeing Napoleon's invasion of the Iberian Peninsula, the Portuguese court moves to Rio de Janeiro. Prince regent Dom João opens Brazilian ports to foreign trade.

1821   The Cisplatine Province is annexed to Brazil.

1822   Portuguese *Cortes* convene and decide that Pedro must return to Brazil. Brazilian independence is proclaimed and Dom Pedro I, the prince regent, is crowned Emperor of Brazil.

1824   Emperor Pedro I enacts Brazil's first constitution, strongly influenced by British models.

1826   Brazil signs a treaty with Great Britain promising to ban the transatlantic slave trade and declares any slave who enters the country after November 7, 1831, free. The treaty has the opposite effect of curbing the slave trade, as more slaves are imported in the 20 years following the law's passage than during any other period in Brazilian history.

1828   As a result of a British-negotiated treaty, Uruguay is an independent nation following a three-year conflict between Brazil and Argentina over ownership of the Cisplatine Province.

1835   Yoruba slaves in Bahia mount the largest slave rebellion in Brazilian history. The Malê Revolt (from the Yoruba word for "Muslim," referring to the Muslim identity of many of the leaders) frightens Brazilian planters and helps generate popular support for the abolition of the slave trade.

1850   Eusébio de Quierós Law definitively outlaws the slave trade.

1865–1870   Paraguayan War/War of Triple Alliance. Brazil, Argentina, and Uruguay form an alliance against Paraguayan leader Francisco de Solorzano López in response to a threat to navigation rights to the Paraná River. The protracted conflict results in heavy casualties and generates resentment of Emperor Pedro.

1871   Lei do Ventre Livre, also known as the Rio Branco Law, emancipates children born to slave mothers. However, the mother's owner maintains the right to the child's labor until age 21.

1881   Brazilian positivist church founded.

1888   Princess Isabel signs the Lei Aurea ("Golden Law") abolishing slavery in Brazil.

1889   A joint civilian–military coalition forces Pedro's abdication and proclaims the Brazilian Republic, headed by Manuel Deodoro da Fonseca.

1891   Brazil's republican constitution creates a dual federalist system and replaces the monarchy with an elected president.

1897   Government troops occupy and destroy the settlement of Canudos in the Bahian *sertão* (interior or backlands), believing that Canudos's spiritual leader, Antônio Conselheiro, seeks to overthrow the Republic. It is one of the deadliest civil conflicts in Brazil's history.

1904   Residents of Rio de Janeiro rebel against mandatory yellow fever vaccines and slum eradication in the "Vaccine Revolt."

1917   Brazil joins the Allied powers in World War I, becoming the only nation in Latin America to actively enter the conflict.

1922	Brazilian Communist Party (PCB) founded.
	Modern Art Week in São Paulo debuts Brazilian *modernistas* (modernists), artists and intellectuals, who celebrate Amerindian and African contributions to Brazilian culture.
1924	*Tenentes* in Rio Grande do Sul, led by Luis Carlos Prestes, march throughout Brazil. The "Prestes Column" is not dissolved for three years.
1929	Worldwide economic collapse plunges the Brazilian agricultural export-oriented economy into crisis.
1930	A military coup brings Getúlio Vargas to power, ousting sitting president Washington Luís.
1932	In July, São Paulo politicians stage a revolt against Vargas but are put down by federal troops by October. The so-called Constitutionalist Revolt tarnishes the public image of *paulistas* (natives of São Paulo), though to appease them Vargas calls for a Constituent Assembly to draft a new constitution.
1933	Gilberto Freyre publishes *The Masters and the Slaves.*
1934	Constitution of 1934 scraps the laissez-faire liberalism of the First Republic and makes the government responsible for social welfare and economic growth.
1935	Communist officers in Recife, Natal, and Rio de Janeiro stage barracks revolts. The revolts are quickly crushed and give Vargas the congressional backing he needs to suspend civil liberties and presidential elections.
1937	Getúlio Vargas stages a coup within his own government and proclaims the "Estado Novo," reading a new constitution on the radio. The Vargas presidency becomes a dictatorship.
1942	Nazi attacks on Brazilian shipping lines prompt Brazil to enter World War II on the Allied side.
1945	The army forces Vargas's resignation. Democratic elections are held for the first time in 15 years and Eurico Dutra is elected as president.
1950	Getúlio Vargas is democratically elected president of Brazil for the first time in his political career.
1953	Petrobrás, a public–private corporation, is created after two years of heated debate.
1954	Getúlio Vargas commits suicide in advance of a military coup that would remove him from office.
1960	Brasília inaugurated as the new national capital.
1964	A military coup ousts sitting president João Goulart, inaugurating over two decades of military dictatorship in Brazil.
1968	The military government passes Institutional Act #5, suspending civil liberties under the guise of maintaining order. Labor unions are banned; university faculty are purged; political opponents are tortured and "disappeared"; and the media are subject to censorship.
1972	Trans-Amazonian Highway is inaugurated.
1978	Congress passes an amendment ending Institutional Act #5.

1979	Congress passes an amnesty law for all political crimes, leading many exiles to return to Brazil.
1985	Tancredo Neves, Brazil's first elected civilian president since 1961, dies from an intestinal infection before he is able to assume office. José Sarney, his vice president, assumes office.
1986	In an effort to stem growing foreign and domestic debt, the *cruzado* replaces the *cruzeiro* as Brazil's currency.
1988	Brazil's seventh constitution, nicknamed the "Citizen's Constitution" for its detailed enumeration of individual rights, is ratified.
1989	Fernando Collor de Melo becomes Brazil's first directly elected president since 1960.
1992	Following up on accusations of political corruption, the Brazilian Senate impeaches President Fernando Collor de Melo. Vice president Itamar Franco assumes office.
1994	The Real Plan, instituted to curb rampant inflation, takes effect. After creating a new value system (the "Unit of Real Value"), it introduces a new currency, the *real*.
1997	An amendment to the Brazilian Constitution allows presidents to run for a second term.
1998	Fernando Henrique Cardoso completes his term in office, becoming the second president of Brazil since 1930 to do so. He is reelected for a second term.
2002	President Cardoso issues a decree to create affirmative action quotas in government agencies.
2002	Luiz Inácio Lula da Silva ("Lula") of the Workers' Party (Partido dos Trabalhadores [PT]) is elected as the 35th president of Brazil.
2003	The Bolsa Família (Family Allowance), Lula's hallmark social welfare policy, is debuted.
2010	Dilma Rousseff, Brazil's first female president and a fellow member of the PT, is inaugurated.
2013	Large street protests break out in major Brazilian cities in June. The protests began in São Paulo in response to proposed transportation fare hikes to help finance the 2014 World Cup, but soon evolve into protests against inefficiency and corruption in a variety of government sectors, including education and health care.
2014	Dilma Rousseff is reelected president of Brazil, narrowly defeating her adversary Aécio Neves.

COLONIZATION AND EARLY HISTORY

On April 22, 1500, a Portuguese fleet that had veered drastically off course on its way to India landed on the coast of the modern-day state of Bahia. The expedition's leader, Pedro Álvares Cabral, claimed the "Land of the True Cross" for the Portuguese Crown and sent the fleet's supply ship back to Lisbon to report on his discovery. Cabral and his men were probably not the first Europeans to visit the land. It is likely that Portuguese and Spanish navigators had explored parts of the South American coast prior to 1500 (thus helping determine placement of the Treaty of Tordesillas meridian line) and that Cabral planned the detour to assert Portugal's rights to the territory.

TREATY OF TORDESILLAS

The Treaty of Tordesillas was a 1494 treaty that revised an earlier papal bull dividing the non-European world between Spain and Portugal. The Treaty of Tordesillas drew a meridian line 370 leagues west of the Cape Verde Islands and established that all land east of the line belonged to Portugal, while all land west of the line belonged to Spain. The treaty ceded the yet-undiscovered Brazil to Portugal, though Brazil's borders with Spanish America remained in dispute for several centuries.

Regardless of whether Cabral's landing was accidental, April 1500 marked the beginning of a Portuguese presence in Brazil. After 10 days spent replenishing food-stuffs and trading with the Tupinambá Indians, Cabral left two *degredados* (criminal exiles) behind and his fleet weighed anchor for India. Though the letter to King Dom Manuel I written by Pero Vaz de Caminha, the scribe on board the expedition, described the fertile land and "handsome" indigenous people ripe for Christian evangelization, sparsely populated Portugal boasted few of the resources needed to found a settler colony in Brazil. As the Portuguese had done in Asia and Africa, they planned to establish *feitorias*, which were fortified trading posts that contained a warehouse, fort, church, and houses where Portuguese agents lived and conducted trade with the indigenous population for local goods, a strategy that promised maximum profits at a minimal human cost. Beginning in 1511, Portuguese merchants with royal contracts set up *feitorias* along the coast to trade with the Tupinambá for brazilwood, a tropical hardwood used to make a valuable reddish-purple dye. For nearly three decades after its "discovery," the Crown's primary interest in Brazil was as a way station for Portuguese fleets en route to India. Spanish and French incursions as well as diminishing returns from the hitherto more profitable Asian trade led the Crown to make a more pronounced effort to colonize Brazil. In 1534, Dom João III divided the Brazilian coast, from the Amazon River to the Río de la Plata, into 15 hereditary captaincies (*capitanias*) and distributed them to "loyal servants of the Crown," a group that included soldiers, merchants, bureaucrats, and lesser members of the Portuguese nobility. The expectation was that the recipients of these land grants (called *donatários*) would encourage Portuguese migration and stimulate economic activity in each region. Although the captaincy system did not flourish (only two captaincies, Pernambuco and São Vicente, were economically profitable), it accelerated sugar cultivation and led to the settlement of new coastal areas.

The land and labor that sugar production required further aggravated Portuguese relations with indigenous populations. At the time of the Portuguese arrival in 1500, an estimated 2–5 million indigenous people inhabited the land encompassing the modern nation-state of Brazil. Although European diseases killed many Amerindians, particularly those living along the coast, enslavement and mistreatment at the hands of the Portuguese probably killed many more. To resist enslavement or resettlement on a Jesuit mission, many Amerindians left coastal regions and headed to the interior. Royal laws declared indigenous slavery illegal in 1570, but exceptions were made for Amerindians who had been captured in a "just war." Brazilian settlers had begun to turn to African slavery, long in use on the Portuguese Atlantic islands, but African slaves were more expensive to purchase than *negros da terra* ("blacks of the earth," a term used in early colonial Brazil to refer to indigenous people). *Bandeirantes* led numerous expeditions into Brazil's interior in search of Amerindian captives, invoking just war on shaky legal grounds. From the Portuguese word for flag, "bandeira," the word *bandeirante* refers to small armed bands separate from a larger company. During the colonial period, *bandeirantes* explored Brazil's interior, searching for gold, runaway slaves, and Amerindian captives to sell as slaves. *Bandeirantes* helped ensure the survival of indigenous slavery well into the 18th century. When the captaincy system showed signs of struggling, the Portuguese Crown sent a royal governor, Tomé de Sousa, along with a largely male contingent of some 1,000 colonists, many of whom were *degredados*, and six Jesuits to the captaincy of Bahia in 1549. Sousa founded the city of Salvador, which served as the capital of the Portuguese colony of Brazil for more than 200 years.

Throughout the 16th and 17th centuries, the French, the Spanish, and the Dutch challenged Portuguese sovereignty over Brazil. The Treaty of Tordesillas, signed in 1494, drew a north–south line through the yet-largely unexplored lands of the "New World" and divided them between the Spanish and the Portuguese. Portugal got all land east of the line, which was some 370 leagues west of the Cape Verde Islands, thus placing Brazil in Portuguese hands. But neither the Spanish nor any other European power placed much stock in the line. As early as 1504, French ships landed on the Brazilian coast to participate in the brazilwood trade. In 1550, French Huguenots escaping religious persecution founded a Protestant settlement on an island in the Bay of Guanabara that they named "France Antarctique." An expedition led by Mem de Sá pushed the French from the area in 1560 and, a few years later, the city of Rio de Janeiro was founded on the site of the former French colony. However, a second French settlement on the northern coast of Maranhão, "France Equinoctal," remained until 1615. The basin of the Río de la Plata was one of several sites of contestation between the Portuguese and the Spanish. The present political borders separating Brazil from its South American neighbors were not resolved until the 20th century, and many remain unclear.

The Netherlands was an important market for Portuguese commodities, particularly Brazilian sugar. When the Portuguese were forced to end trade relations with the Netherlands as a consequence of Portugal's union with the Spanish Crown (1580–1640), the Dutch West India Company sponsored an invasion of the northeastern coast of Brazil. The Dutch occupation of Bahia lasted just one year (1624–1625), but in 1630 the Dutch launched a second, successful invasion of the

A slave market in 19th-century Pernambuco, as drawn by Augustus Earle and engraved by Edward Finden in 1834. Though enslaved Africans worked in a variety of occupations, the demand for African slaves was constant in sugar-growing regions of Brazil, where harsh plantation slavery produced high mortality rates among laborers. (Library of Congress)

leading sugar-producing captaincy of Pernambuco, from which they ruled from the capital city of Recife for nearly 25 years. The blows to Brazil's sugar economy were magnified by contemporaneous Dutch attacks on Portuguese centers for the slave trade on the African coast. Though Portuguese forces, aided by a significant contingent of indigenous and African troops, expelled the Dutch from Angola in 1648 and from Brazil in 1654, the Brazilian sugar industry would never recover its former glory. In the quarter century since the Dutch occupied Pernambuco, the Dutch, along with the English, had set up thriving sugar colonies in the Caribbean. The abundance of sugar in global markets drove down the price of Brazilian sugar and ousted smaller producers.

THE MATURE COLONY

In the late 17th and early 18th centuries, the discovery of gold and diamond deposits in what is today the state of Minas Gerais, as well as limited gold reserves in Mato Grosso and Goiás, drastically altered Brazil's demography. Droves of people left coastal regions for the interior of the country, joined by some 4,000 Portuguese immigrants and thousands of African slaves each year. By the second half of the 18th century, Minas Gerais not only was Brazil's most populous captaincy but also had the colony's largest enslaved population and its largest population of free people of color. Few could aspire to the fabulous wealth of those who held gold and diamond contracts; nevertheless, the region developed a mixed economy that supported urbanization. Though the gold boom ended once the placer deposits had been exhausted in the 1760s, the move of the colonial capital from Salvador to Rio de Janeiro in 1763 underscored the shift of economic, political, and social power from the north of the colony to the south.

Gold and diamonds provided the capital that the Crown needed to pay for the goods Portugal and its colonies could not produce. But when mineral production began to decline in the second half of the 18th century, Portugal no longer had the ability to settle its foreign debts. Sebastião José de Carvalho (the "Marquis of Pombal"), the de facto head of Portugal from 1750 to 1777, enacted a variety of policies throughout the Portuguese Empire to increase royal revenue. In Brazil, these included expelling the Jesuits (so that the Crown could control Amerindian labor), settling Brazil's frontiers with Spanish America, and more efficiently taxing Brazilian products. This increased Crown vigilance generated some resentment on the part of colonists, but after Pombal's fall from power in 1777, Portuguese oversight relaxed.

Two late-18th-century events have traditionally been pointed out to highlight settlers' discontent with colonial rule. Although Brazil did not have any universities and the Catholic Church dominated education, colonial elites who studied in Portugal were exposed to Enlightenment ideas. In 1788 and 1789, residents of Ouro Preto in Minas Gerais plotted to assassinate the governor and declare independence from Portugal in response to a proposed tax hike. The chief conspirators of the so-called Inconfidência Mineira, nearly all wealthy white elites with the exception of José Joaquim da Silva Xavier (also known as Tiradentes, or "teeth-puller" for his sometimes dentistry work), were discovered before their plan could come to fruition. The plotters were banished to Angola, save Tiradentes, who was hanged, drawn, and quartered on April 21, 1792.

An independence plot known as the "Tailor's Revolt" developed 10 years later in Salvador better illustrates how Enlightenment ideas and news of Atlantic revolutions spread throughout the colony. In 1798, a group of tailors and artisans in the port city were inspired by the French and Haitian Revolutions to proclaim Bahian independence. Unlike the Minas conspirators, this group (a majority of which was nonwhite and included slaves) added social and racial equality to their demands for a "democratic, free, and independent" government. Like the Inconfidência Mineira six years earlier, the plot was discovered and the chief conspirators were punished by death or exile. Although neither the Tailor's Revolt nor the Inconfidência Mineira heralded the "coming of independence," they reveal how colonists participated in the political and intellectual currents of the time.

The relocation of the Portuguese court to Rio de Janeiro transformed colonists' relationship to the Crown. Fleeing Napoleon's invasion of the Iberian Peninsula, the Portuguese prince regent Dom João and a fleet of ships containing the royal family, advisors, court members, religious leaders, important manuscripts, and part of the royal treasury set sail for Brazil in late 1807. One of the first actions of the court in its new home was to open Brazilian ports to foreign trade and end prohibitions on manufacturing in the colony. In addition to establishing the Bank of Brazil, a national library, and institutions of higher learning, Dom João brought Brazil's first printing press. The court's transfer helped create scores of middle-sector jobs in public services, transportation, and construction. In little more than a decade, Rio de Janeiro's population doubled.

Once Napoleon was driven from the Iberian Peninsula in 1814, the Portuguese urged the court to return to Portugal. Ignoring the calls for his homecoming, Prince João created the United Kingdom of Portugal, Brazil, and the Algarves in 1815. The elevation of Brazil to the status of a "kingdom" made the colony Portugal's equal and justified the monarch's continued presence in the Americas. In 1820, liberal reformers in Porto, Portugal, launched a political revolution, demanding a new constitution and the end of absolute monarchy. The new constitution was to be drawn up by a representative body (*Cortes*) that would meet in Lisbon and include representatives from Portugal and its empire. The *Cortes* convened in January 1821 without any representatives from Brazil and decided that Dom João (now King Dom João VI) must return to Portugal. In April 1821, the king departed for Lisbon. Prince regent Pedro stayed behind in Brazil, possibly with instructions from his father to collaborate with Brazilians seeking independence, and refused a series of Portuguese orders that he also returned. On September 7, 1822, Pedro rejected a final decree to return to Portugal and proclaimed Brazilian independence. On December 1, 1822, Pedro was crowned emperor of Brazil.

THE BRAZILIAN EMPIRE (1822–1889)

Given Brazil's nearly bloodless break with Portugal, relative political stability, retention of the monarchy, and the persistence of slavery, it is tempting to see post-colonial Brazil as fundamentally conservative, particularly when compared with the rest of Latin America. Such a characterization belies both the new political consciousnesses that came to the fore and the new interest groups that emerged during the empire.

The 1824 Constitution created a bicameral legislature that consisted of a Senate, whose members were appointed by the emperor to serve life terms, and a popularly elected Chamber of Deputies. The emperor had the power to veto legislation and dissolve the legislature. The franchise was limited to men over age 25 who met a certain income requirement, but Brazil's 1824 Constitution was unique among American constitutions of the era in that manumitted Brazilian-born slaves who met the income requirements were considered citizens and could exercise political rights. Independence also helped loosen planters' monopoly on the military, shifting the militia toward officers of color.

Although these small, yet significant concessions helped defuse potential racial conflict, imperial Brazil had the largest slave population in the Americas. Over the course of the Atlantic slave trade, over 4 million Africans were brought to Brazil as slaves. Given the economy's continued dependence on slave labor, now directed toward the production of coffee, indigo, and rice, the abolition of slavery proceeded slowly. In 1826, Great Britain, Brazil's largest trading partner, pressured Brazil into signing a treaty to abolish the African slave trade within a few years, largely because it feared slave-produced Brazilian sugar would have an advantage over Caribbean sugar on the world market. The provisions of the treaty were mostly ignored, and slave imports to Brazil increased to record levels. Slave rebellions like the 1835 Malé Revolt in Salvador, student activism, and the outbreak of disease traced to over-crowded slave ships all goaded Brazilian legislators into passing a series of laws that provided for slavery's gradual end. The War of the Triple Alliance (1865–1870), also known as the Paraguayan War, was crucial in generating popular support for abolition. Although the Río de la Plata basin had long been a site of contention be-tween Spanish and Portuguese America (and the subsequent independent nations of Paraguay, Uruguay, Argentina, and Brazil), the conflict began when a Paraguayan warship seized a Brazilian steamship on the Paraná River in November 1864. Soon after, Paraguayan troops invaded the province of Mato Grosso. Brazil, Argentina, and Uruguay formed an alliance to counter the Paraguayan dictator, Francisco So-lano López. The internecine conflict, characterized by guerrilla fighting mostly on Paraguayan soil, lasted five years and resulted in massive casualties on all sides. The Brazilian military depended heavily on slave soldiers, who were promised freedom in exchange for their service. Following the war's end, many soldiers questioned their obligation to capture runaway slaves. Moreover, Brazilian soldiers' ruthless fighting against the much smaller and more poorly equipped Paraguayan forces generated bad international publicity for Brazil, which was the only American territory (besides Cuba) in which slavery was legal. In 1871, the Brazilian legislature passed the Lei do Ventre Livre (Free Womb Law), which emancipated all children born to slave mothers, while an 1885 law freed slaves over the age of 60. Princess Isabel signed the Lei Aurea, or "Golden Law," abolishing slavery in 1888. Brazil was the last country in the Americas to abolish slavery.

In addition to highlighting the racial dimensions of Brazilian society, the War of the Triple Alliance ushered in a new era in which military officers played an important role in Brazilian politics. Support for Pedro II waned during the pro-tracted conflict, particularly when he refused the United States' 1867 offer to negotiate a peace settlement. In 1868, Pedro further alienated his opponents by dismissing the Liberal cabinet and calling for a Conservative government, even though Liberals held the elected majority in the Chamber of Deputies. Pedro's dissenters formed the Republican Party in 1871 and called for an end to the monarchy. Although Republicans were in the political minority, they enjoyed particular support from junior military officers. On November 15, 1889, a group of junior officers convinced their commander, Marshal Manuel Deodoro da Fon-seca, to lead a coup against the emperor. On November 16, 1889, Brazil became a free republic.

THE FIRST REPUBLIC (1889–1930)

The republican constitution of 1891 created a highly decentralized federalist system that granted broad powers to individual states. Each state had its own government and legislature and was responsible for maintaining its own militia, raising funds by taxing exports to other states, and contracting foreign loans. A directly elected president replaced the emperor. Suffrage was limited to literate males, who, prior to 1930, comprised just 3.5 percent of the population. Under such a decentralized system, the wealthiest states profited at the expense of the poorer states. Politicians from the chief coffee-producing states of Rio de Janeiro and São Paulo and the dairy-producing state of Minas Gerais formed a political coalition known as *café com leite* ("coffee with milk") that dominated national politics, agreeing to alternate the presidency between them.

Republicans sought to "modernize" Brazil in several ways. They believed whitening the population would help align Brazil with the industrialized Western nations they admired, such as Argentina, France, and the United States. The government as well as private industries encouraged immigration from Western Europe, the Middle East, and Japan: between 1870 and 1930, an estimated 2–3 million immigrants arrived in Brazil. Today, Brazil is home to the largest population of Japanese immigrants and Japanese descendants outside of Japan as well as one of Latin America's largest Jewish populations. Industrialists insisted that labor shortages in the southern states justified massive immigration, even though Brazil already had a large reserve of laborers—thousands of former slaves and their descendants—in the north and center of the country.

In the capital city of Rio de Janeiro, "modernization" entailed the demolition of several hundred downtown structures to make way for wide avenues and new public buildings like an opera house and a national library. Many of these buildings had provided housing for lower-income families, and their destruction gave birth to the first *favelas* (slums). In the early 1900s, Dr. Oswaldo Cruz spearheaded a massive public health campaign to eradicate endemic diseases like yellow fever, bubonic plague, and smallpox by exterminating mosquitoes, improving the city's trash collection and sewage systems, and making vaccines for smallpox compulsory. When news of the vaccine legislation, inflated by rumors, spread, a large riot known as the "Vaccine Revolt" broke out in the center of Rio in November 1904. Protesters were not so much opposed to the mandatory vaccinations as they were to the authoritarian measures used by public health inspectors, police, and politicians to "clean up" the city. Officials entered private homes at will for health inspection and destroyed many they deemed unsanitary, without providing an alternative home for the displaced residents. The weeklong confrontation with the police and military resulted in many deaths, arrests, and the deportation of those involved.

Brazilians living in rural areas also challenged elite's visions for Brazil. In the 1890s, thousands of impoverished, largely mixed-race followers of a religious mystic known as Antônio Conselheiro left their homes for a small village in the Bahian interior called Canudos. Conselheiro preached opposition to the Republic, and when several Canudos residents clashed with and killed local police following a dispute over a shipment of wood, and state militia failed to suppress the community, in 1897 government officials in Bahia and Rio de Janeiro sent in federal troops with orders to destroy Canudos and its monarchist residents. The "Canudos War" lasted for months and ultimately killed thousands of troops and civilians.

Hoping to bolster Brazil's international image, Republicans supported Brazil's entry into World War I on the Allied side. Though Brazil was the only Latin American nation that did not remain neutral during the conflict, its participation did little to change Brazil's role in world affairs. Meanwhile, critiques of Republicans' policies grew in various arenas. Junior army officers (*tenentes*) denounced the Republicans' failure to modernize the military or provide social legislation and staged barracks revolts on military bases across Brazil. The most famous such revolt was led by Captain Luís Carlos Prestes in Rio Grande do Sul. Beginning in 1924, the "Prestes Column" marched through the interior of Brazil for three years. A new generation of Brazilian artists and intellectuals challenged Republicans' valorization of white European/U.S. culture above all. The so-called *modernistas* celebrated African and Indian contributions to Brazilian culture, debuting their vision of *brasilidade* (Brazilianness) on a national stage at the Modern Art Week festival held in São Paulo in February 1922.

THE VARGAS ERA (1930–1945) AND THE SECOND REPUBLIC

When world markets collapsed in 1929, Brazil's coffee-dependent economy was plunged into crisis. Between 1929 and 1931, the price of coffee dropped from 22.5 cents a pound to 8 cents a pound (Skidmore 2009, p. 97). In the 1929 presidential elections, Júlio Prestes, the governor of São Paulo, defeated the governor of Rio Grande do Sul, Getúlio Vargas. The murder of Vargas's running mate, João Pessoa, though unrelated to the election outcome, helped Vargas assemble a coalition that counted on the support of *tenentes*, disgruntled coffee planters, and political leaders from Rio Grande do Sul and Minas Gerais who were upset that sitting president Washington Luís had violated the *café com leite* agreement by choosing a fellow politician from São Paulo (Prestes) as his successor. As Vargas and his supporters headed for Rio de Janeiro, Washington Luís went into exile and a military junta temporarily seized power. When Vargas and his supporters arrived in Rio, the junta peacefully handed over power to Vargas. Vargas worked quickly to centralize Brazilian politics by replacing state governors with appointed officials and dissolving the Congress. But Vargas left major political institutions, such as the bicameral legislature, intact and even extended the right to vote to women. Political elites in São Paulo mounted a revolution in July 1932 but were suppressed by the federal army by October. To appease the rebels, Vargas agreed to hold elections for a Constituent Assembly, which would write a new Brazilian Constitution. The Constitution of 1934 made the government the guarantor of social welfare and economic growth. The assembly also elected Vargas to a four-year term as president.

In 1935, Vargas seized his opportunity to take direct control of the Brazilian state. During the 1930s, two national ideological parties had been growing in influence—the National Liberation Alliance (ALN) and the Ação Integralista Brasileira (AIB, Brazilian Integralist Action). The ALN, created in 1935, was a coalition of communists, socialists, and other members of the Brazilian left headed by Luís Carlos Prestes (of

General Getúlio Vargas, head of the military junta and provisional president of Brazil (seated, right), arrives in Rio de Janeiro to take charge of the new government on November 19, 1930. (Bettmann/Corbis)

"Prestes Column" fame) and directed by the Comintern in Moscow. The AIB was a paramilitary organization that provided Brazil's answer to European fascism, and enjoyed the support of some members of the clergy as well as among some of the middle and upper classes. The ALN and AIB clashed openly in the streets of Brazil's major cities throughout the 1930s, but the ALN bore the brunt of government repression, and its leaders were often jailed. In 1935, Prestes returned from a lengthy sojourn in Moscow with a Comintern-formulated plan to overthrow the Brazilian government. Responding to the new directive, communist officers on military bases in Rio de Janeiro, Recife, and Natal staged barracks revolts in November 1935. The revolts were quickly quashed, and Vargas used the 90-day "state of emergency" he was given by Congress to suspend civil liberties and arrest, imprison, and torture leaders of leftist movements.

For the next two years, the Brazilian Congress renewed the state of emergency. In November, Vargas and his administrators fabricated evidence of a new communist threat and staged their own coup. On November 10, 1937, Vargas read the constitution for Brazil's "Estado Novo" ("New State") on the radio. The new constitution converted the Vargas presidency into a dictatorship. Vargas claimed all political power for himself, disbanded the Congress, and canceled the scheduled 1938 presidential election. The ritual burning of state flags a few weeks later symbolized Brazil's turn away from federalism. The AIB had hoped to gain from Vargas's turn to the right, but when it staged its own coup in March 1938, Vargas responded by banning all paramilitary organizations. Vargas had thus suppressed his opposition on both the political left and right.

The Estado Novo blended fascist with populist political strategies. All media were censored, while the police force readily deployed violence against the regime's detractors. Yet Vargas also worked to cultivate an image of himself as "Father of the Poor," instituting a host of social welfare programs for urban workers. He validated the authentic "Brazilian" culture that the 1920s *modernistas* propagated and used government money to patronize the national soccer team as well as Rio's Carnival celebrations and samba schools.

As World War II brewed, Vargas courted both Allied and Axis support. Brazil ultimately entered the war in 1942 on the side of the Allies following Nazi attacks on Brazilian shipping lines and sent 25,000 combat troops to Italy. Brazil grew closer to the United States, supplying rubber, quartz, and other raw materials as well as offering its coastline for naval and air bases in exchange for U.S. military equipment, training, and financing for a Brazilian steel mill at Volta Redonda. The United States took full advantage of the war to increase goodwill toward the United States in Brazil (and the rest of Latin America) by sending "cultural missions" of U.S. writers, artists, and actors to Brazil. Once the war ended, U.S. intervention in Brazilian affairs only increased as it sought to create a bulwark against communism in Latin America.

Vargas used the war as an excuse for delaying the elections scheduled to take place in 1943, but once the war ended in 1945 he could no longer ignore the calls for democratic elections from within Brazil. A new political party, the União Democrática Nacional (UDN, National Democratic Union), was organized in opposition to the Vargas regime, comprising liberal constitutionalists and some former members of the PCB. Vargas also supported the creation of two new political parties, the Partido Social Democrático (PSD, Social Democratic Party) and the Partido Trabalhista Brasileiro (Brazilian Labor Party). As the election drew near, Vargas did not declare whether he was going to run. The army, with UDN and U.S. backing, forced Vargas's resignation in October 1945, and democratic elections were held with record popular participation. The winner was Eurico Dutra of the PSD, who had served as minister of war under Vargas. A new constitution was promulgated yet again. But Dutra soon resorted to the repressionist techniques of Vargas (with U.S. support) in the wake of massive workers' strikes in major Brazilian cities. In 1947, Dutra outlawed the Communist Party, which had received a significant number of votes in the 1945 elections, and ousted its members from the Congress.

In 1950, Vargas returned to office, this time as a democratically elected president. He sought to recommence the economic development measures he had left unfinished, including expanding urban labor unionization and nationalizing the oil and electricity industries. Following nearly two years of heated debate, congress created a state oil corporation, Petróleo Brasileiro (Petrobras). Yet by 1954, scandals and accusations of corruption plagued Vargas's administration. When an assassination attempt on a prominent anti-Vargas journalist, Carlos Lacerda, was traced back to Vargas's bodyguard/chauffeur, his detractors had a field day. On August 24, 1954, Vargas committed suicide. Following his death, public opinion swung decisively pro-Vargas.

Elected in October 1955, President Juscelino Kubitschek de Oliveira launched a plan for a new, modernist Brazilian capital city in the center of the country, to be designed by Oscar Niemeyer and Lúcio Costa. The city of Brasília was constructed on a desert plateau from the ground up in about four years, echoing Kubitschek's campaign slogan of "fifty years of progress in five." Construction of the new capital bankrupted the state and worsened inflation. Kubitschek's successor in the 1960 elections, Jânio Quadros, lasted just seven months in office before resigning. Quadros had hoped that threatening to resign would encourage Congress to grant him emergency powers. Much to his surprise, Congress readily accepted his resignation.

Vargas's former labor minister and Quadros's vice president, João Goulart, took office in 1961. The Brazilian right, including the military, as well as the U.S. government, believed that Goulart was a dangerous communist radical, a view seemingly given credibility by his presence in the People's Republic of China during Quadros's resignation. As Goulart returned to Brazil, Congress docked his presidential authority by creating a parliamentary system with a prime minister (a 1963 plebiscite restored the presidential system). Goulart's plans to unionize long-forgotten rural workers in 1963 agitated landowners, who joined the voices of opposition to Goulart. Beginning on March 31, 1964, military units occupied government offices in Rio de Janeiro and Brasília. Within 24 hours, the U.S. government had recognized the new military regime and Goulart had fled to his native state of Rio Grande do Sul.

MILITARY DICTATORSHIP (1964–1985)

Although every leader of Brazil from 1964 to 1985 was a four-star general, the military dictatorship was hardly homogeneous. From the outset of the coup, the military was divided over the direction Brazilian politics should take. Hard-liners felt that Brazil was not ready for democracy and that the Brazilian government needed a complete overhaul. A more moderate wing viewed military rule as a temporary measure, believing that Brazil would be ready for democracy after some economic and administrative reorganization. Soon after taking power in April 1964, military ministers issued an "Institutional Act" that gave the president the power to suppress political rights and amend the Constitution. The military began to get rid of all traces of the left as well as thousands of civil servants. A second Institutional Act, issued by the administration of General Humberto Castelo Branco, created two new political parties, the pro-government Aliança Renovadora Nacional (National Renewal Alliance) and the "opposition" Movimento Democrático Brasileiro (MDB, Brazilian Democratic Movement).

In the wake of Castelo Branco's death in a plane crash, industrial strikes, and massive student protests throughout Brazil in 1968, hard-liners within the military seized power, passing Institutional Act 5 in December 1968. Under this act, labor unions were banned; subversive activity was monitored (and punished) by the Departamento da Ordem e Estado; political opponents were arrested, tortured, and "disappeared"; and many university faculty were fired. Military repression did not go unopposed, and guerrilla organizations that counted on significant participation from Brazilian university students secured the release of a number of political prisoners by kidnapping prominent public officials, particularly foreign diplomats. One of them was US Ambassador Charles Burke Elbrick, who was kidnapped in 1969. Eventually, however, military officials infiltrated such guerrilla organizations and rendered them ineffective.

The "opening" to democracy began in 1974–1975, when the more moderate General Ernesto Geisel assumed the presidency. Geisel eased up censorship and overturned a campaign law that had blocked the opposition party from television access. The opposition party, the MDB, won major victories. Further easing of the dictatorship came in 1979, when Congress passed an amnesty law that applied to all political crimes. Artists, politicians, and activists who had gone into exile during the military regime began to return to Brazil. In this more open political climate, in 1985, millions mobilized in support of direct presidential elections. Though the government-controlled Congress upheld indirect elections, the democratic outlook seemed brighter when Congress elected Brazil's first civilian leader in over two decades, MDB candidate Tancredo Neves, to the presidency. But on March 15, 1985, the day he was to be inaugurated in Brasília, Neves succumbed to an intestinal infection. Vice president José Sarney was sworn in as the provisional president while Neves underwent treatment. Following multiple surgeries in Brasília and São Paulo, Neves died on April 21, 1985. The tasks of overseeing Brazil's transition to democracy and controlling its ever-worsening inflation would fall on José Sarney.

DEMOCRATIC BRAZIL (1985–PRESENT)

In the years following the end of the military dictatorship, poor public services, weak infrastructure, and rising levels of violent crime in major cities—all problems tied to the persistence of social and economic inequality and worsened by political corruption—mediated the impact of the democratic transition on the lives of ordinary Brazilians. Through the adversities of the past three decades, the commitment of the Brazilian people to democracy and a respect for the rule of law have ensured fluid transitions of power even in the worst of economic times.

Aware of the monumental tasks that lay before him, upon taking office, José Sarney secured a congressional mandate to extend his term by one year (from four years to five years). Tackling Brazil's annual inflation rate of nearly 300 percent was a priority for the Sarney administration. In February 1986, government economists rolled out the "Cruzado Plan." The plan replaced Brazil's currency, the *cruzeiro*, with the *cruzado*, while adjusting prices, exchange rates, and wages. The Cruzado Plan initially achieved a modest reduction in inflation, but, by 1988, inflation skyrocketed to record levels.

As economic troubles awaited a more effective solution, a Constituent Assembly began meeting in February 1987 to draft Brazil's new constitution. Following some 19 months of deliberations, on October 5, 1988, Congress ratified Brazil's seventh constitution. Nicknamed the "Citizen's Constitution" for its detailed enumeration of individual rights, the 1988 Constitution reflected the lobbying efforts of countless interest groups and institutions, including labor unions, the Catholic Church, and human rights activists. The Constitution guaranteed direct, secret elections for public office and reduced the minimum voting age from 18 to 16 years. In addition, the Constitution opened numerous avenues for popular participation in the political process, whether through referenda, plebiscites, or proposing new laws. Beyond ensuring political rights, the Constitution protected civil liberties that had been revoked or severely curtailed during the dictatorship, including the right to strike, the abolition of censorship, and guaranteed job tenure for federal civil servants. Significantly, the Constitution classified racism as a crime without the possibility for bail, paving the way for further antidiscrimination legislation at the federal and state levels.

The end of the dictatorship and the commitment to a democratic society called for revitalized political parties. During the military regime, the Brazilian left, particularly the PCB and militant guerrilla groups, had all but disappeared as major political actors. New parties filled the void left by these older parties, of which the Partido dos Trabalhadores ("Workers' Party," or PT), formed in São Paulo in the late 1970s, was among the most dynamic. The PT drew its strongest support from middle-class professionals, teachers, and government workers, steadily increasing its representation in Congress throughout the 1980s. In the 1989 presidential elections, the PT put forth one of the top contenders for office, Luiz Inácio Lula da Silva, or "Lula." Lula's personal trajectory was markedly different from those of the elites who had governed Brazil since the founding of the republic. The son of sharecroppers from the Brazilian northeast, Lula had toiled in factories for several decades and rose to fame as the leader of an autoworker's strike in the late 1970s. Lula seemed an ideal representative of Brazil's democratic aspirations. Lula's chief opponent was Fernando Collor de Mello, the telegenic governor of the state of Alagoas who hailed from an elite background. Collor narrowly defeated Lula, aided by his monopolization of television airtime and the apparent success of his accusations that Lula was a dangerous radical.

As Brazil's first directly elected president since 1960, Collor promised to crack down on political corruption and "bring Brazil into the first world" via a host of neoliberal economic policies that included privatizing many state-owned industries and lowering tariffs. Collor promulgated many of his less-popular policies by submitting presidential decrees (*medidas provisórias*, similar to the executive orders wielded by U.S. presidents) rather than negotiating with Congress. If Congress did not ratify his *medidas provisórias* within 30 days, Collor reissued them, a practice that led his detractors to accuse him of abusing presidential power. Collor and his administration were soon mired in their own corruption scandals. A congressional investigation revealed the existence of a multimillion-dollar influence trafficking ring. Huge crowds took to the streets to demand Collor's impeachment. On December 29, 1992, a few hours before the release of the Senate's impeachment vote, Collor resigned from office. He was found guilty of corruption and banned from politics until the year 2000 (he has since returned to the political scene). For the second time in less than a decade, the vice president assumed the presidency.

Itamar Franco, a former senator from Minas Gerais, was relatively unknown on the national stage prior to taking over for Collor. Franco did not support most of the neoliberal economic policies Collor had pursued, but was at a loss for a solution to the ever-worsening inflation crisis. Franco's presidency and Brazil's economy were salvaged by Franco's appointment of Fernando Henrique Cardoso, a sociologist who had been serving in the Ministry of Foreign Relations, to head the Ministry of Finance in May 1993. Along with a team of economists from the Pontifical Catholic University of Rio de Janeiro (PUC-Rio), Cardoso drafted a new plan to tame inflation, the Plano Real. The first step of the Plano Real, launched in March 1994, involved the conversion of all prices to the "Unit of Real Value," an index calculated by Brazil's Central Bank that corresponded to the value of the U.S. dollar. Once Brazilian consumers had had time to adjust to the new system and (so the plan's architects hoped) regained confidence in the market, a new currency, the *real*, was introduced on July 1, 1994. Other aspects of the Plano Real involved balancing the federal budget, raising taxes, and further privatizing industries. Franco, like his predecessor, made liberal use of the *medida provisória* to push forth some of the plan's less popular measures. By all accounts, the Plano Real was a success—by December 1995, inflation had slowed to just 1 percent to 2 percent per month.

The 1994 presidential elections pitted Cardoso, the candidate for the center-right Partido da Frente Liberal (Liberal Front Party, known as "Democratas" since 2007) against Lula. Bolstered by the fortune of the Plano Real, Cardoso handily won the election, though the PT also gained many congressional seats.

Much of Cardoso's term was spent securing the economic stability ushered in by the Plano Real, and as a result, much-needed reforms in education, health care, and other public services were, for the most part, placed on the back burner. In January 1997, Cardoso secured a constitutional amendment to allow for consecutive presidential terms, enabling him to run for president in 1998. Once again running against Lula, Cardoso was elected to a second term in November 1998.

During Cardoso's second term, his administration devoted greater attention to issues of social welfare. In 1999, Cardoso launched a program to provide free medical treatment to HIV-positive Brazilians; his minister of health, José Serra, spearheaded a campaign that succeeded in breaking the patents on many medications considered essential to public health. Cardoso's administration also brought race relations to the forefront of Brazilian politics, an issue he had written about as a university professor. Early in his first term, Cardoso made a public statement acknowledging the existence of racism in Brazil and convened an Advisory Council on Race Questions. In 2002, Cardoso issued a decree to establish quotas for the recruitment of Afro-Brazilians, women, and other underrepresented groups in government agencies and entities operating with government contracts. Bolstered by a more visible black movement and international support, affirmative action programs have since spread to the majority of Brazil's federal and state universities. The Cardoso years saw major changes in the conversation about race among white Brazilian elites, who for generations had denied the existence of racism in Brazil simply because there were no legal barriers to one's access to education or jobs as in the pre-1964 United States.

If racial issues began to be addressed, class remained a significant predictor of one's access to education, future job prospects, and life expectancy. Lula, elected president by a landslide on his fourth try in 2002, was the first president since 1964 to espouse an openly socialist agenda. Lula's candidacy had raised concern among some international observers, who feared that his "radical" politics would drive Brazil into bankruptcy. In his 2002 "Letter to the Brazilian People," Lula pledged that, if elected, he would tackle the PT agenda of reducing socioeconomic inequality while pursuing fiscally responsible policies and honoring the International Monetary Fund agreements signed during Cardoso's administration. By some accounts, Lula's economic policies were even more conservative than those of his predecessor, reflecting how the PT had shifted toward the center of the Brazilian political spectrum to appeal to a wider audience.

One of Lula's major domestic policy initiatives was the Bolsa Família, introduced in 2003. The Bolsa Família was designed to eliminate extreme poverty by providing cash grants to families living below the poverty line. The receipt of funds was conditional on the basis that children attend school and receive required vaccinations. In its first decade, the Bolsa Família brought some 36 million Brazilians out of extreme poverty and increased school graduation rates in Brazil's poorest states. The program is also said to have helped empower women, the recipients of the checks, by providing them with a guaranteed income. The success of the Bolsa Família has inspired the creation of similar programs elsewhere in Latin America. Lula left office in 2010 with an approval rating of nearly 80 percent, the highest of any president in Brazilian history.

Lula's fellow PT member and successor in office, Dilma Rousseff, has continued his commitment to eradicating poverty. Rousseff's "Brasil sem Miséria" ("Brazil without Poverty") program, debuted in 2011, expanded many aspects of the Bolsa Família. Though a growing economy and social welfare programs have helped many Brazilians attain a better standard of living in recent years, many feel that it is not enough. Government spending on the 2014 World Cup and planned 2016 Olympics has generated particular resentment. During the Confederations Cup, a soccer tournament held in Brazil in June 2013, street protests broke out in São Paulo in response to a bus fare hike instituted to help finance the World Cup. Though the fare increases were reversed, citizens in many Brazilian cities took to the streets to protest inefficiency and corruption in a variety of government sectors. Following the publicization of various corruption scandals within her administration and slowing economic growth, in October of 2014, Dilma was narrowly reelected in a run-off election that pitted her against Brazilian senator and former governor of Minas Gerais, Aécio Neves.

Despite the challenges of the past three decades, Brazil has emerged as a vibrant democracy, with a stronger economy, improved public services, and commitment to building a more inclusive society. Credit for these triumphs is due in large part to the Brazilian people, who have not been content with the status quo, but rather have worked tirelessly to realize change.

REFERENCES

Alberto, Paulina L. 2011. *Terms of Inclusion: Black Intellectuals in Twentieth-Century Brazil.* Chapel Hill, NC: University of North Carolina Press.

Beattie, Peter M. 2001. *The Tribute of Blood: Army, Honor, Race, and Nation in Brazil, 1864–1945.* Durham, NC: Duke University Press.

Bethell, Leslie, ed. 1987. *Colonial Brazil*. Cambridge: Cambridge University Press.

Carvalho, José Murilho de. 2012. *The Formation of Souls: Imagery of the Republic in Brazil*. Translated by Clifford E. Landers. Notre Dame, IN: University of Notre Dame Press.

Dávila, Jerry. 2003. *Diploma of Whiteness: Race and Social Policy in Brazil, 1917–1945*. Durham, NC: Duke University Press.

Eakin, Marshall C. 1997. *Brazil: The Once and Future Country*. New York: St. Martin's Press.

Fausto, Boris, and Sergio Fausto. 2014. *A Concise History of Brazil*, 2nd ed. Cambridge: Cambridge University Press.

Ferreira Furtado, Júnia. 2008. *Chica da Silva: A Brazilian Slave of the Eighteenth Century*. Cambridge: Cambridge University Press.

Frank, Zephyr L. 2004. *Dutra's World: Wealth and Family in Nineteenth-Century Rio de Janeiro*. Albuquerque: University of New Mexico Press.

Haddad, Mônica A. 2009. "A Spatial Analysis of Bolsa Família: Is Allocation Targeting the Needy?" In Joseph L. Love and Werner Baer (eds.), *Brazil under Lula: Economy, Politics, and Society under the Worker-President*, pp. 187–203. New York: Palgrave Macmillan.

Hemming, John. 1978. *Red Gold: The Conquest of the Brazilian Indians*. Cambridge, MA: Harvard University Press.

Klein, Herbert S., and Francisco Vidal Luna. 2009. *Slavery in Brazil*. Cambridge: Cambridge University Press.

Langland, Victoria. 2013. *Speaking of Flowers: Student Movements and the Making and Remembering of 1968 in Military Brazil*. Durham, NC: Duke University Press.

Lesser, Jeffrey. 2013. *Immigration, Ethnicity, and National Identity in Brazil, 1808 to the Present*. Cambridge: Cambridge University Press.

Levine, Robert M. 1998. *Father of the Poor? Vargas and His Era*. Cambridge: Cambridge University Press.

Lockhart, James, and Stuart B. Schwartz. 1983. *Early Latin America: A History of Colonial Spanish America and Brazil*. Cambridge: Cambridge University Press.

Maxwell, Kenneth E. 2004. *Conflicts and Conspiracies: Brazil and Portugal, 1750–1808*. New York: Routledge. First published 1973 by Cambridge University Press.

Metcalf, Allida C. 2005. *Go-Betweens and the Colonization of Brazil, 1500–1600*. Austin, TX: University of Texas Press.

Monteiro, John M. 1999. "The Crises and Transformations of Invaded Societies: Coastal Brazil in the Sixteenth Century." In Frank Saloman and Stuart B. Schwartz (eds.), *The Cambridge History of the Native Peoples of the Americas, Volume III: South America, Part I*, pp. 973–1024. Cambridge: Cambridge University Press.

Moritz Schwartz, Lilia. 1999. *The Spectacle of the Races: Scientists, Institutions, and the Race Question in Brazil, 1870–1930*. Translated by Lilia Guyer. New York: Hill and Wang.

Russell-Wood, A. J. R. 1998. *The Portuguese Empire, 1415–1808: A World on the Move*. Baltimore, MD: Johns Hopkins University Press.

Schultz, Kirsten. 2001. *Tropical Versailles: Empire, Monarchy, and the Portuguese Royal Court in Rio de Janeiro, 1808–1821*. New York: Routledge.

Schwartz, Stuart B. 1986. *Sugar Plantations in the Formation of Brazilian Society: Bahia, 1550–1835*. Cambridge: Cambridge University Press.

Skidmore, Thomas E. 1988. *The Politics of Military Rule in Brazil, 1964–1985*. New York: Oxford University Press.

Skidmore, Thomas E. 2009. *Brazil: Five Centuries of Change*, 2nd ed. New York: Oxford University Press.

Smith, Joseph, and Francisco Vinhosa. 2002. *A History of Brazil, 1500–2000: Politics, Economy, Society, Diplomacy*. London: Pearson Education.

Williams, Daryle. 2001. *Culture Wars in Brazil: The First Vargas Regime, 1930–1945*. Durham, NC: Duke University Press.

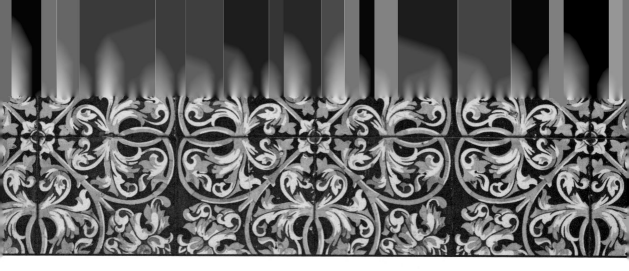

Government and Politics

Renato Lima de Oliveira

INTRODUCTION: CONSOLIDATING A YOUNG AND VIBRANT DEMOCRACY

Since 1985, when José Sarney became the first civilian president after more than 20 years of military rule, Brazil has been struggling to consolidate its vibrant and young democracy. It has now over 140 million voters who cast their ballots on electronic voting machines that allow the results to be known with certainty only few hours after the polls are closed. Contested elections occur regularly every two years and are closely monitored by a free press and numerous stakeholders, including several nongovernmental organizations (NGOs). The military does not play a political role anymore—and past human rights violations are increasingly exposed and condemned. Inclusionary social policies are expanding educational levels and income redistribution, slowly reducing the high levels of inequality that long characterized Brazil. All of these achievements highlight the fact that Brazil has been consolidating its democracy without institutional breaks or the level of polarization observed in neighboring countries in South America, including Argentina, Bolivia, and Venezuela. From one point of view, Brazilian politics after redemocratization of the mid-1980s is a success story.

However, as underscored by a number of widespread popular protests that took place in June 2013 and in opinion polls, there is much complaint about the quality of public services and political representation. Corruption scandals are common and the judicial system is slow and permissive, thus not acting as a deterrent of malfeasance. The quest for accountability, from absentee civil servants to corrupt politicians, is still far from complete. Part of the problem lies in the complex political regime that governs the electoral system in Brazil, which favors the formation of large governing coalitions sustained by parties attracted by pork-barrel and clientelistic practices. There are more than two dozen parties registered and entitled to receive public funds, but very few of them have any ideological cohesion. In sum, as in the classic metaphor of the glass half full or half empty, an assessment of Brazil's political achievements can be optimistic or pessimistic, depending on what the analyst wants to stress. However, taking a broader historical and comparative perspective, a moderate optimism may be warranted. Democracy has been consolidating, and public pressure forces politicians to be more accountable.

This chapter addresses the complex Brazilian party system and its electoral structure, which differs substantially from the rules existent in the United States. It introduces the main political parties and explains how the federal system works in Brazil and the role assigned to municipalities, states, and the federal government in managing the public administration of the country. It concludes by addressing the main challenges and recent achievements in the pursuit of more accountable and efficient governments. To appreciate the latest accomplishments and put the current challenges into perspective, a historical detour is necessary. Therefore, before focusing on contemporary events, this chapter briefly describes the main events in the Brazilian political system since independence from Portugal in 1822.

HISTORY: FROM THE EMPIRE TO THE NEW REPUBLIC

The Empire

Since independence from Portugal, Brazil has been a parliamentary monarchy, a presidential republic, then for a brief period of time a parliamentary republic, and finally back to presidentialism—all that with alternations between periods of democratic regime and authoritarian rule.

In stark contrast to its neighbors in South America, Brazil adopted a monarchical form of government between 1822 and 1889. The recently independent country was first ruled by Pedro I, the son of Dom João VI, the king of Portugal. Dom João VI and the rest of the royal family returned to Europe in 1822, after the end of the European Napoleonic wars that led them to move to Brazil in the first place. Portuguese elites also pressured Pedro I to return to his motherland. As he refused to do so, he declared Brazil independent of Portugal and became Brazil's first emperor. His mercurial and authoritarian personality and a low skill to deal with the politics of a constitutional monarchy led him to abdicate the throne in 1831 in favor of his son, Pedro II. At the time, Pedro II was just five years old, which is why the country was ruled by a series of regents until Pedro II turned of legal age. Successive crisis marked the regency period, and Pedro II was allowed by the politicians of the time to be crowned king at the age of 14, putting an end to the tumultuous years of the regency period.

Pedro II reigned from 1840 up to the collapse of the empire, in 1889. He consolidated Brazil's territory and the agriculturally based economy, grounded on sugarcane and coffee production dependent on slave labor. During the empire, two parties alternated in power: the Conservative and the Liberal parties. Franchise was restricted (including by income levels), and elections were dominated by local bosses who used government positions to secure votes. As in a parliamentary regime, the executive power was awarded to the leaders of the party that succeed electing the higher number of representatives to the parliament. However, the emperor had a moderating power, which gave him prerogatives far beyond what exists today in a constitutional monarchy. More specifically, the emperor could appoint Senate members (who would serve with lifetime tenure) and dissolve the Chamber of Deputies, calling for new elections. Between the dissolution of the chamber and the results of new elections, the emperor appointed a new government. Because of the high centralization of powers, the party of the temporary government appointed by the emperor always succeeded in fabricating the votes necessary to win elections and stay longer in power. Regardless of any ideological difference, both Conservative and Liberals played the patronage game—the use of government largesse to reward allies. Prime ministers had the prerogative of nominating political allies to virtually all public positions that could influence elections, which allowed the party in power to secure the ways to win elections. As Richard Graham shows, the right to appoint loyalists to public positions guaranteed whoever was occupying the prime ministership to "almost determine the results of elections, and do so entirely within the law" (Graham 1990, p. 85). When necessary, coercion of voters by police chiefs and judges was used. Consequently, patronage was the most important "asset" in the political competition of the two parties. The practice of appointing or promoting civil servants because of political connections rather than merit was frequently criticized by parties while in the opposition. However, once in power, that was how they governed. A bureaucracy staffed by political connections contributes to low state capacity, and this practice, while severely reduced with the years, is still present in pockets of the Brazilian state.

By late 19th century, the emperor was an old man without a male heir that could succeed him. In addition, the support base of the monarchy was the declining economic sectors of the country, like the slaveholders of large plantations. On the other hand, considerable part of the rising middle class, the military, and politicians from the state of São Paulo were agitating for republican ideals and an end of the centralization of resources in the court, the city of Rio de Janeiro.

The Proclamation of the Republic

In 1889 a military coup initiated the republican form of government that has continued ever since. The first president of the new republic of Brazil was the field marshal Deodoro da Fonseca, who became well-known during the Paraguayan War (1864–1970) and rose up to be a leading officer in the Brazilian Army and in the proclamation of the republic. His tenure was marked by an ineffective government and successive crises. A growing civilian and military opposition forced Fonseca to resign in 1891. He was replaced by another marshal, Floriano Peixoto. Of authoritarian inclination, historians dub Peixoto as "The Iron Marshall," as he dealt with insurgencies like the Naval Revolt of 1893–1894. The early years of the republic were marked by instability and authoritarianism; the latter used to crush opponents such as the small monarchist resistance and separatist movements.

By late 1894, Peixoto left office, handing power to a traditional politician, Prudente de Moraes, a lawyer from São Paulo. Moraes initiated a period of an oligarchical regime where politicians from São Paulo (a big coffee producer) and leaders from Minas Gerais (at the time the most populous state and also host of an economically powerful elite) governed in agreement, and elections were basically riddled with fraud and controlled by local strongmen. In addition, only literate males had the right to vote—in a country where illiteracy prevailed. In the so-called Brazil's Old Republic, the central government was relatively weak with respect to state governments. However, the central government was crucial in consolidating the republic (in contrast to the monarchical option) and the territorial unity of the country (against regional insurgencies). There were no national parties, and states had extensive tax and defense power. In fact, national politics was an extension of the politics within the state (Soares 2001). Whoever was supported for president by the local strongman in each state got a majority of votes in that state—either real or through fraud.

The oligarchical arrangement passed through some cracks but was only directly challenged in the 1930s with the rise of Getúlio Vargas, a skilled politician from the state of Rio Grande do Sul. He had already been the minister of finance in 1926 and later president (analogous to governor) of his home state, in 1928. In 1930, he started a presidential campaign as a reformist candidate, criticizing the electoral corruption that was widespread at the time. After an unsuccessful bid for presidency, he took part in a coup d'état that removed president Washington Luis from power and prevented the inauguration of the president-elect Júlio Prestes. The Revolution of 1930, as how it came to be known in history, profoundly influenced the Brazilian state and ended the decentralized oligarchical regime that had prevailed since the end of the empire. Among other tools to achieve such goal was the direct nomination of political appointees to rule by force (without elections) in key states and municipalities.

Regardless of the merit of his actions, Vargas left a lasting legacy with his multiple years in power, first between 1930 and 1945 and later from 1951 to 1954. He was the most important Brazilian politician of the 20th century and shaped much of the national politics while alive and dead, as he committed suicide in a move that was his ultimate political card.

Getúlio Vargas and His Legacy

A good indication of the lasting influence of Getúlio Vargas is the fact that *getulismo*, *varguismo*, or the associated *trabalhismo* are still political meaningful terms in Brazilian politics. All three refer to movements and ideologies based on the legacy of Getúlio Vargas and a specific type of labor organization (*trabalhismo*). Yet these terms represent a mixed bag of goods, a blending of concession of workers' rights, nationalism, paternalism, corporatism, and authoritarianism.

After seizing power in 1930, Vargas governed as a provisional president up to 1934, when the Congress elected him as president under a new constitution. He was supposed to govern as a constitutional president from 1934 to 1938, but by 1937 he seized total power, declaring the Estado Novo (New State). The new regime was of fascist inspiration and Vargas became a full-fledged dictator, closing the Congress, jailing political enemies, and censoring the press. However, that is not the full story. Throughout his tenure, Vargas also promoted economic nationalism and extension of workers' right, which was politically popular and at the time helped to counterbalance the impacts of the Great Depression in Brazil. The Brazilian state started to intervene heavily in the economy, just like other governments in the rest of the world, including the United States during the New Deal. That kind of intervention required from the Brazilian state a bureaucratic capacity that was largely absent hitherto. Through a number of actions, including the creation of state-owned enterprises (SOEs) and civil service reforms, Vargas modernized the Brazilian public sector. The Departamento Administrativo de Serviço Público (DASP, Administrative Department of Public Service) was a key agency that aimed to increase efficiency and merit-based recruitment in the public sector, at a time when public jobs were considered a sinecure awarded to political allies (just like it was during the empire, as discussed earlier). As Barbara Geddes (1994) shows, while DASP was not able to extend the modern and rational civil service model to the whole administration, it succeeded in upgrading bureaucratic performance and creating an elite civil service that the emerging developmental Brazilian state could rely on. The Brazilian state continued far from being a professional bureaucracy without political ties, but at least part of it was organized among modern lines.

The adoption of social legislations and extension of workers' right made Vargas widely popular among the working poor. This was also helped by the government propaganda efforts, which included the censorship of the press and the promotion of a positive image of the dictator. The regime propagated the image of Vargas as "Father of the Poor," an epithet that did not disappear despite subsequent political opening and intense opposition.

By 1942, after a period of neutrality during World War II, Brazil entered in the war effort in the side of the allies. Vargas negotiated with President Roosevelt of the United States a package of compensation for taking their side during the war. Brazil entered in the war effort supplying raw materials, offering its costal line to host naval and air bases, and even sending troops to fight in Europe during the last couple of years of the war. In exchange, Brazil received U.S. armaments, investments, and technical assistance.

As the war was coming to an end, the contradiction between fighting fascist governments in Europe while having a dictatorship back at home became increasingly clear. By the end of the war, Brazil had another bloodless coup. Vargas was forced to resign the government by his military aides, who pressured for elections.

National Parties and Competitive Elections

The end of the New State authoritarianism also brought new parties to the political arena. Opponents of Vargas formed the União Democrática Nacional (UDN, National Democratic Union). UDN was initially born as an ideologically plural gathering of opponents of Vargas, but later it was characterized as having as its core constituency businessmen and the middle class as well as for preaching an anticorruption and free enterprise agenda. UDN would play an important role in the opposition against Vargas, but the party would never be able to establish itself as the main political force in the country. In fact, it progressively lost seats in the national legislature from 1945 to 1962.

Vargas sponsored the creation of two parties, one that represented the traditional (oligarchic) interests that governed with him and the other a party that embodied the union movement that had close ties with the government during his tenure. The former was the Partido Social Democrático (PSD, Social Democratic Party), of strong rural base, and the latter was Partido Trabalhista Brasileiro (PTB, Brazilian Labor Party).

While PSD and UDN had nationwide representation and were able to compete in most municipalities in the country already by 1945, PTB initially did not have the same structure of the other two parties. However, structural changes such as the growing urbanization and industrialization of Brazil made PTB a party poised to conquer its own space. By 1962 it was already bigger than UDN in terms of elected federal deputies, closely following PSD, the then biggest party in the country. PTB would conquer votes by pressuring for the extension of labor legislation. Internally the party was far from being democratic—party affairs were managed by the iron fist of Vargas and his family. Together, PSD, UDN, and PTB would be the most important parties in the period between the end of World War II and the military coup of 1964. The Brazilian Communist Party (PCB) was another important political force of that period; however, it had a short life as a regular party disputing regular elections. PCB had a leader of national expression, Luiz Carlos Prestes, and a significant nationwide vote share of close to 10 percent. However, in the context of the Cold War and due to the close ties that the PCB had with the totalitarian government of the Soviet Union, its status as a national party was questioned. After a legal battle that was decided by the Supreme Court in 1947, the PCB was banned from the political life. Members of PCB, including Prestes, lost their mandates. That decision indirectly benefited PTB, as members of PCB would frequently use PTB as a vessel to dispute elections. Other smaller parties competed and had representation at the Congress, but they had to gravitate around and make coalitions with PSD, UDN, or PTB during the 1945–1964 period.

Vargas Returns to Power as an Elected President

With Vargas out of the federal government, democracy returned in 1945 with presidential elections where the two main contenders were military officers, general Eurico Gaspar Dutra (who was elected, from PSD) and air force brigadier Eduardo Gomes (UDN), with a third candidate from the Communist Party, Yeddo Fiúza, with a substantive amount of votes. Enfranchisement was significantly extended in comparison to the period of the Old Republic—for the first time it reached 10 percent of the population, while in the period of the Old Republic it stayed below 5 percent. In raw numbers, registered voters amounted to 1.5 million in 1933 and went up to 7.5 million in 1945. Enfranchisement would keep expanding, and by 1962 it reached 18.5 million people (Soares 2001).

Vargas was out of the presidency but not of politics. He returned to his home state Rio Grande do Sul and was elected senator. However, he would not be away from the presidency for much longer. In 1950 he ran in the ticket of PTB to succeed Gaspar Dutra and was elected by a wide margin. However, the political conditions this time were very different. Now, elected in a democratic environment, Vargas would have to govern with an open Congress with an aggressive opposition and an unfettered press—in other words, a much harder environment to rule than what he was used to during the New State.

His period as democratically elected president promoted nationalistic policies, including the creation of Petrobras, an oil state-run company. The period was marked by economic and political crisis, including corruption scandals. The situation became critical after one of Vargas's bodyguards hired a gunman to murder the main political opponent of the president, Carlos Lacerda, a journalist and politician from UDN. The assassination attempt failed, and authorities uncovered that the intellectual author of the crime was a close aid of the president. This complicated even further the political situation of Vargas. This time, however, he did not want to accept resignation, despite growing opposition. On August 24, 1954, he took his own life, leaving a suicide note that was a vigorous defense of his government's nationalistic policies and an attack on the opposition. As Thomas Skidmore (1999) puts it, with his suicide Vargas turned the table on his enemies. Masses went to the streets to cry the death of the "Father of the Poor" and protest against the political enemies of *varguismo*. With his death, Vargas disappeared from the political scene, but his influence remained. The PTB, which in 1945 elected only 22 federal deputies, was growing steadily and by the early 1960s was already a major political force in the country rivalling with the conservative PSD.

Developmentalism and the Creation of Brasília

Developmentalism, or the use of state activism to structurally transform the economy, was an idea that got more traction in Brazil than in other neighboring countries like Argentina. In this regard, the president who succeeded Vargas, the politician from Minas Gerais Juscelino Kubitschek (from PSD), would deepen the model largely initiated by his predecessor. However, Kubitschek was more open to foreign investments, and Brazil economically boomed during his tenure.

Whoever visits the capital of Brazil, Brasília, will see the fingerprints of Kubitschek. It was one of his main political promises to move the capital of the country from Rio de Janeiro to a new city to be built in the geographical center of the country. Few thought he would do it, but Kubitschek successfully erected a new city in about four years out of a virtual desert area. The political core of the country moved from the coastal area of Rio de Janeiro to the middle of the country, moving politicians but also a growing bureaucracy.

Building Brasília was a strong show of optimism and state capacity. In fact, Kubitschek promised progress equivalent to 50 years in just 5. However, his strategy had high fiscal costs, and by the end of his tenure an economic crisis was approaching, with rising foreign debt and inflation.

Kubitschek was succeeded by Jânio Quadros, a politician from the state of São Paulo and former governor. Quadros campaigned with a broom, claiming that he would sweep corruption from the country, and he had support from the UDN. His tenure, however, was brief: it lasted only seven months. In what is interpreted as a political maneuver aimed at accumulating more political power and trump over opposition in the Congress, Quadros sent a resignation letter to the legislature, confident that deputies would refuse to accept it. At the time, the population voted separately for president and vice president. His vice president was João Goulart, from PTB, who had run in an opposition ticket to Quadros. Goulart was a leftist politician and former labor minister of Vargas. Quadros calculated that the political establishment would prefer to give more powers to him rather than see the populist Goulart as president. Not only was Goulart already known for his leftist inclinations, but also at the time of Quadros's resignation the vice president was in an official visit to communist China. Quadros's political maneuver proved to be perhaps the biggest miscalculation in the history of Brazilian politics.

The Political Crisis of the 1960s

Quadros's resignation created a stalemate. The Congress was willing to accept the resignation of the president, but the military command and some politicians were reluctant to accept Goulart as his replacement. The solution came with an unusual compromise: Goulart was allowed to become president, but Brazil would switch from presidentialism to parliamentarism through a constitutional amendment. In other words, he could be president as long as he was powerless. The executive power was transferred to a prime minister and his cabinet.

The compromise solution did not last. Goulart unwillingly accepted the arrangement, but, once in office, he struggled to gain full powers and switch back the regime to presidentialism. That was exactly what happened after a referendum approved the return to presidentialism in early 1963. This time, Goulart needed to prove that he was able to govern and approve the social reforms that he defended so much, including the land reform. A great deal of the political and economic crisis that happened between the resignation of Quadros and the reversal back to presidentialism was blamed on the nature of the parliamentarian regime. However, the return to presidentialism did not make things easier to the government. The administration was fragile on all fronts: economically, politically, and on foreign affairs.

Inflation was rampant and accelerated to 100 percent a year by the end of Goulart's days in power. The government was also slow and unable to react with a coherent economic plan to fight inflation and the growing external debt. In exchange of political support, Goulart sacrificed bureaucratic capacity by the use of widespread appointments to government positions using political criteria. The president had a weak political base and was pushing for social reforms that polarized the country. If all that was not enough, Goulart supported acts that directly confronted the top echelons of the military and the pillars that they defended, such as the respect for hierarchy and order.

Finally, on the external front, Brazil was caught in the middle of the cold war ideological battle. The Cuban Revolution of 1959 brought a communist regime to the neighborhood of the United States. Cuba actively tried to promote socialist revolutions in other countries, and the United States was equally forceful in preventing them. The majority of the Brazilian military was decidedly anticommunist, and some key military members had ties to the United States. In fact, both Presidents Kennedy and Johnson followed closely the political crisis in Brazil and allowed operations to support coup-plotters.

Goulart was presiding over a country in economic crisis and frequent strikes, with a weak legislative and military support, and facing the open opposition of the Church and other conservative groups and classes of the society. He proved to be

João Goulart, popularly known as Jango, had his political roots in the Brazilian Labor Party and became president of Brazil in 1961. He was ousted from power in 1964 by a military coup. (Library of Congress)

an unskilled president, unable to diffuse the crisis that accumulated on all fronts. By the end of March 1964, the general Olimpio Mourão and his troops marched from Minas Gerais toward Rio de Janeiro (the then state of Guanabara), thus initiating a coup d'état. Goulart decided to not confront the military coup and eventually left the country to an exile in Uruguay. The democratic experience of 1945–1964 came to an end. For the next 21 years, Brazil would be ruled by generals.

Brazil's Military Dictatorship (1964–1985)

In Brazilian political history, the military dictatorship is perhaps the period most talked about in nonfiction books, novels, movies, television shows, and documentaries. Unfortunately, the abundance does not go hand in hand with quality. The depiction of the period frequently lacks the nuance necessary to understand a complex regime that had points of contact with other military dictatorships in the continent (e.g., Argentina and Chile) but was also markedly different. In this section, we can only make a cursory analysis of the period, but hopefully it will stir the interest of the reader in seeking more information.

Brazil's military came to power in 1964 in another bloodless coup and with support from key politicians and the majority of the press. To the much surprise of key political figures who supported the coup, like Carlos Lacerda from UDN, the military was not willing this time to return to the barracks easily. Unlike other dictatorships that had military members as rulers, like Augusto Pinochet in Chile, in

Brazil the military as an institution stayed in power for 21 years. Generals would rotate in power with fixed terms as if they were regularly elected. The Congress in fact indirectly elected military presidents; however, there was no room for surprises or real contestation, what makes such elections just a mock ritual. The military also set the principles of the economic policy, which, with the years, became increasingly of a right-wing nationalistic color instead of a free market orientation. Politicians who thought that they would use the military to conquer the presidential power after the removal of Goulart had to conform in playing a minor role—either supporting the government or picking a side in the opposition.

The military was quick to purge politicians, union leaders, and supporters of Goulart once it set into power. However, repression itself was not constant—it markedly increased after 1968 when a decree, the Institutional Act number 5, effectively turned Brazil into a full-fledged dictatorship. The AI-5 revoked civil liberties; allowed the president to close the National Congress at will and legislate by decree whenever the Congress was in recess; imposed the censorship of the press, films, theater, and television, among other restrictions. AI-5 also gave the government the right to sack any civil servant or force his or her premature retirement. Only 10 years later, in 1978, was the AI-5 revoked, following a slow and gradual process of political opening that initiated in 1974.

Throughout the authoritarian regime, elections were not completely abolished. This is just one facet that makes the Brazilian military dictatorship far from typical. It kept for the most part an open Congress and elections for federal deputies and senators. The press was censored but not totally blocked. The leaders of the regime wanted to convey an image of democracy, although ruled by generals indirectly elected after having removed a civilian president in power. The contradiction about the image that the regime wanted to portray (especially for foreigners) and the reality was constant throughout the Brazilian military regime.

In 1965, the military abolished the existing political parties and imposed a bipartisan regime that lasted until the return to multipartism in 1979. One party would support the government and the other would (mostly) offer a token opposition to the regime. The former was the Aliança Renovadora Nacional, ARENA (National Renewal Alliance Party), and the latter the Movimento Democratico Brasileiro, MDB (the Brazilian Democratic Movement). Whenever things threatened to get out of control, the wizards of the military regime would twist the electoral rules in their favor, fabricating favorable results. When the opposition surprised the government with important electoral victories or the prospect of getting them, what would put in danger the majority of the regime in the Congress, the government would step in and change the rules of the game. A major example was in 1977 when President Ernesto Geisel closed the Congress and approved unilaterally changes in the constitution and in the electoral rules. That included, for instance, creating senatorial positions filled by direct appointment of the president.

Political opposition was exercised on several fronts. In fact, the majority of today's political leaders of the country forged their careers while in the opposition to the military dictatorship in different roles. That could be running in the MDB ticket, organizing workers' strikes, taking leadership roles in student movements, or even going clandestine in guerrilla groups. Because traditional political channels were blocked or were not truly representative, resistance took many forms, and cultural expressions were a major one. Popular artists had to fight censorship whenever their lyrics criticized social conditions, the political restrictions, or even moral costumes. Some artists opted to live in exile, waiting for the return of democracy. At the other extreme, urban and rural guerrilla groups tried to overthrow the right-wing military dictatorship to impose a left-wing dictatorship of the proletariat. With more democratic credentials, a new union movement started to grow by the late 1970s and joined forces with the opposition in complaining about political repression and calling for different economic policies.

The top military command planned its way out of power before it was forced to leave. The transition started in 1974 through slow and gradual movements toward political opening. By that time, Brazil had been growing by an average of 10.9 percent since 1968, what came to be known as an economic miracle, and the military enjoyed popular support—which was also helped by the fact that the press was censored and therefore negative news was limited. Moreover, by 1974, most of the guerrilla activities—comparably smaller than what happened in neighboring countries like Argentina and Chile—had already been defeated. The military in Brazil handed power through a careful and controlled process that ultimately took 11 years. The military not only negotiated with the civilian leadership but also kept in check the hard-liners within the armed forces—officers contrary to the political opening. The military establishment sidelined radicals both on the left and on the right, which changed the strategy of civilians from direct confrontation with the military to a push in the direction of faster and wider reforms. This controlled liberalization was also supported by the business elite, who at times voiced criticisms of the scale of the government's economic intervention but would ally with the administration on issues of wage controls and limits to strikes. At the end, the timing was the one chosen by the military leadership and only after a return to the barracks free of retribution was guaranteed (Couto 1999).

A major landmark in the opening process was taken in 1979 when the government-controlled Congress passed an Amnesty Law that protected from prosecution both government officials and opponents—a law that was then called for by civil society groups. One of the intended aims of the law was to ease the transition to civilian power and to promote national reconciliation. It allowed the return of activists, politicians, and artists from exile—and the protection of officers involved in the repressive apparatus. The Amnesty Law is currently under question because it prevents the criminal prosecution of human right violations committed by former military officers.

The 1980s started in a dire financial situation for the military regime. Brazil was hit hard by the sudden rise of oil prices in the late 1970s after the Iranian Revolution. Brazil's fast growth years were fueled by foreign debt and imported oil. The political economy of the military regime focused on replacing imported manufactured goods by local production, a model known as import-substitution industrialization that relied heavily on less efficient SOEs. Foreign debt and fiscal pressures led to restrictions on import capacity and increasing inflation, leading the economy of the country to the brink of collapse. This also meant that the repression necessary to control workers' demand in a stagnating economy with rising living cost was becoming unbearable. The economic crisis became a liability to the supporters of the regime and a window of political opportunity to oppositionists. Among the oppositionists were rising stars like the sociologist Fernando Henrique Cardoso and the union leader Luiz Inácio Lula da Silva. The former was already a renewed intellectual, who was forced by the dictatorship to retire from a professorship at the University of São Paulo, which led him to the exile and an international career. The latter did not even finish high school but showed remarkable leadership capacity leading strikes in São Paulo. Despite different backgrounds, both of them were leaders in the opposition and helped to found new political parties (the social-democrat PSDB and the Workers' Party PT, respectively) and succeeded in reaching the presidency with the return of democracy.

Democratization and a New Constitution

Formal relinquishment of the military from power was accomplished only in 1985 by an indirect election of the moderate oppositionist Tancredo Neves by the Congress—after the same Congress rejected an amendment that would allow direct popular vote for president. Neves, a skilled politician from Minas Gerais, was a historical leader of MDB known for his ability to compromise. Neves died from natural causes before taking office and the position was then assumed by his running mate, José Sarney, a conservative politician and former collaborator of the authoritarian government.

Sarney's tenure was mediocre in economic terms and marked by constant political crises. However, it helped in consolidating democracy after years of military rule. It was also under his tenure that the country approved a new constitution in 1988—one that put heavy emphasis on the expansion of social rights and decentralization of resources.

THE 1988 CONSTITUTION AND ITS AMENDMENTS

Brazil has had eight constitutions since independence. The current one was promulgated in 1988 and encapsulated the high hopes of a country that had recently transitioned from a centralized military dictatorship to a democracy. It was dubbed at the time as the "Citizens' Constitution" due to its progressive character. Nevertheless, the text produced by the constituents was overly detailed and economically backward, thus incompatible with the challenges that the nation had to face to upgrade its economy and control inflation. The consequence was that every subsequent government pushed for constitutional amendments, resulting in a young constitution that has passed through more than 70 amendments so far.

Sarney was the last president elected by indirect vote. His successor was to be selected by direct popular vote in 1989. In that year, 22 politicians had run for office, with the runoff elections being disputed by the union leader Lula, on the left, and a young governor of the state of Alagoas, Fernando Collor, on the center-right. By a tight margin Collor was elected. He came to office with high hopes of modernizing Brazil's economy, opening the country for imports, and dealing with the inflation that was already beyond 1,000 percent in a year. Collor's economic plan failed, and he was caught on a major corruption scandal that ultimately cost his mandate through impeachment. Once again, Brazil was ruled by someone elected as vice president, this time Itamar Franco, a relatively unknown politician from Minas Gerais. After failed attempts to deal with the inflation, Franco selected Fernando Henrique Cardoso as finance minister. That was Cardoso's political jackpot: he handpicked a group of ingenious economists who finally succeeded in curbing inflation to civilized levels. Suddenly, he was propelled from the ministry of finance to a presidential run. His PSDB faced Lula and the PT, but the overwhelming success of the economic plan was enough for Cardoso to win the election right on the first round.

While in government, Cardoso sponsored a polemical amendment to the Constitution allowing for one reelection of incumbents. In 1998, he one more time faced and defeated Lula in a presidential campaign. Cardoso's second term was less fortunate, and basically he had to manage one crisis after another. By the end of his administration, the popularity of the government was reaching low levels and he was unable to help elect José Serra, a former minister of health and candidate running at the PSDB ticket. Instead, voters selected Lula and his PT to the presidency. However, Lula and PT had changed. The former union leader moved to the political center, broadening his coalition base, and had run a campaign as a moderate reformer, abandoning the blue-collar image that characterized his first run for the presidency in 1989. Despite a major corruption scandal in his first term, Lula managed to be reelected in 2006 due to a fast-growing economy and the expansion of social programs, including the very popular Bolsa Família, a conditional cash-transfer aimed at poverty alleviation. In 2010, he handpicked Dilma Rousseff from PT as his successor. Rousseff is an economist and former guerrilla leader during the days of the dictatorship, who later went to occupy positions in left-leaning administrations during the country's democratization. She had never run for office before 2010 but served as minister of mines and energy and later chief of staff in the Lula administration. With the blessing of Lula, who became the most popular politician in the contemporary history of Brazil, Rousseff was elected in 2010 against Serra (PSDB) in a runoff election and reelected in 2014 beating Aécio Neves (PSDB) by a tight margin.

Nowadays, the fear of a military coup has vanished, and there is a virtual consensus that democracy has taken deep roots in Brazil. However, it does not mean that voters are satisfied with the political system. Some of the problems are as old as the empire, such as the practices of clientelism and patronage. Others are more recent, like issues of campaign finance and unrepresentative parties. To understand the nature of the grievances, it is necessary to delve into Brazil's complex political system. That is what we will do in the next section, introducing the federative organization of the country, how leaders are selected, and the problems brought by the electoral system.

GOVERNMENT TODAY

A Multiparty Presidential Democracy in a Federative System

The Federative Republic of Brazil currently comprises 26 states, one Federal District (Brasília, the country's capital), and more than 5,500 municipalities. The legislative power at the federal level is exercised by the National Congress, which is formed by two chambers. The Chamber of Deputies is the lower house and comprises 513 elected deputies. The Senate is the upper house and is composed of 81 members, where each state and the Federal District are represented by three members.

Since the return to democracy in the mid-1980s, elections occur regularly for city councilors, mayors, state deputies, federal deputies, senators, governors, and president. State and federal elections occur on the same year and day, while municipal elections occur two years later. Mandates are of four years for all positions with the exception of senators who serve for eight years. Beginning in 1997 when the Congress approved reelection for executive positions, incumbents are allowed one consecutive reelection. Legislators can be reelected indefinitely. This section will introduce the different levels of government in Brazil and the powers, as well as the main parties. It will first address Brazil's electoral system and then it will describe the different levels of executive government, followed by the legislative and the judicial power.

The Electoral System

Voting in Brazil is mandatory for citizens in the age range of 18 to 70, and voluntary for citizens between 16 and 18, as well as above 70 and illiterates at any age. Those who do not vote have to justify their absence to the Electoral Justice or pay a small fine. Failure to do so results in severe civil penalties, including denial of a passport and receiving wages and benefits from the federal government. Election Day occurs on Sunday, with ample mobilization, and is preceded by widespread campaigns. That includes television and radio ads during prime time in all channels. Airtime is allocated to parties in proportion to their representation in the Congress—parties that elected more deputies in the previous election get more free airtime in the subsequent campaign. Turnout is constantly above 80 percent. Popular street concerts and free distribution of t-shirts and all types of gifts used to be common until the 2006 campaign, when a law prohibited this practice in order to reduce the undue influence of money in politics and reduce campaign costs. Notwithstanding that, campaign costs continued to rise every year, driven by fancy television ads, shot by professional staff and actors and directed by *marqueteiros*, political consultants that run the entire campaign. Television and radio airtime is free—during election time, airwaves televisions and radios are mandated to broadcast campaign material following prespecified time slots.

BRAZIL'S ELECTORAL JUSTICE

It is a considerable challenge to organize elections that are held on the same day throughout one of the largest countries in the world and for so many positions. This task is done by the Electoral Justice, which is in charge of all the electoral process, from enrollment of voters to vote count and proclamation of the elects. It is a branch of the judiciary, independent from the executive and legislative, and has a well-deserved fame for ensuring well-run elections without fraud incidents.

In the 2012 election, more than 115 million citizens went to the polls and voted for the positions of city council and mayor in the more than 5,500 municipalities. In 2014, citizens were able to choose state and federal deputies, governors, one-third of the Senate, and the president. Since 1988, elections for the president, governors, and mayors of municipalities where the electorate is larger than 200,000 voters are decided by a two-round system if in the first round no candidate achieved an absolute majority.

Candidates in Brazil cannot run as independent—they have to be affiliated with one party. This is not a problem, as the permissive rules allow the proliferation of numerous parties. Brazil currently has over 30 different parties (see Table 3.1), with many more groups attempting to form new ones. The electoral institutions in place is of open list with proportional representation (PR), which allows even very small parties to be part of ruling coalitions. A comparison with other electoral institutions is worth in order to highlight the peculiarities of the Brazilian system. In single-member legislative districts, such as the system mostly associated with that existent in the United Kingdom and former colonies (also known as first-past-the-post), voters select one candidate to represent the district. Whoever gets the majority is selected—all other votes are "wasted." In a hypothetical scenario, if a party gets 51 percent of the votes in all districts that it were running, it will get 100 percent of the contested seats. All other voters, the 49 percent of citizens who voted for other parties, end up not being represented in Congress. In PR systems, seats are allocated in proportion to the number of votes received. In perfect proportionality, a party that receives 51 percent of votes would have 51 percent of the seats, with the remaining 49 percent divided among competing parties according to their vote share. In practice, most countries adopt some form of disproportionality such as upper chambers, where even small rural states can elect the same number of senators with much lower absolute votes. In addition, in Brazil, seats for the lower chamber are assigned disproportionately, with populous states limited to a maximum of 70 deputies and small states to a minimum of 8. What this means is that it is much easier to be elected as a federal deputy or senator in a small state such as Rondônia or Alagoas than in Minas Gerais or São Paulo. Having said that, PR even when moderated by such specific rules favors the representation of small parties. Assume that 10 percent of the voters in a state care about the environment and are willing to vote for a candidate of the Green Party. In addition, let us assume that these voters are spread more or less equally throughout the different municipalities of the state. Because each citizen will have his or her votes counted and seats will be proportionally allocated, a candidate may have a small number of votes per municipality, but the sum of all of them will help the candidate get elected as a deputy. In a majoritarian system, if only 10 percent of voters in a district want to vote for the Green Party (or, e.g., the Libertarian Party), they do not form the majority, and their votes are wasted.

Proportional representation is the most common formula for translating votes into seats—80 percent of democracies adopt one form of PR. However, Brazil mixes PR with the less common feature of open-list voting. This means that parties can present multiple candidates to voters and the order of election is entirely decided by the voters themselves. This means that even candidates of the same party compete among themselves for votes—and during elections, not primaries, which is almost absent in Brazil. This open-list procedure stands in contrast to closed-list systems,

TABLE 3.1   Parties in Brazil

Acronym	Name	Translation	Number of Elected Federal Deputies, 2014
PMDB	Partido do Movimento Democrático Brasileiro	Party of the Brazilian Democratic Movement	66
PTB	Partido Trabalhista Brasileiro	Brazilian Labor Party	25
PDT	Partido Democrático Trabalhista	Democratic Labor Party	19
PT	Partido dos Trabalhadores	Labor Party	70
DEM	Democratas	Democrats	22
PCdoB	Partido Comunista do Brasil	Communist Party of Brazil	10
PSB	Partido Socialista Brasileiro	Socialist Party of Brazil	34
PSDB	Partido da Social Democracia Brasileira	Brazilian Social Democratic Party	54
PTC	Partido Trabalhista Cristão	Christian Labour Party	2
PSC	Partido Social Cristão	Christian Social Party	12
PMN	Partido da Mobilização Nacional	National Mobilization Party	3
PRP	Partido Republicano Progressista	Progressive Republican Party	3
PPS	Partido Popular Socialista	Popular Socialist Party	10
PV	Partido Verde	Green Party	8
PTdoB	Partido Trabalhista do Brasil	Labor Party of Brazil	1
PP	Partido Progressista	Progressive Party	36
PSTU	Partido Socialista dos Trabalhadores Unificado	Unified Socialist Workers Party	
PCB	Partido Comunista Brasileiro	Brazilian Communist Party	
PRTB	Partido Renovador Trabalhista Brasileiro	Brazilian Labor Renewal Party	1
PHS	Partido Humanista da Solidariedade	Humanist Party of Solidarity	5
PSDC	Partido Social Democrata Cristão	Christian Social Democratic Party	2
PCO	Partido da Causa Operária	Workers Cause Party	
PTN	Partido Trabalhista Nacional	National Labor Party	4

Acronym	Name	Translation	Number of Elected Federal Deputies, 2014
PSL	Partido Social Liberal	Social Liberal Party	1
PRB	Partido Republicano Brasileiro	Brazilian Republican Party	21
PSOL	Partido Socialismo e Liberdade	Socialism and Freedom Party	5
PR	Partido da República	Party of the Republican	34
PSD	Partido Social Democrático	Social Democratic Party	37
PPL	Partido Pátria Livre	Free Nation Party	
PEN	Partido Ecológico Nacional	National Ecological Party	2
PROS	Partido Republicano da Ordem Social	Republican Party of the Social Order	11
SD	Solidariedade	Solidarity	15
			513

where party officials decide the order of the list and the amount of votes the party receives determines who gets elected. Closed list favors party identification and discipline, while open list reinforces personalistic votes and independence from party positions. Therefore, by design, the electoral system of Brazil favors the existence of multiple parties (because of PR) but with candidates relatively independent of party platforms (due to the open-list vote). To sum up, parties are many, but they can rarely be ideologically distinguished from each other, despite the names that they carry.

Without ideological cohesion, what glues together different parties in a governing coalition in Brazil is patronage—spread of governmental favors to political allies. Governments in order to secure a majority in the Congress attract different parties by offering cabinet positions, thousands of public jobs for political appointees to be filled by recommendation of party leaders, and budget resources for public investments in the allied congressman's political base (which in the United States is known as pork barrel). In such a system, to be an oppositionist politician is of high cost—it means that you do not have the same kind of visibility that comes from heading a cabinet position or is able to divert resources to public projects in your home base. No wonder few presidents and state governors have to face an oppositionist majority in the legislature. If there is a need to accommodate a new party in the governing coalition, the federal government may then create a new administrative structure. With the current historical record of close to 40 ministries in the federal government, there are many such examples: the Ministry of Fishing and Aquaculture and the Secretariat of Civil Aviation are two recently created bureaucratic structures formed in a quid pro quo to accommodate leaders of allied parties in exchange of favorable votes in the Congress.

Given the incentives existent to join the ruling coalition, real deadlocks in the Congress are rare (as long as there is money to buy political support through the public budget). However, while this system can lead to comfortable political majorities, it brings its own destructive seeds: political criteria in determining appointments and investment decisions tend to inefficient spending and is more likely associated with corrupt practices. Hence, we have come full circle in partly explaining how Brazil can have both a very institutionalized and stable politics and, at the same time, widespread complaint about the quality of public services and of corruption practices.

Government Levels: The Federal Government, States, and Municipalities

The president of the republic is both the head of government and state, and the commander in chief. The power of the presidency is heightened by the relative weakness of the mechanisms of checks and balances. The president can freely appoint his or her cabinet of various ministers without the need for legislative confirmation. The president also has control over the appointment of the head of state agencies and the various SOEs, thus virtually being able to reshape the public administration in key positions.

In addition to its bureaucratic command, the executive in Brazil has extensive agenda powers in the Congress. It can send bills directly to the legislative and can even issue temporary decrees that have the force of law, and can be converted into permanent law after congressional vote (an instrument called *Medida Provisória*, a temporary executive law). Another important prerogative of the presidency is the veto of legislation: the executive can veto a law in its entirety or only the parts that it does not want to see enacted (partial veto). Finally, the legislative approves the annual budget, but the executive may choose to retain the expenditures of areas that it does not consider a priority.

In smaller scale, governors and mayors also have far-reaching influence over the local legislatures. However, regional and local governments are much more limited in terms of powers and administrative capacity. The Constitution of 1988 had a strong bias in favor of decentralization, but since then the federal government has been increasingly recentralizing resources and policy initiatives. This makes Brazil almost a federation in name only, as states and municipalities are relatively weak vis-à-vis the federal government. Of all the taxes collected in the country, the federal government keeps 60 percent, while states 33 percent and municipalities 7 percent, in approximate numbers. As a general rule, the poorest the state and the municipality, the more dependent they are from transfers from the federal government to fund infrastructure projects. Both the federal and state governments have considerable freedom to redirect transfers to places of their choice, thus easily co-opting regional leaders to support the executive. On the other hand, states that are rich in industries like São Paulo and Minas Gerais enjoy much more autonomy and can actually become strongholds for the opposition.

Each state has a constitution, a justice system, and an elected unicameral legislative body. Municipalities have an organic law (that is akin to a constitution), an elected mayor, and a legislative body comprised of city councilors. The number of city councilors are determined in proportion to the population of the city, with a minimum of 9 and a maximum of 55. The number of state deputies follows a specific rule related to the number of federal deputies that each state has. At minimum, each state legislature is composed of three times the number of federal deputies assigned to the same state.

Main Parties

The center-left Partido dos Trabalhadores, PT (Workers' Party), is currently the most important party of Brazil. Since 2003, with the election of the union leader Luiz Inácio Lula da Silva to the presidency after three unsuccessful bids, PT has occupied the presidency and heads the governing coalition with different parties. PT was formed at the end of the military dictatorship gathering activists from all walks of life, including union members, former guerrilla fighters, and community organizers of grassroots movements. This heterogeneous origin still affects the works of the party: PT is known for accommodating several different factions, from leftist radicals to more or less free marketers who push for European-style welfare redistribution. As the party consolidated its grip on the governmental machine, it has loosened its ties to grassroots movements and became more bureaucratic and centralized. Despite that, no other party in Brazil has a comparable presence in union movements and NGOs.

The social-democrat Brazilian Social Democracy Party (PSDB) is the main political adversary of PT. PSDB is a party that was mainly formed by intellectuals and politicians who opposed the dictatorship at the MDB ticket, as well as from various movements from the civil society. The PSDB was formed from a leftist faction of the Party of the Brazilian Democratic Movement (PMDB) in the late 1980s. The party ruled the country from 1995 to 2002 under the presidency of Fernando Henrique Cardoso and has occupied important state governments, including São Paulo, Minas Gerais, Rio Grande do Sul, and Ceará. PSDB would fit as a typical center-left party in Europe due to its belief in the capacity of the state to regulate the economy and a duty to redistribute wealth and reduce inequalities through governmental action. However, in Brazilian politics, PSDB is commonly associated with the center-right, partly by its role in privatizing state companies in the 1990s and partly due to the strong association that the left in Brazil already has with the PT.

PMDB is the main swing party in Brazilian politics. It has its roots in the legal electoral dispute against the dictatorship but nowadays is more characterized as a party that gathers different regional leaders who can join governmental coalitions to enjoy the benefits of being in power. Despite its size (see Table 3.1), PMDB does not have any coherent leadership or unity to present a strong candidate to the presidency of the republic—a task that has been occupied by either PT or PSDB. However, political analysts and pundits in Brazil will say that every government will need the PMDB at its side in order to have legislative majority.

PT, PSDB, and PMDB are by far the most important parties in Brazil, with others playing minor positions as members of the ruling coalition or joining the ranks of the opposition. Two other parties are worth addressing due to their different trajectories: the Democrats (DEM) and the Socialist Party (PSB). The former has its origins in ARENA, the party that supported the dictatorship. With the political opening and the return of the multiparty system, former ARENA members created the Partido da Frente Liberal (PFL, Party of the Liberal Front), a center-right party that had important national leaders, and ruled several municipalities and states. PFL was the main partner of PSDB during Fernando Henrique Cardoso's government, having appointed the vice president and several ministers. However, political struggles led PFL leaders to leave the governing coalition with the PSDB before the presidential race of 2002. During Lula's first term, PFL played an important role as opposition in the Congress. Nonetheless, several members of PFL left the party to join the governmental coalition through smaller allied parties. As addressed earlier, congressional representatives who support the government are rewarded with transfers to their constituencies and the right to appoint allies to key positions in the bureaucracy. This gravity power exerted by the government was fatal to PFL that in the 2000s continually lost members and potential allies. Furthermore, the party was still tarnished by the old image of having supported a right-wing dictatorship in a country that was now ruled by the leftist PT. In 2007, PFL tried to rebrand itself and picked the name of Democrats. The effort so far has failed, and the party is still decreasing in numbers and national importance.

The recent trajectory of PFL/DEM goes in opposite direction to the socialist PSB. From a minor member of the PT government, PSB has been steadily increasing its size and is threatening to break the polarization between PSDB and PT in the national politics. Under the leadership of the governor of Pernambuco and former minister of science and technology during Lula's first term, Eduardo Campos, PSB built a party machine that skillfully played its cards in local elections, at times making coalitions with the PT, and others with the PSDB. In late 2013, PSB left the national coalition led by PT and prepared a presidential run. However, Campos died in August 2014 in a plane crash while going to attend a campaign event in the state of São Paulo. He was succeeded in the campaign by his running mate, Marina Silva, herself a founder of PT and a former environmental minister. Marina left PT to found her own party, Rede, which at the time of the campaign could not get the approval of the Electoral Justice, and she was invited to join PSB. It was the first significant crack in the large governmental coalition that PT has been leading since 2003, when Lula was elected. Despite a close race where opposition candidates headed the polls many times, the opposition was defeated again in 2014 with the reelection of Dilma Rousseff (PT). However, it was by a tight margin and with significant losses to the PT in the Congress—from the then 88 seats that the party occupied to 70 in the current legislature. Given a weak economy and corruption scandals directly linked to her campaign, party, and allies, Dilma Rousseff's second term started very fragile in terms of political support.

It may be odd, but in Brazil parties tend to heavily occupy the left in terms of the ideological spectrum. Individual candidates sometimes can voice right-wing positions, but parties and their leaders rarely voice support for proposals that could be associated with right-wing positions, such as free trade, and reduction of governmental intervention in the economy. A rising political force in terms of social conservatism is the growing evangelical caucus. This caucus unites politicians from different parties but that together tends to campaign against socially progressive policies, like same-sex marriage and abortion.

Corruption

According to the international NGO Transparency International, Brazil figures out in the 69th position in the 2014 edition of the Corruption Perception Index (CPI), where the lowest score is the best. The CPI measures perceptions of corruption in the public sector, and Brazil ranks far behind even to countries in South America, like Chile and Uruguay (both tied in the 21st position). Corruption is an important issue in Brazilian politics, and virtually no year goes by without the revelation of major corruption scandals in the different branches and levels of the government. In national polls, Brazilians constantly complain about the widespread corruption that is perceived to dominate the political system. Notwithstanding a frustration that does not seem to wane with the succession of governmental administrations, the struggle for accountability has produced important landmarks in the institutional framework of Brazil.

Since redemocratization, Brazil has built or continuously strengthened the role of government agencies that audit and expose malfeasance with public money, like the Federal Court of Accounts and the Office of the Comptroller General. Their role goes beyond saving money and preventing overinvoice in public construction projects: by exposing corruption, they help to disseminate information to voters. Scholars have shown that politicians caught in association with corrupt deeds have lower rates of reelection and amount of campaign donation. There is no lack of known corrupt politicians still occupying senior positions, but that is not the whole story.

A major landmark in the fight against corruption was the approval of the Law No. 135/2010, known as Ficha Limpa (Clean Record), which tightens eligibility criteria for running for office in the country. This law originated from public demand, as 1.5 million voters signed a petition calling for its approval by the Congress. Previously, only individuals convicted at final decisions—without possibility of recourse to higher courts—were ineligible. However, this was not very effective because Bra-

A Brazilian demonstrates in favor of the "ficha limpa" law, which prohibits people who have been sentenced in court to run for public office, in front of opponents in Brasília, Brazil, on September 22, 2010. (AP Photo/Eraldo Peres)

zil's judicial system is notoriously slow and allows multiple appellations procedures. The result was that, by 2010, over one-quarter of the members of the lower house of Congress were facing charges of wrongdoing. While some politicians would be punished at the ballot after the discovery of a corruption scandal, others could get away from charges of wrongdoings and convictions by lower courts and be reelected. Clientelistic practices—like granting individual favors for voters and job appointments in the government—are used by corrupt politicians to survive in power, despite reputation damages. The new law disqualifies politicians from running for office for eight years in case of conviction of serious crime by collegiate decisions (made by a group of judges). The aim of the law is twofold: (1) to prevent corrupt acts by discouraging politicians from committing wrongdoings due to future electoral consequences and (2) to avoid having politicians with known criminal record continue to hold office.

Finally, even the long-standing popular belief that the wealthy could always get away without serving time is now under challenge. In the last decade, a few politicians occupying positions as high as governorship were jailed for the first time—if not by a final court decision, at least temporarily. In 2013, the Supreme Court finally issued jail sentences to senior members of the PT and key political allies convicted in the Mensalão scandal. For the first time in its history, senior Brazilian lawmakers and former ministers went to jail for using public money to buy votes in the Congress and pay illegal campaign expenses. In another scandal involving the state oil company, Petrobras, senior executives, politicians, and contractors were jailed in 2014 and 2015 for running a corruption scheme that transferred kickbacks from contracts to fund parties, with PT being the main beneficiary. The full extent of this latest corruption scandal is still to be determined by investigations of the police and the Public Ministry, but Petrobras recognized in 2015 losses of over $2 billion due to corruption.

THE MENSALÃO SCANDAL

The Mensalão, or the big monthly allowance, was a scandal that broke out in 2005 during Lula's first term. Roberto Jefferson, then president of an allied party (PTB), defended himself from charges of corruption by attacking the PT and the government. He accused Lula's chief of staff, José Dirceu, of managing a scheme to buy the support of individual politicians by offering monthly cash allowances. Subsequent investigations confirmed some of the allegations and revealed other malfeasances involving the PT and allied parties. Because it involved lawmakers, the Brazilian law required the criminal case to be tried by the Supreme Court. The court found 25 people guilty, including Jefferson himself and Dirceu.

Foreign Relations

Brazil's landmass occupies half of the total area of South America. This makes Brazil to share borders with 10 different countries: Argentina, Paraguay, Uruguay, Bolivia, Peru, Colombia, Venezuela, Guyana, French Guiana, and Suriname; the exceptions are Chile and Ecuador. Due to its size and economic prowess, Brazil naturally occupies a prominent place in terms of foreign relations in the Southern Cone. In addition to this natural vocation, the Ministry of External Relations, also known as Itamaraty, is staffed by a cadre of civil servants recruited through meritocratic means. Many government positions in Brazil have been given in exchange of political support, but rarely was the case of politicians or wealthy donors receiving ambassadorships through political connections. Foreign affairs are professionally run as a state business and out of clientelistic exchanges. Even the minister of external relations, which is a political position, tends to come from civil servants who made their career at Itamaraty, regarded as highly professional.

During the military dictatorship, Brazil kept for the most part strong links with the United States, which was natural for an anticommunist military government in an ideologically polarized world. In the 1980s, Brazil's foreign policy was focused on addressing the serious economic crisis and renegotiating the external debt. The economic stability achieved in the mid-1990s coupled with the charisma and international respect enjoyed by the presidents Fernando Henrique Cardoso and Luiz Inácio Lula da Silva helped to consolidate Brazil as an important player in the world stage.

The international status of Brazil is of a middle power country, leading important negotiations on behalf of developing countries and seeking to occupy top positions at multilateral organizations. Throughout the government of Fernando Henrique Cardoso (1995–2002), Brazil's foreign relations were focused on opening markets for its products and strengthening regional integration with its neighbors. During Lula's tenure (2003–2009), Brazil's foreign relations took a more pronounced left turn, emphasizing South–South cooperation and strengthening relations with left-leaning regimes (like Cuba and Venezuela) and African countries. Under Dilma Rousseff, analysts have pointed out a more pragmatic foreign policy, but no significant ideological changes have occurred. In fact, Rousseff is criticized for not having too much appetite for foreign policy, quietly neglecting initiatives in this area.

One of the top international priorities of Brazil is to occupy a permanent seat in the United Nations Security Council. To achieve that, the country has been articulating a coalition of nations in support of the reform and enlargement of the Security Council. In economic negotiations, Brazil has kept a strategy of leading the regional economic bloc of Mercosul and of privileging multilateral trade negotiations at the World Trade Organizations (WTO) instead of signing bilateral trade and investments agreements. This strategy has received criticism at home, as the Mercosul is a rather small market with severe disagreements among its members and the WTO has been slow in concluding multilateral trade negotiations. There is domestic hopes that the deadlock at the multilateral trading system may be resolved with the rise of WTO's new director, the Brazilian career diplomat Roberto Azevêdo. He became the head of WTO in September 2013 for a four-year term.

CONCLUSION

Brazil came a long way to consolidate its current democratic regime. After having passed through a European-like empire to an oligarchic republic and later a military dictatorship, Brazil enters the 21st century as one of the largest mass democracies of the world. Despite various criticism, it is an inclusive, multiparty democracy, where political disagreements are solved through institutional solutions, not coups. Civil society is actively engaged in public debates and voices its concerns and disagreements. Clearly, the challenges are many, and the debates about institutional reforms of the political system are intense. However, it is definitely a country where voters are heard, the suffrage is universal, and politicians have been increasingly held accountable.

REFERENCES

Couto, Ronaldo Costa. 1999. *História indiscreta da ditadura e da abertura—Brasil: 1964–1985*, 2nd ed. Rio de Janeiro: Editora Record.

Geddes, Barbara. 1994. *Politician's Dilemma: Building State Capacity in Latin America.* Berkeley: University of California Press.

Graham, Richard. 1990. *Patronage and Politics in Nineteenth-Century Brazil.* Stanford, CA: Stanford University Press.

Skidmore, Thomas. 1999. *Brazil: Five Centuries of Change.* New York: Oxford University Press.

Soares, Glaucio Ary Dillon. 2001. *A democracia interrompida.* Rio de Janeiro: Editora FGV.

Suggested Reading

Ames, Barry. *The Deadlock of Democracy in Brazil: Interests, Identities, and Institutions in Comparative Politics.* Ann Arbor: University of Michigan Press, 2001.

Cheibub, José. 2007. *Presidentialism, Parliamentarism, and Democracy.* New York: Cambridge University Press.

Goertzel, Ted, and Paulo Roberto Almeida (eds.). 2014. *The Drama of Brazilian Politics: From 1814–2015.* Amazon Digital Service.

Hagopian, Frances. 1996. *Traditional Politics and Regime Change in Brazil.* New York: Cambridge University Press.

Isbester, Katherine (ed.). 2010. *The Paradox of Democracy in Latin America: Ten Country Studies of Division and Resilience.* North York : University of Toronto Press.

McCann, Bryan. 2008. *The Throes of Democracy: Brazil since 1989.* Black Point: Fernwood Publishing.

Melo, Marcus André, and Carlos Pereira. 2013. *Making Brazil Work: Checking the President in a Multiparty System.* New York: Palgrave Macmillan.

Power, Timothy, and Matthew Taylor (eds.). 2011. *Corruption and Democracy in Brazil: The Struggle for Accountability.* Notre Dame: University of Notre Dame Press.

Samuels, David. 2003. *Ambition, Federalism, and Legislative Politics in Brazil.* Cambridge: Cambridge University Press.

Schneider, Ben Ross. 2013. *Hierarchical Capitalism in Latin America: Business, Labor, and the Challenges of Equitable Development.* New York: Cambridge University Press.

Economy

Rodrigo R. Coutinho

INTRODUCTION

Since 2009, many of the most important newspapers and magazines worldwide have repeatedly reported on Brazil. Topics have included its recent and increasing economic growth, its rapid recovery from the world crisis initiated in late 2008, the stabilization of its democracy, and its increasing internationalization. There have also been references to Brazil's increasing exports of agricultural goods, commodities, and industrialized goods, and to its increasing role in world organizations, such as the World Trade Organization (WTO). Its increasing leadership role in regional Latin American and South American organizations, such as the Mercosul (Southern Common Market, which comprises five member states: Brazil, Argentina, Paraguay, Uruguay, and, more recently, Venezuela), and Unasul (União de Nações Sul-Americanas, or Union of South American Nations, which includes 12 of the 13 South American countries, except for French Guiana, as member states), has been on the news as well. Some of these reports have also focused on the country's preparation to hold big events, such as the 2014 FIFA World Cup and the 2016 Olympic and Para-Olympic Games in Rio de Janeiro, and on other events recently held in Brazil, including the 2011 Miss Universe contest, the 2013 Confederation Cup, the World Youth Day of 2013, the Rio+20 (the UN Conference for Sustainable Development) in 2012, the 2011 World Military Games, and the 2007 Pan American Games.

However, not everything that international media have shown about the largest South American country is good news. Brazil has continuously been reported because of its violence, human rights abuses, and destruction of the environment, especially when it is related to the Amazon forests, a topic of preference of world magazines and newspapers. The country was on the news due to its "equivocal" involvement in negotiations with Iran and Honduras, after the ousting of its president

José Manuel Zelaya, because of its leadership in the United Nations' MINUSTAH Mission (Mission des Nations Unies pour la Stabilisation en Haïti, or United Nations Stabilization Mission in Haiti) in Haiti, because of its "bid" to hold a permanent seat at the UN Security Council (it had been a rotating member 10 times), for its combats against trafficking in the *favelas* (slums) of Rio de Janeiro, and for its continuous dialogues with political figures the developed Western world frequently excoriates, such as Venezuelan presidents Hugo Chávez and Nicolás Maduro, Bolivian president Evo Morales, Ecuadorian president Rafael Correa, and Cuban presidents Fidel and Raul Castro.

Some of these reports were and are frequently published in/from venues such as the Council on Foreign Relations, *Foreign Policy* magazine, *The Economist*, *The New York Times*, *The Wall Street Journal*, *The Financial Times*, and *60 Minutes* (CBS). A fine example is the famous cover of *The Economist* from November 12, 2009, in which the statue of Christ seems to be taking off from the Corcovado Mountain, on which it stands, followed by a 14-page special on Brazil titled "Brazil Takes Off." On September 28, 2013, however, *The Economist* published another 14-page report, in which it contests if Brazil has blown it, titled "Has Brazil Blown It?" It is interesting to observe the difference between the two covers in a period of four years: one is very optimistic and the other extremely pessimistic. Such diverse viewpoints about Brazil make it of extreme importance to look closer at the country and its numbers, so as to have a better understanding of the causes for the country's success, eventual failures, and most likely future and economic growth, as well as to look at some challenges Brazil now faces and if and how the country can (further) foster its development.

Data from the U.S. Department of State ("U.S. Relations with Brazil") and the Central Intelligence Agency (CIA, The World Factbook) show that Brazil is the world's largest wetland area (with 14% of the world's renewable freshwater). The country has an annual population growth rate of 1.02 percent, has a mostly Roman Catholic population (65%), a literacy rate of 90.3 percent among the adult population, a life expectancy of 73.1 years, and has an infant mortality rate of 21/1,000. The same sources show that the country's national identity is strong, and Brazil prides itself on being open to all races; according to the CIA, 47.7 percent of the population is white, 43.1 percent mulatto, 7.6 percent black, 1.1 percent Asian, and 0.4 percent indigenous (2010 estimate). More than 130 million people are eligible to vote in Brazil, a mandatory civic duty. It is a federal republic, with 26 states and a federal district, 81 senators, and 513 deputies. Among its population, 85.7 percent lived in urban areas in 2015, and six major group origins prevail: Portuguese, Africans, other Europeans (especially Italians, Spanish, and Germans), Middle Eastern, Japanese, and other Asian immigrant groups.

According to *The Economist* (Pocket World in Figures, 2013 Edition), Brazil is the world's fifth-largest country (8,512,000 square kilometers), and ranks fifth in world population (195 million). It has the world's seventh-largest economy (following the United States, China, Japan, Germany, France, and the United Kingdom), and is soon expected to become the world's fifth. In 2015, the country was the world's fifth largest in industrial output ($478 billion), sixth in manufacturing ($281 billion), eighth in services ($1,201 billion), and fifth in agriculture ($103 billion). However, according to the same report, Brazil was ranked 37th in global competitiveness for doing business, and 40th in business environment. This ranking is a contrast, considering that, according to the World Bank ("Foreign Direct Investment, Net Inflows"), since 2009 Brazil has been, on average, among the four most attractive countries for foreign direct investments (following China and Hong Kong, the United States and Belgium).

Brazil's nominal GDP (gross domestic product) is $2.2 trillion (2013 estimate), and its per capita GDP is of $12,100 (2013 estimate). Its major foreign markets are China (17%), the United States (11%), and Argentina (7%), and its major suppliers are China (15%), the United States (15%), and Argentina (7%). In 2013, however, according to the United Nations Development Programme (UNDP, Human Development Reports), Brazil was ranked 79th among 194 countries in the Human Development Index, with 0.74. Nevertheless, in the past few years (especially during President Lula's term in office), it has lifted around 35 million people out of poverty. Its current unemployment rate, according to the CIA, is 6.9 percent (2015 estimate).

Today, Brazil is one of the creditors of the International Monetary Fund (IMF). In December 2011, the country received a visit from the IMF chief Christine Lagarde. Brazil's stand during that time, according to finance minister Guido Mantega, was not to borrow money from the IMF (as it had mostly been the case), but to lend it. The Brazilian Development Bank currently has more money than the World Bank. The country has one of the most balanced and diversified economies in the world. It is the world's largest producer of coffee, sugarcane, tropical fruits, and frozen concentrated orange juice, and is the world leader in the production of soybeans, cotton, tobacco, cocoa, oilseeds, corn, pork, poultry, beef, milk, and seafood, among other items. Brazil also has abundant sources of energy. It is self-sufficient in gas and oil (it is to become among the top 10 oil exporters in the years to come, after gigantic recent discoveries offshore), and has an established and well-developed ethanol

A coffee plantation in the city of Carmo, south of Minas Gerais State, in 2014. Minas Gerais is presently responsible for more than 50 percent of the coffee production in Brazil. (Adrianoreis/ Dreamstime.com)

(sugarcane-based) industry, which has enabled the country to be "a superpower in renewable energy in the twenty-first century" (Rohter 2010, p. 182).

According to Larry Rohter, ethanol from sugarcane "is by far the most attractive: for every unit of energy that is expended to produce sugarcane ethanol, the final product generates more than eight units of energy. In contrast, the energy ratio of ethanol made from corn, the favored source in the United States, is less than two to one" (2010, p. 183). Ethanol is described by its advocates as a wonder fuel—it is renewable, cheaper to produce, and more environment friendly than gasoline. Furthermore, Brazil has an enormous hydropower capacity, and barely a quarter of it has been tapped (though it represents more than 70 percent of the electricity consumed in Brazil). Water in the country is not only abundant, but also renewable, and the cost of hydroelectric power is very low.

As Rohter (2010, p. 199) points out, "Brazil's energy system today is one of the least carbon-intensive in the world, with nearly half of total consumption provided by less-polluting renewable fuels." He remarks that many other countries would like to be in a similar position, which has placed Brazil in a unique spot for world climate mediations (2010, p. 199). Brazil leads biofuels technology, and was the first country to achieve sustainable use of biofuels. Most of its fleet of vehicles are flexible fuel vehicles (FFVs), which work with any mix of gasoline and ethanol, at any time. Not for less, Brazil is a leading player in international climate change negotiations. The country also has vast deposits of iron ore, bauxite, copper, manganese, nickel, lead, tin, chromite, beryllium, copper, tungsten, uranium, zinc, and gold, among others.

Brazilian industries range from pulp and paper to cellulose, mining, steel, cement, automotive, aluminum, heavy machinery, electric equipment, chemicals and petrochemicals, plastic, textiles, shoes, computers, aircrafts, and consumer durables. Brazil is also a leading player in deep-sea oil production. It has a diverse and sophisticated services industry, which includes telecommunications, banking, energy, commerce, transportation, and computing sectors.

Brazilian companies, once faced with an environment with monetary reforms, foreign debts negotiations, and privatizations, are now not only well prepared for growing but have also been integrating more and more into the world economy, and can already be seen worldwide. Some examples include, but are not limited to, Vale (among the three world-leading mining companies), Inbev (now the owner of Anheuser-Busch), Petrobras, Embraer (the world's third-largest aircraft manufacturer, following Boeing and Airbus), JBS Friboi (world's largest meatpacking company, which bought Swift Foods Company), Perdigão, WEG (electric motors), Marcopolo, Bradesco, and Itaú-Unibanco.

Brazil has large and well-developed agricultural, mining, manufacturing, and service sectors, and the country's economy outweighs that of all other South American countries (CIA, "The World Factbook"). Besides, Brazil is expanding its presence in world markets, and since 2003, it has steadily improved its macroeconomic stability, building up foreign reserves, and reducing its debt profile by shifting its debt burden toward real denominated and domestically held instruments. In 2008, Brazil became a net external creditor, and two ratings agencies awarded investment grade status to its debt. The CIA also highlights Brazil's emergence from the 2008 economic crisis, with a GDP growth of 7.5 percent in 2010, a historic high. To illustrate the diversity of the Brazilian economy mentioned earlier, and still according to the CIA data, the

Brazilian GDP composition, by end use, as of 2013 (estimate), was 5.5 percent in agriculture, 26.5 percent in industry, and 68.1 percent in services, levels similar to those of developed nations.

However, since 2010, the economic growth has slowed down. According to the IPEA (Instituto de Pesquisa Econômica Aplicada, or Institute of Applied Economic Research), Brazilian GDP has grown, between 2004 and 2013, as seen in Table 4.1.

The higher growth rates verified in the 2004–2008 period reflect, among other elements, the high prices of commodities, of which Brazil is a major exporter. However, after a rapid recovery from the 2008–2009 crisis, in 2010, the GDP growth slowed down, and most economists predicted an estimated GDP growth of no more than 1.7 percent for 2014, and negative growth for 2015.

In addition to that, after a period of lower interest rates—the Sistema Especial de Liquidação e Custódia (Special Clearance and Escrow System) rate (the Brazilian Bank's overnight rate) went down from 12.5 percent per year in July 2011 to 7.35 percent per year in October 2012. It has increased again and is now, as of April 2014, at 11.0 percent per year. The lower interest rates also followed the financial crisis, during which the rates went from 13.75 percent in December 2008 to 8.75 percent in July 2009 and up again to 12.5 percent in July 2011. The lower interest rates favored consumption following the crisis, and again after July 2011, but as inflation increased, the rates followed and increased too.

The annual rate of inflation in Brazil, according to the IPEA and Índice Nacional de Preços ao Consumidor Amplo (Amplified Consumer Prices Index), between 2004 and 2013, is shown in Table 4.2 (per year). It is interesting to note that, for 2004, according to the Banco Central do Brasil (Central Bank of Brazil), the inflation target was 5.5 percent per year, with a tolerance of 2.5 percent; and, since 2005, the target has been 4.5 percent per year, with a tolerance of 2 percent. Therefore, one can note that, since 2010, inflation has been closer to the top of the tolerance range than to the target. Inflation reached 6.41 percent in 2014 and it is expected to hit close to 10 percent in 2015.

According to IPEA data, trade went from a deficit of US$698 million in 2000 to a record surplus of US$46,456 million in 2006, mostly stimulated by the commodities

TABLE 4.1   Brazilian GDP, 2004–2013

2004	2005	2006	2007	2008	2009	2010	2011	2012	2013
5.71%	3.16%	3.96%	6.09%	5.17%	–0.33%	7.53%	2.73%	1.03%	2.28%

Source: IPEA. IPEA Data. Retrieved from http://www.ipeadata.gov.br.

TABLE 4.2   Inflation in Brazil, 2004–2013

2004	2005	2006	2007	2008	2009	2010	2011	2012	2013
7.60%	5.69%	3.14%	4.46%	5.90%	4.31%	5.91%	6.50%	5.84%	5.91%

Source: IPEA. IPEA Data. Retrieved from http://www.ipeadata.gov.br.

boom of the year 2000 and a high demand from China; however, since 2006, it has been decreasing, having reached a surplus of US$2,258 in 2013, mainly due to decreasing demand for commodities and higher imports. Still, according to the Instituto Brasileiro de Geografia e Estatística (Brazilian Institute of Geography and Statistics), the savings rate closed at 13.9 percent in 2013, whereas the investment rate closed at 18.4 percent. These levels are considered low in comparison to developed nations and even lower in comparison to developing nations. The former—low savings rates—reflects an increase in consumption by both families and government—by families through government policies to stimulate consumption, through lower interest rates, and lower taxes for specific goods, such as automotives and white goods, and by the government through higher expenditures in social security and loosened fiscal policies. The latter—low investment rates—reflects not only lower savings rates but, by and larger, higher government expenditures.

With all that in mind, it is important to understand how Brazil got to its current condition, since colonization, through industrialization and stabilization, in order to establish an eventual path toward the future.

COLONIZATION: AN EXPORT-LED ECONOMY

The legacy of colonialism, namely arbitrary territorial borders, authoritarian political institutions, and infrastructure for extraction and production of primary products, prevailed in Brazil, as in the other countries of the Latin American region. As Halperín Donghi (1993) pointed out, Mexico was a major silver exporter, whereas agriculture remained important due to local consumption in the colonial period. The Greater Antilles (Cuba, Santo Domingo, and Puerto Rico) had intensive cultivation of commercial crops, such as sugar. Colombia was a major exporter of gold, and Venezuela was a major exporter of cacao, indigo, coffee, and cotton. Bolivia was a major exporter of silver (p. 25). However, even after the Spanish Crown's reforms, the interregional trade in Spanish America was altered and, according to Halperín Donghi (1993, p. 25), "encouraged the growth of export economies based on agriculture and ranching alongside the older ones based on mining." According to him, "Only Venezuela and Cuba had developed export economies completely unassociated with mining. Meanwhile, mining retained its pride of place within the growing economies of Mexico, New Granada [Colombia], and the Río de la Plata [Bolivia, Argentina, Paraguay and Uruguay] by dominating their exports" (1993, p. 25). Halperín Donghi explains that normally the mining sector produced for the overseas market and the agricultural sector basically for local or regional markets (1993, p. 25). He mentions the exceptions, which were tobacco and sugar from Cuba, cacao from Venezuela and Ecuador, and hides from the Río de la Plata (p. 25). The reform also included the addition of governors and *corregidores*, *intendencies*, and *subdelegados* (mayors, intendancies, and deputies) to the viceroyalties, aiming at strengthening control over the region (p. 25).

The transformation that took place in Brazil was more substantial than that in Spanish America in the course of the 18th century, as it moved from essentially an important sugar-producing area to the extraction of gold and diamond. The exploration of its north and south territories, extended Portuguese control to tropical forest regions in Amazonia and temperate grasslands in Rio Grande do Sul (Halperín Donghi 1993, p. 25). The "Gold cycle" centered around Minas Gerais, motivated the transfer of the capital from Salvador, Bahia, to Rio de Janeiro, the closest port to Minas Gerais, in 1763. However, one might say that the relations between Brazilian producers and the Portuguese merchants of port cities like Bahia, Recife, and Rio de Janeiro contrasted with those between their counterparts in Spanish America. According to Halperín Donghi (1993), the Portuguese colonizing power was weaker than that of Spain, and "not until the eighteenth century did tiny Portugal succeed in creating a colonial administration comparable to Spain's of the sixteenth century, and the Portuguese ability to impose a cohesive system of mercantile exploitation on its colonies was correspondingly weaker" (p. 39).

For two centuries, the colonization in Brazil was based on the exploration of wood (*Pau-Brasil*, or Brazilwood) on the coast and on sugar plantation in the northeast (Galeano 1976, p. 73). However, during the 18th century, the Brazilian gold extraction in Minas Gerais was higher than that of the Hispanic colonies. The country's population grew 11 times (from 300,000) during that period. According to Galeano (1976, p. 73), the gold from Brazil ended up in England, facilitated by the Treaty of Methuen, signed in 1703 by Portugal and England, and this was one of the main facilitators for the industrialization of England. The Treaty of Methuen opened the Portuguese market to England in exchange for the English market for Portuguese wines. It also opened the Brazilian market for English products. Due to commercial inequalities, though, Portugal ended up by paying its deficits to England with Brazilian gold.

In sum, most Latin American economies were export-led economies then. For example, Mexico and Bolivia were major exporters of silver, Colombia was a major exporter of gold, and Venezuela was a major exporter of cacao. Brazil in the 18th century went from being a sugar-producing area to also being an exporter of gold and diamond, coffee, and, later, rubber. It is also interesting to point out the importance of the *ciclo da borracha* (or the rubber cycle) in Brazil. Around 1770, rubber was used only as an eraser. However, 70 years later, Charles Goodyear and Hancock discovered the vulcanization process of rubber, which, in 1850, was already used to produce tires. In 1890, the *seringueira* (the rubber tree) was responsible for 10 percent of Brazilian exports and, in 1910, for 40 percent, most of it coming from Acre, a former Bolivian territory. Manaus was the world's rubber capital at that time, and rubber generated so much money that the rich people who lived there built mansions of extravagant architecture and sumptuous décor, with wood from the Far East, tiles from Portugal, and columns of Carrara marble and French furniture. Expensive food came from Rio de Janeiro, dressmakers and tailors from Europe made dresses and suits, and the children went to study in England. The Amazonas Theater, a monument in eclectic style, is the greatest symbol of vertigo of those fortunes in the turn of the century (Galeano 1976).

By 1808, with the occupation of Spain by Napoleon and the imprisonment of the Spanish king, Ferdinand VII, the imperial power of Spain over its colonies began to decrease, and, within two years, the process of political independence was developed and spread throughout most of Spanish America (Halperín Donghi 1993, p. 47). This process, which occurred through violent struggles, lasted until 1825, with the independence of the Republic of Bolivia. In the case of Brazil, however, the looser colonial bonds with Portugal had prepared the way for a less disruptive separation. With Napoleonic expansion, the Portuguese court moved to Rio de Janeiro in 1807, backed up by the British naval forces. Following this event, Great Britain received the status of most-favored nation. In 1820, with the liberal revolution in Portugal, the king returned to Portugal, but his son Pedro stayed in Brazil. In 1822, Pedro, the prince regent, proclaimed the independence of the country on September 7, 1822, thus becoming the emperor of Brazil.

According to Halperín Donghi (1993), "In 1825, only ruins remained of the former Spanish empire. Its trading networks had been undone" (p. 74). In his words again, "The years of warfare had widened the circles of political power, and, in many places, they had put it within reach of new social groups. The people who had struggled for independence now looked ahead to a brighter future, but the legacy of colonialism was a heavy one, and the new order did not quickly emerge" (p. 74). Halperín Donghi explains that similar problems happened in Brazil, "despite the fact that the country's shift from colony to nation was more peaceful" (p. 74).

The author also highlights the struggle for political and administrative power in the period, the use of political connections as routes to landownership, the lack of European investments in the region during the first half of the 19th century, and the deterioration of artisan production, especially in the Andean highlights, due to imports of cotton cloth from New England. He mentions the generalized growth in imports in the region, the increasing British dominance, and, following the Mexican-American War (1846–1848), the appearance of the United States as a growing "force," starting with Central America and Cuba (1993, p. 74). Particularly in the case of Brazil, Halperín Donghi stresses the rise of the coffee economy in the 1840s, which brought some political stability following conflicts in the 1820s and 1830s, and the divergence between Brazil and England regarding slave trading, which ended in 1851. These aspects call attention to the "unification" and consolidation of Brazil, opposite to the increasing political fragmentation of Spanish America, in spite of the attempts from Simon Bolívar to maintain the unity of Spanish America (1993, p. 91).

After 1850, Latin America's economies began to develop in a way that matched the elites' old aspirations. Gas lighting started spreading over the region, and the increasing availability of credit enabled national governments to consolidate their power. However, after 1870, a neocolonial order emerged in the region, based on the production of primary products for European markets, while importing manufactured goods from Europe in return. This new order was in line with the interest of the local ruling classes and landowners, who could not only protect themselves from the eventual harms of the process but also actually benefit from it, to the detriment of the ordinary people, especially in the rural areas. Immigration played an important role in this order, as with the end of slave trade, and increasing masses of immigrants, especially from Italy, started to work in the haciendas and rural areas of Brazil, in the cultivation of coffee, among other products.

The neocolonial order, or British imperialism, that lasted until shortly after the crisis of 1929, was characterized by the presence of positivism in Brazil (e.g., the Brazilian flag has two positivist motto written on it—*Ordem e Progresso*, or order and progress), and, in many cases (e.g., Argentina, Paraguay, Uruguay, Mexico, Venezuela, and most of Central America), military dictatorship took over. In the case of Ecuador, a progressive dictatorship not rooted in military basis but in the

Catholic Church dominated the country. Colombia, Chile, and Peru were ruled by a liberal progressivism. In Brazil, the Crown began to lose support from the Church, the army, and the liberals, and the country entered a period of rapid social and economic change. Between 1870 and 1885, for example, the country's export structure displayed a decisive shift, and in 1889, the landowners, discontent with the actions and decisions of the imperial order, led to a military coup that ended up by replacing the empire with the establishment of a republic.

That order lasted for half a century. The successes and failures of the region's export economies by the 1920s began to produce social realities that no longer fit the political forms inherited from the previous period of economic expansion. The oligarchical republics and progressive dictatorships that had been erected on narrow social bases in the Latin American states needed more popular support in order to ensure their stability.

INDUSTRIALIZATION

After the crisis of 1929, the decline in trade affected the countries in Latin America, and a shift of resources to the manufacturing sector occurred. However, not all countries in the region had enough market size, as Brazil, Argentina, and Mexico did to support industrialization. Therefore, in the longer term, results from the import-substitution industrialization (ISI) of the 1930s varied across the region. Actually, instead of that, what happened, according to Halperín Donghi (1993), was that "this process of partial industrialization further accentuated the regional inequalities that had appeared during the earlier period of export expansion" (p. 212). Countries with larger markets, higher-skilled labor, and larger urban areas, thus, benefited more from the process than others with lower-skilled, rural population.

World War II further boosted the process of ISI, as trade with European countries and the United States, after its engagement in the conflict, was further restricted, since most of the efforts were being directed to the war. Therefore, the economies in Latin America were somehow "isolated." The impact, though, was harder on imports than on exports, which increased buying power in the region, something that ended up by stimulating local factories, which enjoyed a period of "dizzying growth" (Halperín Donghi 1993, p. 213). The *bonanza*, however, was not equal within the region. Countries "that barely produced supplies of food sufficient to feed their populations (and there were a number of these, from Mexico to Chile) suffered with particular intensity" (Halperín Donghi 1993, p. 212), while "the industries of the larger Latin American countries were able not only to dominate their domestic markets but also to initiate exportation of manufactured good. Brazilian industrial production soon reached markets in Spanish America and Africa" (Halperín Donghi 1993, p. 213), and the increasing importance of Brazil in the region was able to attract more investments to the country. According to Halperín Donghi, "The strategic importance of Brazil convinced U.S. policymakers to subsidize the creation [in 1941] of a state-owned steel industry at Volta Redonda in the state of Rio de Janeiro despite their long-standing hostility to such enterprises" (1993, p. 234).

After the war, industrialization became a "question mark" in the region, as returning to an export-led economy again became a possibility. Demand for Latin

America's primary exports rose, but the supply of capital goods fell, and the capital accumulated in the region was directed to nationalizing foreign-owned public services, repatriating the public debt, and importing manufactures not destined preferentially for European reconstruction. Gradually, technical modernization was no longer a priority in the region.

By the mid-1950s the postwar export boom had ended, inflation risen, the trade balance deteriorated, and devaluations of currencies spread in the region. However, according to Halperín Donghi (1993, p. 250), developmentalist policies in the region, though, promoted the expansion of basic industries, such as steel and automotive—a Volkswagen industry was created in São Paulo, later followed by other car industries in the state—which brought back dynamism to industrializing initiatives in the region and attracted foreign investment. Once more, though, Latin American industrialization was unequal, and whereas countries such as Brazil and Mexico, with larger markets, could take this step, others like Chile and Peru were not able to transform their economies in the same way.

With increasing inequalities among, but also within, countries in the region, especially between urban and rural areas, social tension rose, and the ideas of communism and socialism spread in the region as an alternative to liberalism, especially after the Cuban Revolution of 1959. The 1945–1960 period was also characterized by the spread, in the region, of what later came to be known as populism. As examples of populism, one can mention Getúlio Vargas in Brazil and Juan Domingo Perón in Argentina.

In the postwar period, as the impetus on industrialization slowed, the Soviet Union gained importance, the Cuban Revolution occurred, and several Latin American countries "appeared" to be moving to the left, as some achievements in Cuba rose in the imagination of many in Latin America. In the 1960s, following the Cuban Revolution, the concepts of First, Second, and Third Worlds emerged, and Latin American countries saw themselves apart from the prosperous economies of both the First and Second Worlds.

The period was also marked by the rise of ideas known as "dependency theories," first developed by Raul Prebisch but later followed by others such as Enzo Faletto and Fernando Henrique Cardoso (Cardoso 2010), who in 1967 wrote a book called *Dependência e Desenvolvimento na América Latina* (*Dependency and Development in Latin America*), in which the underdevelopment of Latin America was characterized as structural. The authors followed the analysis of Hans Singer and the dependency theory—according to which underdeveloped countries would be "trapped" by a continuous loss in trade, due to imports of industrialized goods and exports of commodities—which segments countries between center and periphery, being the underdeveloped countries in the latter category. This theory also applies to countries whose regions can be segmented between center and periphery—and this is somehow the case of Brazil.

According to Cardoso (2010), it would be necessary for the countries in the periphery (the underdeveloped) to escape from the trap of producing agricultural goods and exporting minerals. It would be, thus, necessary to accumulate capital through taxes to invest in technology and to orient production to the internal market, to keep exchange rates under control, and to attract capital for investment. In order to accomplish that, it would be necessary to improve government efficiency and public policies. Cardoso and Faletto (in Cardoso 2010) also argued that there are differences among the underdeveloped countries as regards the opportunities of growth and integration to the world markets. This is in agreement with Harrod-Domar's theory of industrial development, according to which economic development equals an increase in production. Therefore, to increase production, a country must invest in the factors that go into production, and, so as to invest, a country must divert resources from consumption toward savings.

Still according to the dependency theory, to break that subordinate role, radical economic restructuring would be necessary, and socialist revolutions, as many *dependencistas* (dependency theorists) called, would be a recommended solution. Therefore, if on the one side one could argue that the Cuban Revolution, right "next door" of these subordinate economies, could serve as an example for other economies in the region, on the other side one could also understand why the United States suddenly increased its activism through a highly complex and differentiated presence in the region. U.S. activism in the region can be exemplified by its "cooperation" in the 1964 military coup in Brazil, which overthrew João Goulart from the presidency, installing a military government that endured until 1985, and following participations on later struggles in Honduras (late 1980s), Nicaragua ("Contras," after 1981), Guatemala (1982), and Panama (1984–1987), just to name a few.

Bulmer-Thomas (2003) highlights some of the strategies adopted by the Latin American countries, including Brazil, since the end of World War II. He makes some remarks on the positive and negative effects some of the strategies adopted have had over the Latin American economies. Some of the strategies, he discusses, are ISI, export promotion, primary-export development, and debt-led growth (pp. 296–390). In the case of Brazil, the ISI strategy was adopted through high increases on taxes—taxes on industrialized goods, on circulation of goods and services, on social security, among others. Somehow it worked. Through high barriers to trade (in the early 1960s tariffs on manufactured imports averaged 184 percent in Brazil), Brazil was able to develop, sometimes with the establishment of SOEs (state-owned enterprises), a broad industrial base—mining and steel, aluminum, automotive and auto parts, chemicals and petrochemicals, oil, consumer durables (or "white goods"), agribusiness, cellulose, pulp, paper, electric equipment, plastic, textiles, shoes, computers, livestock, aircraft, heavy machinery, among others. Brazil has also been able to diversify its services industry, which includes telecommunications, banking, energy, commerce, and transportation.

DEMOCRATIZATION

In the early 1970s, several changes in the international context created a substantially new situation in Latin America. Among these changes, one can mention the suspension by Nixon, in 1971, of the parity of the dollar to the gold, established since the Bretton Woods Accords of 1944, and the "consecutive" oil shocks of 1973 and 1979. These facts, among others, led to increasing economic difficulties in Latin America, especially to the countries with high levels of debt in dollar. In addition to that, according to Halperín Donghi (1993), a nonincreasing interest from the Soviet Union in the region discouraged further "engagement" from the United States in the region (p. 338). Therefore, a "new isolationism," economic difficulties, high levels of debt, high inflation, all together with difficult, albeit peaceful, transitions from dictatorships to democracies in the 1970s and 1980s, especially in Uruguay, Argentina, and Brazil, led to what was later called the "lost decade," the 1980s, especially, but not limited to, in Brazil and Argentina, where successive attempts to restore the economy through successive economic plans failed. In Colombia, Peru, and Bolivia, drug trafficking escalated, and violence followed, with the emergence of paramilitary groups such as Fuerzas Armadas Revolucionarias de Colombia (Revolutionary Armed Forces of Colombia) in Colombia and terrorist groups such as Sendero Luminoso (Shining Path) in Peru.

In Brazil, transition to democracy was particularly turbulent. In 1984, following popular protests, the so-called *Diretas Já!* (a claim for direct elections), a civil president took power, ending a 21-year period of military dictatorship. However, the civilian president, Tancredo Neves, elected by the majority of the Electoral College never took office—he became severely ill the day before his inauguration on March 15, 1985, and died on April 21, 1985. Thus, the vice president, José Sarney, took office. In 1988, Brazil approved a new constitution, and in 1989, direct elections occurred.

However, the president popularly elected in 1989, Fernando Collor de Mello, who took office on March 15, 1990, resigned on December 29, 1992 to prevent his trial of impeachment, economic crisis, mass demonstrations, and allegations of corruption. His vice president, Itamar Franco, then, took office, on the same day, December 29, 1992. Finally, democracy stabilized in Brazil, with presidential elections taking place every four years. Following Itamar Franco, Brazil had, in office, Fernando Henrique Cardoso, from 1995 to 2002, Luiz Inácio "Lula" da Silva, from 2003 to 2010, and Dilma Rousseff, currently in office, since January 1, 2011.

In 2002, however, Brazil went through a turbulent moment. According to Lourdes Sola (in Larry Diamond et al. 2008), "The market disciplines adopted by President Luiz Inácio Lula da Silva [from the left-of-center Workers' Party (PT)] upon his 2002 election were meant to address the crisis of confidence that had accompanied his win, as investors feared a break with the macroeconomic policies and the moderately market-friendly reforms of his predecessor, Fernando Henrique Cardoso of the Brazilian Social Democratic Party (PSDB)" (p. 125).

In short, Brazilian democracy is young yet stable, and its democratic institutions are still consolidating. The debt crisis of 1982 was devastating and generated regressive outcomes everywhere. However, since the debt crisis, still according to Sola, the country has been witnessing a shift toward the fuller social and political incorporation of those who had long been left on the system's edges, and the "broad social reforms that began in the 1980s gathered momentum and moved toward a tipping point in 1993, even as the embrace of market-friendly changes in economic policy was growing stronger. Cardoso's administration oversaw the expansion of universal public programs in health, education, and assistance to the elderly and disabled, as well as the implementation of conditional cash transfers programs such as Bolsa Escola, Bolsa Alimentação, and Projeto Alvorada, which would be, during the Lula administration, consolidated under the name Bolsa Família. The Comunidade Solidária program targeted the poorest communities in all Brazilian municipalities with basic social programs to help provide more and better health care, primary schooling, and (later) job training for the people living there. Other targeted land-reform and poverty-relief efforts sought to assist rural dwellers." According to Sola (2008, p. 127), this transformation of the welfare system is even more remarkable if one considers the external economic shocks in Brazil as a consequence of the Mexican peso crisis of 1994–1995, the Asian currency crisis of 1997, the Russian bond default of 1998, and the crisis of confidence surrounding Lula's election in the country. Still, the social reforms were followed by trade liberalization, the end of restrictions on capital mobility, privatization of public enterprises, and fiscal responsibility, many of these in accordance to the Washington Consensus.

WASHINGTON CONSENSUS

In the 1980s, a new set of reforms, which became known as the Washington Consensus, were designed to promote economic growth in underdeveloped countries. According to Dani Rodrik (2007, p. 17), these rules of good behavior are: fiscal discipline, reorientation of public expenditures, tax reforms, interest rate liberalization, unified and competitive exchange rates, trade liberalization, openness to direct foreign investment, privatization, deregulation, and secure property rights. In the end of the 1990s, however, the list was changed so as to include a series of so-called second-generation reforms. They were corporate governance, anticorruption, flexible labor markets, adherence to WTO disciplines, adherence to international financial codes and standards, "prudent" capital-account opening, non-intermediate exchange rate regimes, independent central banks/inflation targeting, social safety nets, and targeted poverty reduction. It is interesting to notice the shift in recommendations. While the original Washington Consensus favored liberalization (of trade, rates, and investments), deregulation, and a lesser, but a more responsible state, the second-generation reforms favored a more participative state, yet responsible (good governance, anticorruption, adherence to institutions), more geared toward social welfare (social safety nets, poverty reduction, inflation targeting) but still open.

Brazil went through several attempts to control inflation and to stabilize its currency in the 1980s and early 1990s, with several economic plans and currency exchanges, such as the so-called Plano Cruzado, Plano Cruzado II, Plano Collor, and Plano Bresser, until, finally, in 1994, Plano Real was developed under the direction of Cardoso as finance minister, during the presidency of Itamar Franco. The success of the plan is considered by many to be the main reason that led Cardoso to be elected president later in 1994.

During his term in office, Cardoso further expanded the liberalization started during Fernando Collor's term in office, further integrated Brazil into the world's economy, and expanded the privatization of public companies (e.g., in the telecommunications, energy, and mining sectors), which also started during Collor's term (with the privatization of Usiminas, a steel industry, in October 1991). He started and expanded social programs and aimed at institutional harmonization, with, for example, the approval of the Lei de Responsabilidade Fiscal (or fiscal responsibility law).

In addition to that, Brazil has successfully avoided what, according to William Easterly (2002), are "actions that create poor incentives for growth: high inflation, high black market premiums, high budget deficits, strongly negative real interest rates, restrictions on free trade, excessive red tape, and inadequate public services" (p. 237). Brazilian people responded positively to these actions. Most of the measures mentioned earlier are included in the neoliberal "set of reforms" known as the Washington Consensus, mentioned earlier, which were implemented in Latin America during the 1990s. However, once more its effects were felt unequally in the region. For example, countries that have committed to fiscal prudence, trade openness, and market orientation have been able to attract increasing foreign investments and be granted "investment grade" classification.

According to Dani Rodrik (2007), "Macroeconomic policies aim to achieve static and dynamic efficiency in the allocation of resources. Macroeconomic policies aim for macroeconomic and financial stability (p. 31). Rodrik (2007) also points out that "social policies aim at poverty reduction and social protection. . . . Efficiency requires property rights, the rule of law, and appropriate incentives. Macroeconomic and financial stability require sound money, fiscal solvency, and prudential regulation. Social inclusion requires incentive compatibility and appropriate targeting, and macroeconomic stability requires . . . central bank independence, adherence to international financial codes, and sundry 'structural reforms" (p. 219). These would be, according to the author, "universal principles" of sound economic management, and countries that have adhered to these principles have done well. Brazil is included in this group (p. 30-31).

Jeffrey Sachs (2005) believes that the focus on the public sector should be on five kinds of investments: human capital (health, education, nutrition), infrastructure (roads, power, water and sanitation, environmental conservation), natural capital (conservation of biodiversity and ecosystems), public institutional capital (a well-run public administration, judicial system, police force), and parts of knowledge capital (scientific research for health, energy, agriculture, climate, ecology) (p. 251). The private sector, however, must turn to investments in businesses (agriculture, industry), services, and in knowledge capital (new products and technologies building on scientific advances), and "household contributions to health, education, and nutrition that complement the public investments in human capital" (p. 251). In this case, Brazil has not done very well so far.

Following the ISI strategy and the Washington Consensus, and under the government of President Lula (2003–2010), Brazil turns more into a model of nationalism and development, preserving the state while keeping social stability, but without jeopardizing any of the achievements of Lula's predecessor, such as austerity, macroeconomic stability, low inflation, and a stable currency. Brazil moves closer to an example of state capitalism—as one could characterize China today, in which the state keeps control of the economy but also stimulates businesses. This strategy generated extraordinary results until the financial crisis of 2008–2009.

In regard to foreign policy, however, and according to Paulo Roberto de Almeida (2009), the Lula administration brought a new emphasis on preferential alliances. According to the author, "Besides a strong emphasis on political multilateralism traditional to Brazilian diplomacy (but now with an evident 'anti-hegemonic' leaning, i.e., against American unilateralism), the focus fell sharply onto South-South diplomacy, as well as in a great effort to see Mercosul reinforced and broadened as the basis for political integration and of consolidation of a unified economic space in South America" (p. 171). de Almeida argues that, "Lula's administration put in motion all kinds of tools and all forms of foreign policy—multilateralism, bilateral relations, and informal mechanisms of cooperation—in order to promote its new diplomatic priorities" (p. 175). During his administration, President Lula had tried to increase trade with his partners of Mercosul and other South American countries, as well as with Africa, and Asia, not to mention Russia—therefore, all BRICS (Brazil, Russia, India, China, and South Africa) countries were included, and with Europe.

GLOBALIZATION

The process of globalization that enhanced trade and financial integration in the world posed both opportunities and challenges to the national economies. Nowadays, governments actively compete with each other by pursuing policies that they believe will earn more market confidence and, therefore, would be able to attract trade and capital inflows: tight money, small government, low taxes, flexible labor legislation, deregulation, privatization, and market openness all around. According to Dani Rodrik (2007), these policies comprise what Thomas Friedman, in his book *The Lexus and the Olive Tree: Understanding Globalization*, called the Golden Straitjacket (p. 201). The author believes, though, that "reaping the efficiency benefits of complete international economic integration requires the further empowering of multilateral institutions and greater reliance on international standards" (Rodrik 2007, p. 212).

Paul Collier (2007) believes that the effects of globalization on the economies of developing countries come from three distinct processes: the first one is trade in goods, the second one is flows of capital, and the third one is the migration of people (p. 80). In regard to trade, according to Dani Rodrik (2007), "Very few countries have grown over long periods of time without experiencing an increase in the share of foreign trade in their national product (p. 219)." The author believes, though, that "no country has developed by only opening itself up to foreign trade and investment. The trick in the successful cases has been to combine the opportunities offered by world markets with a domestic investment and institution-building strategy to stimulate the animal spirits of domestic entrepreneurs (p. 219)."

Therefore, it seems that if, on the one side, globalization offers opportunities and challenges from the inside out, on the other side, it offers challenges from the outside. Thus, the major challenge for governments would be to turn opportunities and challenges into benefits for the country. One thing that looks certain, though, is the fact that there is a window of opportunities opened and that it is up to countries to figure out how, and to what extent, they want to take advantage of that. To Dani Rodrik (2007), trade can bring development, provided we accept the five simple principles that follow (p. 227):

1. Trade is a means to an end, not an end in itself.
2. Trade rules have to allow for diversity in national institutions and standards.
3. Nondemocratic countries cannot count on the same trade privileges as democratic ones.
4. Countries have the right to protect their own institutions and development priorities.
5. But countries do not have the right to impose their institutional preferences on others.

In reference to globalization, Cardoso (2010) mentions a statement by Giorgio Napolitano, former president of Italy, who said, at the end of the 1970s, when he was the secretary of the Italian Communist Party, that, "either we internationalize ourselves or they will internationalize us." According to Cardoso (2010), there is no *single way* to development, or, as Dani Rodrik (2007) affirms, "no one-size model fits all (p. 112)." Countries can select different paths toward developments, with varying success—after all, there are ways of internationalizing yet safeguarding national interests and better conditions of life for its peoples.

According to Jeffrey Sachs (2005), although antiglobalization movement leaders' motivations are accurate, their assessment of the problems are not: "High rates of foreign direct investment inflows have been associated with rapid economic growth. . . . The same is true with trade. Countries with open trade generally have grown more rapidly than countries with closed trade, and rising per capita incomes in most countries have generally been associated with a rise in the ratio of trade (exports plus imports) to GDP" (p. 356). Therefore, according to Sachs, it is important that the antiglobalization movement mobilizes its members to engage in a proglobalization movement that addresses the needs of the poorest of the poor, the global environment, and the spread of democracy. For Sachs (2005) this "is the kind of globalization championed by the Enlightenment—a globalization of democracies, multilateralism, science and technology, and a global economic system designed to meet human needs. We could call this an Enlightened Globalization" (p. 358).

It is interesting to note that the wealth generated in Brazil through commerce, in the early stages, was not used to foster development, but, rather, to import luxurious goods for the rich, as exemplified earlier during the *ciclo da borracha*. Many people built huge mansions and farms, and imported clothes and other luxurious goods from Europe—to such an extent that there is a saying that some people used to roll cigars with money in Brazil.

According to Carlos Alberto Montaner (2003),

the nationalistic, interventionist, and anti-market mentality, sustained by the great majority of Latin Americans, does not result from an appraisal of the economic realities of the region. What is obvious in Latin America—as opposed to the United States, Canada, Europe, and Japan and other Asian enclaves—is the weakness of the production apparatus [with few exceptions]. National companies, state and private-owned, produce little, and what they produce is generally of very low quality and little added value. At least in the case of state companies, it is done with disregard for the real costs of operation. Distribution methods tend to be inefficient. Management makes no use of modern administrative tools. Banking systems are untrustworthy and the legislation that regulates them poor. Innovation is minimal and originality nonexistent. All these factors lead to high numbers of unemployed, low salaries, and poor work conditions. Capital is chronically low and a great portion of what could be available instead "leaks" to other countries with clearer laws and better legal guarantees, countries where currency is not devalued from an inflation born of tax collection disorder and public spending. (p. 84)

Since 2003, when the author made his argument, many improvements have been made in the region—especially in regard to the banking systems, management, and capital availability, yet a lot has to be done to prove his argument false—mainly in regard to production efficiency, value added, and distribution.

Montaner (2003) also states that, "a society must concentrate its efforts on improving human capital if it wants to gain a prominent place in the world. It must educate more and better, primarily during children's formative years, when character and certain behavioral habits and skills are first developed" (p. 173). Education, though, has never been a priority for colonizers in Latin America (although there are some old universities in the region, such as the Universidad Autónoma de Santo Domingo, Dominican Republic, founded in 1538, the longest continuously operating university in the Americas). Education is a problem especially in Brazil, and mostly in regard to elementary education, although in recent years efforts to change this situation have been made.

DEVELOPMENT

Development can be measured and pursued in many different ways. For Sachs (2005, p. 232), however, "The international development community should speak of the Big Five development interventions that would spell the difference between hunger, disease, and death and health and economic development," which are agricultural inputs (fertilizers, irrigation, improved seeds, etc.), investments in basic health and education, power, transport, and communications services, safe drinking water and sanitation.

According to Dani Rodrik (2007), economic principles do not map into unique policy packages. The author argues that the import-substitution strategies that reigned until the late 1970s and the Washington Consensus of the 1980s did not prove as a single recipe for success. According to him, several developing nations succeeded without following these policies (e.g., South Korea and Taiwan). Not only did they not undertake significant deregulation or liberalization of trade and financial systems, but also they did not privatize, and heavily relied on public enterprises instead (p. 30). On the other hand, some countries in Latin America, which followed the "recipe" closely, reaped little growth benefit out of it. Therefore, Rodrik (2007) concludes that "even the simplest of policy recommendations is contingent on a large number of judgment calls about the economic and political context in which it is to be implemented" (p. 30). However, he points out that "universal principles" of sound economic management, such as productive efficiency, macroeconomic and financial stability, and distributive justice and poverty alleviation are common (p. 30).

There are three points related to economic policies that stand out: fiscal policy, exchange rate, and monetary policy. The first one impacts the fiscal balance (revenues—expenses) and may lead to a surplus, which can be used in investments or kept as reserves, or a deficit, which will have to be financed, and eventually paid for, and may increase interest rates. Exchange rate policies may favor either exports or imports, inhibit or stimulate investments, if undervalued or overvalued. Finally, monetary policy impacts growth and interest rates, as more money in the market means more credit and less interest rates, and less money means less credit and, thus, higher interest rates. The more money in the market, with lower interest rates, the more investments the country should have with higher returns for investing in productive activities rather than in financial activities.

As Dani Rodrik (2007) says, "What stands out in the cases of real success [. . .] is not gradualism per se but an unconventional mix of standard and nonstandard policies well attuned to the reality on the ground" (p. 35). This author argues that "growth spurts are associated with a narrow range of policy reforms" (p. 35). Therefore, not many institutional reforms are needed to cause growth. Rodrik characterizes policy reforms as a combination of elements of orthodoxy with unorthodox institutional practices, which is not to say that all unorthodox remedies work (p. 35). He remarks that one must be careful with institutional innovations because "growth strategies require considerable local knowledge" (Rodrik 2007, p. 42). Therefore, what works in a country may not work in another. Rodrik points out that, it is more of a challenge to maintain growth, due to the need of more extensive institutional reform, than to set it in place. Increases in growth decelerate with time. Therefore, initial experiences of growth, even medium-term ones, are no assurance of long term achievement. For Rodrik, "the key to longer-term prosperity, once growth is launched, is to develop institutions that maintain productive dynamism and generate resilience to external shocks" (Rodrik 2007, p. 43). Therefore, countries must find their own path to development through trial and error. The successful growth strategies, Rodrik argues, are those that "are based on a two-pronged effort: a short-run strategy aimed at stimulating growth, and a medium- to long-run strategy aimed at sustaining growth. [. . .] The key . . . is to develop institutions that maintain productive dynamism and generate resilience to external shocks" (2007, pp. 43–44).

According to Larry Diamond et al. (2008), there is a consensus emerging: the fact that strong, efficient, and lean states are so important as sound policies to ensure properly functioning and free markets. According to him, Latin American States, Brazil included, "must actively invest in infrastructure, education, health, research and development, the environment, and the quality of life—not only to address social needs but also to foster national competitiveness" (Diamond et al. 2008, p. xv). In the case of Brazil, although many of the investments mentioned earlier are being made, a lot has yet to be done in regard to infrastructure—especially in roads and railroads, ports and airports—and toward a more efficient and lean state.

Here, especially, lies the importance of good governance, a sound economic policy, solid democratic institutions, and improved infrastructure, which can lead to economic growth, poverty eradication, and employment creation. Sound and responsible monetary, exchange rate, and fiscal policies give the country credibility, which is enhanced as the more transparent the rules and policies are. A sound and responsible fiscal policy can generate investments, and coherent monetary and exchange rate policies can generate confidence in that the interest rates and the exchange rates will not change suddenly and abruptly. This creates a positive cycle. Brazil has successfully had these three conditions—responsible monetary, exchange rate, and fiscal policies—and the results can already be seen. The country is, for instance, considered investment grade by the three major rating agencies (since April 2008 by Standard & Poor's, May 2008 by Fitch Group, and September 2009 by Moody's).

Solid democratic institutions also give confidence, by lowering the political risk, and Brazil's transition from Cardoso to Lula, in 2002, was a very good example of that. When Lula was to be elected president, many investors "fled" the country, and both the exchange rate and the interest rate went way up. When it became clear that no radical shift would happen, foreign investors returned, and both exchange and interest rates began to fall.

According to Joseph Stiglitz (2006), global warming has become a true challenge of globalization. Other demands have been the necessity of increasing collective action, and collaboration among people and countries toward solving common problems. Stiglitz explains that, "In effect, economic globalization has outpaced political globalization. We have a chaotic, uncoordinated system of global governance without global government. . . . There is a clear need for strong international institutions to deal with the challenges posed by economic globalization; yet today confidence in existing institutions is weak" (p. 21). This is a major opportunity for Brazil, if the country takes advantage of its current leading role in sustainable development.

Stiglitz (2006) points out that the question facing developing countries today is whether their governments will be able to promote development, regulate markets, and provide basic social services (p. 21). The author argues that while the process of globalization has brought new demands on nation-states to address the increasing inequality and insecurity that it presents, globalization has limited, in many ways, the capacity of nation-states to respond, since it unleashed market forces in such a way that governments often cannot control (p. 21).

However, for Stiglitz (2006), "the good news is that economics is not zero-sum. We can restructure globalization so that those in both the developed and the developing world, the current generations and future generations, can all benefit" (p. 24). On the other hand, there has to be a comprehensive approach toward development. Still according to Stiglitz, the government must be involved in providing basic education, legal frameworks, infrastructure, and some elements of a social safety net, besides regulating competition, banks, and environmental impacts. This is in line with the five investments that Sachs (2005) believes the public sector should focus on, and mentioned earlier: human capital, infrastructure, natural capital, public institutional capital, and parts of knowledge capital (p. 251).

According to Montaner (2003), it took medieval Europeans 500 years to produce paper in industrial quantities after slowly adopting a method developed by the Chinese and transferred by the Arabs. Yet the Koreans took just 20 years to build automobiles and transport them to the opposite ends of the earth (p. 175). Today, thanks to the examples of Taiwan, Singapore, Korea, and Hong Kong, among other countries, despite their occasional slips into financial crisis, it is possible to go from poor to rich, from Third World to First, in a single generation. This transition is accomplished by opening nations to the world's great economic and scientific flow, opening economies to competition and collaboration, and accepting the guidance of the great civilization epicenters.

The process of development is long; it is a cycle, and one thing tends to lead to another. For example, the lesser the unemployment rate, the less violence one tends to have; the more transparent the institutions, the less corruption one tends to have; the more one enforces contracts and protects property rights, the more businesses one tends to generate, and again more employment, and so it goes. That is clearly happening in Brazil. There is still violence, corruption, among other problems, but the rates are lowering, and they tend to lower even more with more sound policies, transparency, and development.

According to Paul Collier (2007), there are four traps that may condemn countries to an endless stage of underdevelopment. They are:

1. *conflicts*: violence, conflict situations or the verge of, and especially civil war;
2. *natural resources*: countries dependent on a single natural resource, such as oil, can cause the so-called Dutch disease, a situation in which the exports of a natural resource pressure exchange rates and prevent the development of other industries;
3. *landlocked* (bad neighbors): countries in the midst of regions wrapped by underdeveloped countries, in conflict situations, or without access to the sea, may have their development condemned, as, for example, they may not have access to foreign trade; and
4. *bad governance*: countries without good governance may not reach development.

Jeffrey Sachs (2005, p. 56) points out that countries often fail to achieve economic growth because of circumstances such as "the poverty trap," "physical geography," "fiscal trap," "constant deficits," "governance failures," "cultural barriers," "geopolitics," "lack of innovation," and "the demographic trap." Poverty can cause economic stagnation—no savings, no investment, no consumption; fiscal traps, with constant deficits, impede investments and growth, since the government is always financing its deficits and, therefore, there are not enough resources left for investments; the lack of innovation impedes development of new technologies, and hence, there are no investments and growth; government failures may lead to the nondevelopment of policies toward development.

In the case of Brazil, neither the traps mentioned by Collier nor the causes of failure mentioned by Sachs (2005) are very significant. Brazil does not have significant conflicts, has vast and varied natural resources. It is neither landlocked (Brazil shares 16,886 kilometers of borders with 10 of the 13 countries from South America, the exceptions being Chile and Ecuador—the 13th is Brazil—and has 7,367 kilometers of maritime borders), nor has "bad" neighbors. Furthermore, as mentioned earlier, it applies the concepts of good governance. In addition to that, the country is not under poverty or fiscal or demographic trap. Its geography is favorable. Brazil is not involved in any major geopolitical conflict; it does not present any significant cultural barriers—on the contrary—and does not lack innovation, although in this regard it needs improvement.

CONCLUSION

According to Sebastian Edwards (2010), countries in Latin America "group" themselves in a few clusters. Those comprise, to the author, a populist category, which includes Venezuela, Ecuador, Bolivia, and Nicaragua, and a second group of countries "formed by those that will neither fall for the populist temptation nor move forward in the implementation of the pro-competition policies and institutional reforms needed to spur productivity growth" (Edwards 2010, p. 233). He includes in this group Brazil (which I would not) and most of the Central American countries.

Argentina, according to the author, raises several important questions (p. 234), especially regarding whether the support for the populist measures of President Cristina Kirchner will increase or decrease. The third group, according to him, embraces "the innovative, productivity-based path to development and prosperity" (Edwards 2010, p. 234). Chile is the leader of this group and might be followed by Peru, Colombia, and Costa Rica, and, eventually, Brazil (which I would include in this group) and Mexico. I would add another group, which might include Paraguay and Uruguay, which seems to have an unknown future, as their small size makes them more "dependent" on larger close economies rather than autonomous.

According to a report released in 2011 by the Council on Foreign Relations (2011) ("Global Brazil and U.S.-Brazil Relations"), Brazil "has undertaken a peaceful economic and social transformation to become the cornerstone of South American growth and stability and a significant power and presence on the world stage" (p. 3). The report recognizes Brazil as an integral force in the evolution of a multipolar world, in a leadership position in Latin America and in the world, and mentions similarities with the United States regarding ethnicity, the respect for democratic values and rules of law, individual rights, religious freedom, diversity and equality, which corroborates my including Brazil in the group of countries leading toward development and prosperity.

To some extent, Brazil's success is based on the same five pillars Jeffrey Sachs rested on in his plan to stabilize Poland's economy in 1989 (Sachs 2005, p. 114). They are:

- *stabilization*: control inflation and establish a stable, convertible currency;
- *liberalization*: legalize private activity, end price controls, and establish the necessary commercial laws;
- *privatization*;
- *social safety net*: pensions, health care, and other benefits for the elderly and the poor, especially to help cushion the transition;
- *institutional harmonization*: adopt economic laws, procedures, and institutions of Western Union (in the case of Poland, to join the European Union, or European Community, at that time).

According to Dani Rohter (2010), the main distinction between Brazil and other members of the BRICS group, as well as from those of advanced nations, such as Japan and Germany, is its unequal agricultural potential and production. That is, Brazil has the ability to be the planet's breadbasket, feeding not only itself, but also much of the rest of the world (p. 151). The country has always been a producer of crops, ranging from coffee and sugar to cocoa and tobacco, dating back to colonial times. Over the past generation, though, Brazilian agriculture has dramatically diversified and modernized, with noticeable increases both in yields and in the variety of crops grown. When the U.S. secretary of state Colin Powell visited Brazil in 2004, he described the country as an "agricultural superpower" rivaling the United States and even surpassing it in some areas. Its potential in agriculture is yet enormous, as there are still vast amounts of land available for agriculture (land for crops can more than quintuple, according to *The Economist*, August 26, 2012, "The Miracle of the

Cerrado"), and with a climate that varies little the year round, which allows multiple (up to three) harvests annually. In regard to infrastructure and distribution, however, and as mentioned earlier, Brazil has yet a long way to go.

Much of the recent success of Brazilian agriculture can be credited to Embrapa (the Brazilian Agricultural and Livestock Research Company), a group of scientists and agronomists who have, according to Dani Rohter (2010, p. 154), developed technology and seeds to grow crops not only in places thought as not viable but also in conditions for long thought as not conducive. They have also developed breeds of hogs with lower fat and cholesterol, and higher yield of loin and ham.

In addition to agricultural success, in regard to industry, today, all major multinational corporations are present in Brazil—from food chains (McDonald's) to supermarkets (Casino), beverages (Coca-Cola), oil companies (Shell), IT industries (Cisco), and software (Facebook), among others. In some cases, their operations in Brazil are more important than in their original countries. Brazil is, for example, the second-largest market for Facebook, and among the main markets for Santander, Citibank, and Coca-Cola.

Brazilian companies, such as Vale, Votorantim, InBev, Itaú-Unibanco, Camargo Correa, Gerdau, Perdigão, WEG, and Marcopolo, not only benefited from the strategies adopted by the country but also became huge corporations, some being world leaders in their industries and now making acquisitions abroad. Just to mention a few examples, InBev purchased Anheuser Busch (Budweiser), 3G Capital purchased Burger King, Gerdau bought Ameristeel, Marfrig bought Keystone, Vale (a former SOE) has acquired Inco, Braskem bought Sunoco, Votorantim bought Cimpor, JBS has bought Swift & Company and Pilgrim's Pride.

Other good examples worth mentioning here are Petrobras and Embraer. Petrobras has been financed through taxes and high prices of gasoline, but it is now among the largest oil and gas companies in the world and among the largest corporations in Latin America (was number one in 2008), and has more oil reserves than Chevron Texaco or Royal Dutch Shell. It has also helped Brazil become a leading player in deep-sea oil exploration and has recently been granted authorization by the U.S. government to explore oil in the gulf region. Likewise, Embraer, after privatization (1994), became the third-largest producer of aircraft in the world—following Boeing and Airbus—and the planes it produces are responsible for about 70 percent of the U.S. regional flights.

Brazil is the largest exporter of beef (Brazilian JBS Friboi is the largest meatpacking company in the world, and Brazil is the largest producer, with more than 200 million heads of cattle, more than double that of the United States) and chicken, and the third-largest exporter of pork. It is the largest producer and exporter of coffee (about 40% of world's production), orange and orange juice (Tropicana bottles used to refer to Brazilian oranges), sugar, tropical fruits, ethanol, and tobacco. It is among the largest producers of cotton, wheat, corn, rice, soybeans, cocoa, oilseeds, corn, among other products, not to mention the fact that it is the second-largest exporter of iron ore, and has substantial reserves of manganese, bauxite, copper, lead, zinc, nickel, tungsten, tin, uranium, among others.

Some of the country's success is also due to government investments in research and development. The creation of the Instituto Tecnológico de Aeronáutica (Technological Institute of Aeronautics) made possible the development of aircraft-manufacturing technology and the creation of Embraer, mentioned earlier. The creation of Embrapa (the Brazilian Agricultural Research Corporation), for example, made possible the production of crops never imagined to be produced in large amounts in Brazil, such as wheat. It also allowed the development of better cattle, the development of breeds of hogs with lower fat and cholesterol, and higher yield of loin and ham; and the development of better-producing technology (e.g., in Brazil, eucalyptus, for cellulose, can be grown to be cut in 5 or 6 years, a record, considering it is usually cut in 7, 14, or 21 years), just to mention some examples.

In regard to energy, the oil crisis of 1973 and 1979, together with the deficit imbalance of the 1970s, motivated the development of the ethanol technology in Brazil, previously mentioned, through university research centers supported by the federal government. Nowadays, 80 percent of the cars sold in Brazil are FFVs. Boeing has been developing, with Embraer, ethanol technology for airplanes—in fact, Embraer already has a small plane, for agricultural aviation, which uses ethanol, the Ipanema, a market leader, with 75 percent of Brazil's fleet in this segment. It is also worth mentioning that the increasing use of ethanol as a fuel reduces dependence on fossil fuels, such as oil.

Therefore, Brazil presents several opportunities, which include agribusiness, infrastructure, energy, real estate, health and education, and retail, as the country develops and moves people out of poverty. Some estimate that Brazil has lifted more than 35 million people out of poverty in the last few years through its Conditional Cash Transfer program, Bolsa Família. As these people consume more, they become an increasing market. Regarding infrastructure, just to give an example, according to *Americas Quarterly* (Winter 2012, p. 37), Brazil has 0.35 kilometers per 100 square kilometers of rail lines, whereas the average in South America is 0.51 kilometers. Although Brazil has more roads (21 kilometers per 100 square kilometers) than the South American average (15 kilometers), only 6 percent of its roads are paved, against an average of 22 percent for South America. Still according to the magazine, the average increase in GDP from 2000 to 2010 was 3.7 percent per year, but its investment/GDP average is still very low, at 17.6 percent, whereas in countries such as Indonesia, Malaysia, Singapore, Thailand, and Vietnam, it is close to 30 percent. Trade, though, is increasing at a faster rate—exports grew at an average 7.2 percent in the period and imports grew at an average 9.4 percent in the same period.

Brazil has also been very successful in diversifying its markets, away from the United States and Europe, and toward Latin America, Africa, the Middle East, and Asia Pacific, especially during Lula's term in office, as mentioned earlier. Brazilian diplomacy has largely focused on South America and South–South cooperation, and the Free Trade Area of the Americas has been clearly set aside. Brazil fostered not only IBAS (India, Brazil, and South Africa) Forum, but also the relations among the BRICS nations.

According to Alexander Busch (2010), in 2003, in Cancun, the Brazilian diplomacy conquered its first major "victory" in the international scene when, during the WTO meeting, it organized a coalition involving very heterogeneous nations, the so-called G-21, to challenge the industrialized countries in granting a larger access to their agricultural markets. Negotiations failed and have not resumed yet. Since then, Brazil has had growing relative importance in the G-20 meetings and summits on environment, always based on *soft power* politics, as opposed to *hard power* ones ("soft power" and "hard power" are expressions coined by Joseph Nye, Harvard University).

Still according to Busch (2010), in 2008, Brazil has granted itself a large victory with the establishment of Unasul, which naturally has a Brazilian leadership, a regional leadership whose importance is essential to ensure the Brazilian ascent in international politics. Brazilian leadership in the region has also been guaranteed by its initiatives to integrate the region physically and economically—Brazil is financing (or proposing to finance) the construction of hydroelectric power plants in the region (Peru) and roads to link the Atlantic and Pacific oceans, through Bolivia and Peru. Cardoso (2010, p. 65) argues that, countries considered "beginners" in the globalization process, Brazil included, have learned to use the WTO to protect their interests against the protectionism of the rich countries, or to utilize the rules of the treaties of intellectual property to protect their interests or their peoples. Brazil, for instance, used the WTO to grant access to the U.S. market for its cotton and was given authorization to retaliate the United States against its subsidies.

Regarding infrastructure, however, it is of utmost importance that the government further invest and stimulate private investments, which could be done through either public–private partnerships or privatizations. One of the major problems Brazil currently faces is the so-called Custo Brasil (or Cost Brazil), or, basically, the cost to get the goods from producer to consumer, considerably high in Brazil, due to the poor quality and extension of its railroads and roads, ports, and airports. This is one of the main reasons that prevent Brazil from furthering its exports of agriculture, and lower Brazil's competitiveness in world markets, especially for industrialized commodities. As for agricultural goods, productivity is higher and prices, in international markets, are also higher.

It is also of utmost importance that Brazil conduct structural reforms, such as legal, fiscal, and administrative. Tax reform, for example, is of utmost importance, as tax legislation in Brazil is extremely complex, especially for new entrants—foreign companies willing to enter the Brazilian market, not to mention the high tax burden on businesses (especially on smaller businesses) and the lack of tax incentives for new entrants. The political reform is also fundamental, as it could facilitate approval of legislation in Congress and eventually reduce the number of political parties—Brazil currently has more than 30 political parties. Labor laws should be adapted to a new world reality; that is, they should be more flexible. The current labor laws in Brazil, named Consolidação das Leis do Trabalho (Consolidation of Labor Laws), date from 1943 and are considered by many to be one of the main reasons for the high degree of informality in the Brazilian labor market (which is over 40%), which affects mainly small and medium enterprises. It is also worth mentioning that Brazil should reduce the size of government—it has almost 40 ministries and more than 20,000 government-nominated jobs, in several companies and institutions, and it has almost as many senators and congressmen as the United States, with about half the amount of states. On the other hand, in regard to governance, Brazilian institutions are democratic and trustworthy, with high levels of transparency and accountability, although some still present high levels of corruption. In addition, Brazil adheres to the rule of law, and political risk is low.

As far as poverty is concerned, Brazil has been very successful in eradicating it. The Bolsa Família Program—a cash-transfer program, or the donation of money to families with children enrolled in school—has helped the country to lift more than 35 million people out of poverty since adopted during Cardoso's term in office, and now it affects more than 30 percent of the Brazilian population. Nevertheless, there is still a way to go, and eradicating poverty is one of Dilma Rousseff's main government priorities.

The degree of informality in the economy, as a whole, although decreasing, is still high in Brazil, and the level of savings is low (around 17% of the GDP, while in China and India, it is around 30%–40%). However, the level of credit is low (around 43% of the GDP), the macroeconomic indicators are solid, and the level of foreign reserves is satisfactory (around $375 billion). That tends to attract foreign direct investments—although further deregulation is necessary—mainly in agribusiness, infrastructure, energy, real estate, health and education, and retail, especially as the country has been preparing to hold the major events mentioned earlier. Brazil still needs to further increase trade. According to the CIA ("The World Factbook") and The World Bank (Merchandise Trade), Brazilian exports are $244.8 billion (2013 estimate) and imports are $241.4 billion (2013 estimate), or 11 percent and 11 percent of GDP, respectively, or less than 22 percent combined, a very low percentage when compared to countries such as China (around 47% of GDP), Chile (59%), Colombia (32%), Russia (43%), and India (42%). That makes Brazil not to rank among the world's 20 leading exporters or 20 leading importers, behind nations such as the Netherlands, Belgium, Mexico, Canada, Singapore, and Taiwan in both exports and imports, a very low ranking, therefore, for the world's eighth-largest economy. Still, when compared to BRICS countries—to which it is generally compared— Brazil is very competitive. Unlike China, it is a democracy; unlike India, it has neither ethnic and religious conflicts nor hostile neighbors; and unlike Russia, it exports more than oil and gas, and has a fair business environment.

Dani Rodrik (2007) believes "in the ability of governments to do good and change their societies for the better: Government has a positive role to play in stimulating economic development beyond enabling markets to function well" (p. 4). The Brazilian government has done well, in the last two decades, in stabilizing democracy and the economy, and in guaranteeing a stable growth, especially in the last decade, stimulated by the bonanza in world markets of agricultural goods and commodities and increasing demand from China for both. However, in the past few years, growth has slowed, inflation has risen (it is now close to the upper band established by the government), and new demands arise from the "new middle-class." Therefore, in order to revamp growth, Brazil needs to tighten its fiscal discipline and conduct the necessary structural reforms mentioned earlier, such as legal, fiscal, and administrative. These should have been pursued during the period of bonanza and favorable political support—the executive has had majority in both houses for most of the past decade, which could have favored reforms. Brazil needs to increase productivity, as unemployment is at a record low (around 5%), through investments in education, infrastructure, a more flexible labor market, more openness to foreign direct investments, and further deregulation. It needs to conduct a fiscal reform and simplify the tax burden, guarantee more efficient and reduced public expenditures, combat corruption, and increase savings and investments. It needs to stimulate the production of more value-added products in selected industries (e.g., produce and export furniture instead of wood), establish a more competitive environment, and make it easier for doing business in the country. Finally, it needs to open its economy further and increase trade and business not only with other developing nations (e.g., China, BRICS, and other South–South negotiations mentioned earlier), but also with the major world economies and markets, such as the United States and the European Union, through regional, bilateral, and/or multilateral agreements. Tourism, for instance, is underexplored, given the country's diversity and natural beauty. Put simply, Brazil has achieved significant advances in the past two decades, since stabilization of currency and inflation in 1994, but it still has a long way to go to achieve a new level of development, consistent with the world's seventh-largest economy.

REFERENCES

Almeida, Paulo Roberto de. 2009. "Lula's Foreign Policy: Regional and Global Strategies." In Joseph L. Love and Werner Baer (eds.), *Brazil under Lula: Economy, Politics, and Society under the Worker-President*, pp. 167–183. New York: Palgrave Macmillan.

Americas Quarterly. Winter 2012. Volume 6, Number 1.

Baer, Werner. 2008. *The Brazilian Economy: Growth and Development*, 6th ed. London: Lynne Rienner.

Banco Central do Brasil (BCB). 2013. "Histórico de Metas para a Inflação no Brasil 1999–2015." https://www.bcb.gov.br/Pec/metas/TabelaMetaseResultados.pdf.

Bhagwati, Jagdish. 2004. *In Defense of Globalization.* Oxford, UK: Oxford University Press.

Brown, Chris. 2001. *Understanding International Relations*, 2nd ed. New York: Palgrave.

Bulmer-Thomas, Victor. 2003. *The Economic History of Latin America since Independence*, 2nd ed. Cambridge, MA: Cambridge University Press.

Busch, Alexander. 2010. *Brasil, País do Presente: o Poder Econômico do "Gigante Verde."* São Paulo: Cultrix.

Cardoso, Fernando H. 2010: *Xadrez Internacional e Social Democracia.* São Paulo: Paz e Terra.

Central Intelligence Agency. 2013. "The World Factbook." https://www.cia.gov/library/publications/the-world-factbook/geos/br.html.

Collier, Paul. 2007. *The Bottom Billion: Why the Poorest Countries Are Failing and What Can Be Done about It.* Oxford, UK: Oxford University Press.

Council on Foreign Relations. 2011. "Global Brazil and U.S.-Brazil Relations." Independent Task Force Report No. 66. New York.

Diamond, Larry, Marc F. Plattner, and Diego Abente Brun. 2008. *Latin America's Struggle for Democracy.* Baltimore, MD: The Johns Hopkins University Press.

Easterly, William. 2002. *The Elusive Quest for Growth: Economists' Adventures and Misadventures in the Tropics.* Cambridge, MA: The MIT Press.

The Economist. November 12, 2009. "Brazil Takes Off." http://www.economist.com/node/16886442.

The Economist. 2011 Edition. "Pocket World in Figures."

The Economist. August 26, 2012. "The Miracle of the Cerrado: Brazil Has Revolutionized Its Own Farms. Can It Do the Same for Others?" http://www.economist.com/node/16886442.

The Economist. September 28, 2013. "Has Brazil Blown It?" http://www.economist.com/news/leaders/21586833-stagnant-economy-bloated-state-and-mass-protests-mean-dilma-rousseff-must-change-course-has.

The Economist. 2013 Edition. "Pocket World in Figures."

Edwards, Sebastian. 2010. *Left Behind. Latin America and the False Promise of Populism.* Chicago, IL: University of Chicago Press.

Encyclopedia Britannica.

Fausto, Boris. 2001. *História Concisa do Brasil.* São Paulo: Edusp.

Frieden, Jeffry A. 2006. *Global Capitalism: Its Fall and Rise in the Twentieth Century.* New York: W. W. Norton.

Galeano, Eduardo. 1976. *As Veias Abertas da América Latina.* Rio de Janeiro: Paz e Terra.

Gilpin, Robert. 2001. *Global Political Economy: Understanding the International Economic Order.* Princeton, NJ: Princeton University Press.

Halperín Donghi, Tulio. 1993. *The Contemporary History of Latin America.* J. C. Chasteen, edited and translated. Durham and London: Duke University Press.

Instituto Brasileiro de Geografia e Estatística (IBGE). http://www.ibge.gov.br.

Instituto de Pesquisa Econômica Aplicada (IPEA). "IPEA Data." http://www.ipeadata.gov .br.

Kohli, Atul. 2004. *State-Directed Development: Political Power and Industrialization in the Global Periphery.* New York: Cambridge University Press.

Montaner, Carlos Alberto. 2003. *Twisted Roots: Latin America's Living Past.* New York: Algora Publishing.

Reid, Michael. 2007. *Forgotten Continent: The Battle for Latin America's Soul.* New Haven, CT, and London: Yale University Press.

Rodrik, Dani. 2007. *One Economics, Many Recipes: Globalization, Institutions and Economic Growth.* Princeton, NJ: Princeton University Press.

Rohter, Larry. 2010. *Brazil on the Rise.* New York: Palgrave Macmillan.

Sachs, Jeffrey. 2005. *The End of Poverty: Economic Possibilities for Our Time.* New York: Penguin Books.

Santibañes, Francisco de. 2009. "An End to U.S. Hegemony? The Strategic Implications of China's Growing Presence in Latin America." *Comparative Strategy*, 28: 17–36.

Stiglitz, Joseph E. 2002. *Globalization and Its Discontents.* New York and London: W. W. Norton.

Stiglitz, Joseph E. 2006. *Making Globalization Work.* New York and London: W. W. Norton.

Sweig, Julia E. November/December 2010. "A New Global Player: Brazil's Far-Flung Agenda." Foreign Affairs. http://www.foreignaffairs.com/articles/66868/julia-e-sweig/ a-new-global-player?page=show.

United Nations Development Programme (UNDP). 2013. "Human Development Reports, Brazil." http://hdr.undp.org/en/countries/profiles/BRA.

U.S. Department of State. 2013. "US Relations with Brazil." http://www.state.gov/r/pa/ei/ bgn/35640.htm.

The World Bank. 2013. "Foreign Direct Investment, Net Inflows." http://data.worldbank. org/indicator/BX.KLT.DINV.CD.WD.

The World Bank. 2013. "Merchandise Trade." http://data.worldbank.org/indicator/TG.VAL. TOTL.GD.ZS.

Society

Religion and Thought

Joseph Abraham Levi

RELIGIOUS LIFE, MORAL VALUES, AND BELIEF SYSTEMS IN BRAZIL: AN OVERVIEW

Like in most countries in the world, religion, spirituality, and belief systems based on the human quest for a contact with and a connection to the Creator(s) of the Universe and the spiritual world, on Earth as well as in the Afterlife, have always played an important role in the lives and mores of Brazilians.

Since the beginning of time, history and politics have oftentimes shaped the formation of nations and the mind-set of their future inhabitants. Brazil is no exception. Moreover, as in most places in the Western Hemisphere, in Brazil the coming together of people from different backgrounds, either willingly or unwillingly, as free men and women, as well as slaves, has created a unique and heterogeneous culture and civilization that perforce also had an impact on the way people behave when it comes to express their inner and outer connection(s) with the divine and/or spiritual world(s).

Pre-European Brazil, oftentimes referred to as Pre-Cabraline Brazil—from the name of the Portuguese navigator and explorer Pedro Álvares Cabral (ca. 1457–ca. 1520), under whose leadership the Portuguese "discovered," or rather, "stumbled upon," what is now Brazil, landing on Porto Seguro, in today's Brazilian state of Bahia—was a mosaic of Amerindian peoples and tribes, belonging to different clans, tribes, and nations that were part of an equally diverse mosaic of ethnic, racial, and linguistic groups. Obviously, their religions and spirituality obeyed to their lifestyle and place(s) of residence. The arrival of the Portuguese in 1500 changed the entire equation: from that moment on the doors were thus open for the gradual, yet ever-increasing change in the human and spiritual component of Brazil and its inhabitants.

During the 322 years of colonial presence (1500–1822) and subsequent imperial rule (1808–1822), the Portuguese imposed their own way of life. That included their religion, that is, Catholicism. Evangelization was thus an instrument used by the various religious orders that were sent to Brazil by the Portuguese Crown and/or by the Holy See to "conquer" the souls of the native population as well as to nurture and guide the souls of those Europeans who settled in Brazil. Among the European settlers there were also Sephardic Jews of the Diaspora and their descendants who—fearing persecution, imprisonment, and deportation to Portugal where they eventually would be burned at the stake—for the most part practiced (some form of) their ancestral faith in the privacy of their homes while overtly they boasted "proud and unwavering" adherence to Catholicism. Of course there were exceptions whereby there were cases of overt return to Judaism in some parts of Brazil, particularly in the plantation areas dominated by Crypto-Jewish, or rather, New-Christian families. Dutch occupation (1630–1654) of almost half of a colony then still in expansion (from north/northeast to south/southwest) brought freedom of religion to this area of Brazil, whereby Protestants and Jews, Sephardic as well as from other Jewish ethnic groups (as in the case of the Ashkenazim, at the time still a minority as compared to the Sephardim), could freely practice their faith without fear of retaliation. Things changed when the Portuguese eventually regained control of the entire colony in 1654. Once again, Catholicism became the official religion of the only Portuguese colony in the Americas.

Given the vast territory that the Portuguese gradually explored, conquered, and took control of, it soon became necessary for the Portuguese Crown to devise a system in order to avoid the prohibitive costs of maintaining an empire that spanned from the Atlantic archipelagos of the Azores, Madeira, and Cape Verde to Brazil, the shores of West and East Africa, Hormuz, Muscat, India, Sri Lanka, Macau, Nagasaki, Malacca, and Indonesia. In most parts of the empire, these new lands in Africa, Asia, and Brazil, usually along the coast, were oftentimes given to noblemen who, on payment of certain tributes, were granted full powers to administer and develop the lands for the Portuguese Crown. These *capitanias* (captaincies)—ruled by *capitães donatários* (captaincy generals), as they were known in Brazil and other parts of the Portuguese possessions overseas—were in a sense the backbone of the empire, or rather, they guaranteed Portuguese presence throughout the world, at a time when trade and economy were beginning to expand on an unprecedented and ever-expanding scale, eventually to reach a worldwide dimension.

As for Brazil, the scarcity of European settlers and the inability to "effectively" utilize the Amerindian population for the advancement of the Portuguese colony in the Americas, coupled with the fact that, like in the rest of the Americas, the native population was decimated by diseases and wars, gave the Portuguese, soon followed by other Europeans interested in expanding their political and/or commercial power in the Americas, the idea of transplanting to Brazil an old human practice, yet transforming it for their own advantage: slavery. By 1550, slaves from the western shores of Africa, down to present-day Angola, later on with contingents from the East African Coast (particularly present-day Mozambique), were thus being transported to Brazil in order to compensate for the lack of manual labor needed for the construction, maintenance, and prosperity of a new colony.

Just like the Amerindian population, also the African slaves who were brought to Brazil came from very diverse ethnic, racial, sociopolitical, and linguistic cultural backgrounds. Their everyday outlook on life, their spiritual beliefs, and their intricate religious systems were thus transplanted to the Portuguese colony in the Americas where they soon had to adapt to the dominant religion of their masters: Catholicism. Though overtly Catholic, most African slaves and their descendants found a way of maintaining a connection with their ancestral faith by way of adapting their beliefs and creed to the intricate hierarchical system of the Catholic faith: each Catholic saint, for example, corresponded to a given traditional African deity or spirit; a feast commemorating a Catholic dogma was thus seen as a way of celebrating an old African rite; or praying to the Christian God and celebrating the Christian saints soon became a way of practicing rituals that linked the slaves and their offspring born in Brazil to their native lands. Obviously rituals varied from village to village, from town to town, and from geographical area to geographical area, north/northeast versus south/south central Brazil. Yet, overall, this form of syncretism in time led to a new way of being and a new belief system within normative Catholicism. At times the ancestral beliefs of many Amerindian tribes also entered into the equation, thus contributing to the uniqueness of this religious syncretism in Brazil which, as time went by, was also adopted by non-Afro-Brazilians as a way of asserting their unique "Brazilian" way of worshipping the Almighty. With Independence (1822) and the subsequent establishment of the Brazilian Empire (1822–1899), the abolition of slavery (1888), and the founding of the Brazilian Republic (1889–), Brazil opened up to the rest of world, thus inviting in and/or taking in people from all over the world, people with different religious views, affiliations, and beliefs: for example, different Protestant denominations, Orthodox Christians, Ashkenazi Jews, Muslims, Bahá'ís (as of 1919), Hindus (as of the 1960s), Buddhists, and Shintoists. These new immigrants (mainly from Europe and Asia) brought with them their faith, their values, and their belief systems, thus contributing to the rich diversity of Brazilian religions and spirituality. Today, we can by far ascertain that, unlike its South American neighbors, and perhaps more like its North American counterparts—at least when it comes to adherence to a religious belief and to religious diversity, that is, the United States and Canada—Brazil has an extreme high percentage of believers with a multifaceted variety of religions, creeds, and spiritual beliefs. In addition, though with echoes in other parts of the Americas, particularly the Caribbean, Brazil boasts syncretic forms of religion and spirituality that are indeed unique, thus making it the leading country in the world when it comes to expressing the human desire to connect with the Creator and the spiritual world(s).

RELIGIOUS ORDERS IN COLONIAL BRAZIL

Augustinian, Capuchin, and Franciscan missionaries played a major role in the spiritual and educational history of colonial Brazil, though not all at the same time and not all in the same geographical area. Yet of all the religious orders that were sent to Brazil, the Jesuits were by far the most influential in forging and molding a new religious identity for the autochthonous population and, though to a lesser degree, the Afro-Brazilians who soon became the majority of the population living in the Portuguese colony in the Americas. Despite their relative small number Jesuits missionaries were instrumental in establishing *aldeias* (villages) where, aside Catechism, Amerindian neophytes would learn a trade and, more importantly, were being kept away from the European colonists who for the most part tried to enslave them.

NATIVE BRAZILIAN TRIBES AND NATIONS

The pre-European era of Brazil embraces a period that is hard to pinpoint. The discovery in 1973 of a collection of more than 800 archeological sites and rock paintings at *Pedra Furada* in the Brazilian state of Piauí points at human presence in Brazil between 48,000 and 32,000 before the Common Era, hence suggesting that at least in Brazil and surrounding areas (e.g., Argentina and Chile) there were civilizations that antedated the Clovis culture, in present-day New Mexico, by most considered the first human settlement in the Americas (ca. 13,500–13,000 before the Common Era). Yet there are scholars who question the veracity of these sites, attributing them to geological formations. Nevertheless, the diversity of native languages in the Americas supports the theory of an earlier presence on the entire continent, one that is certainly older than the widely accepted one: 20,000–15,000 before the Common Era.

Unlike the rest of the American continent where, despite the existence of still some puzzling questions, there is a general consensus on how the native populations led their lives, archeologists, anthropologists, and historians have always had difficulties in drawing a definite map of the sociopolitical makeup of most of what is today Brazil before the arrival of the Portuguese in 1500. The lack of written and architectural records, as well as burial and ritual sites, mainly due to the humid climate and the acidic soil, has left scholars with only pottery and war weapons as arrowheads, as their sole guide to aboriginal life in pre-European contact Brazil.

Just like elsewhere in the Americas, mounds are a good indication of Amerindian life and mores in present-day Brazil before the arrival of the Europeans in the 16th century. Coastal *sambaquis* and the Amazon River *terra preta* are "shellfish" and "black earth" deposits left by prehistoric native Brazilian men and women, thus testifying to complex socioeconomic configurations rather than the stereotypical "nomadic cultures" of most of the American continent.

Most linguists and philologists divide native Brazilian languages and dialects into four basic language groups: the Macro-Gê, the Tupi-Guarani, the Arawak, and the Caribe. It is believed that before the arrival of the Portuguese the indigenous population of what is today Brazil preferred living in small nomadic groups oftentimes at war with one another. Given that like most areas in the Americas Brazil was colonized by Europeans in an east-to-west movement, it is understandable that more information is known about the coastal Amerindian groups, the ones physically closer to/who had the most interactions with the Europeans. Inner groups like the Amazonian tribes and nations were the last with whom the Europeans had contact, hence the scarcity of information on their pre-European contact way of life and culture, including their religious and spiritual beliefs.

Generally, the native population of present-day Brazil makes a clear distinction between the sacred and the mundane, between the supernatural and the natural. Yet the supernatural has a way of interfering in unexpected ways with the daily lives of the population. Spirits and the souls of the departed wander through Nature, oftentimes warning against imminent natural and supernatural catastrophes. Dances with elaborate masks, for instance, are thus a way of pleasing the spirits and asking for their intercession with the world beyond (above and below Earth). Myths, rites, and religious beliefs are centered on an intricate pantheon of gods and goddesses and of human and heavenly/infernal spirits, as well as of souls of animals, plants, and inanimate objects. As in most autochthonous populations of the world, also in many areas of present-day Brazil the native population does not have an absolute god ruling the entire Universe. Moreover, it appears that myths and religious/spiritual rites are stronger among agricultural societies than among hunters and gatherers. Given the great distance between the Supreme Being and humans, perforce the Supreme Being cannot or will not want to have a direct role in humans' everyday life. Hence, the Supreme Being is seen as just another of the many deities whom humans have to revere and seek protection from. In fact, most of the time the Supreme Being is not seen as the Creator, since any deity or spirit could generate humans, animals, plants, and inanimate objects. Nevertheless, and though invisible, the Supreme Being is strictly linked to the origins and the mores of all tribes; hence, rituals are very important to their very existence and prosperity.

Followers of Candomblé, *a syncretic Afro-Brazilian religion of oral tradition, dance and pray next to offerings to Yemanjá, a West African water goddess. On Yemanjá's Day, which was made to correspond with the Catholic feast of Our Lady of Navigators, thousands of followers take their flower, perfume, and jewel offerings in procession to a nearby beach. (AP Photo/Renzo Gostoli)*

Almost never the Supreme Being is a mythological entity linked to Earth and its inhabitants, since most native tribes in the Americas appear to favor the idea of the hero seen as a trickster and/or a twin deity. The myth of the two twins is perhaps a variant of the link between the Supreme Being and the hero-trickster. Of a capricious nature, oftentimes the trickster is conceived of as the lord of all animals; hence, he is represented as a migrant or as an errant, changing/transforming the landscape, animals, and humans. For most tribes, for example, the hero-trickster has stolen from the world beyond: the fire, daylight/the Sun, and/or water. In many myths the hero-trickster saves humans from monsters and cannibals, as well as from natural disasters and supernatural forces. Being responsible for their destiny, the hero-trickster is therefore the one who has given humans arts and crafts, skills to perform all tasks, and the laws that govern their society, as well as the mythology, the rites, and the rituals of their respective tribes and nations. It is not surprising then that for most tribes and nations in present-day Brazil the hero-trickster has gone through an animal form phase. For the Pioyes near the Amazon River he is a tapir, whereas for the Pareci (Arawak) and the Bakairi (Caribes) in Mato Grosso, the hero-trickster is a "horse-mouse" and a "spider," respectively; the Mundurucu in the Tapajós River area, instead, believe in Karusakaibo, a Mundurucu hero-trickster. The Supreme Being of the Surara in the Amazon region coincides with the Moon, the abode of the departed, where their lord resides, usually described as an old man with a long beard. The phases of the Moon thus link the Supreme Being with humans and their daily lives on Earth. On the other hand, the Canelas, in eastern Brazil, and the Guarani, of southern Brazil/Paraguay, venerate the Sun, almost always represented as a man. Most tribes of the Tupi ethnolinguistic group worship *Tupã*, the thunder god. Hence, it is no surprise that during the colonial rule (1500–1808) and imperial rule (1808–1822) Jesuit and other missionaries almost always associated this powerful deity with the Christian God. On the other hand, the Pareci people in Mato Grosso believe that everything emanates from *Maiso*, the Earth goddess, while most Mundurucu tribes—in present-day Pará, Amazonas, and Mato Grosso—celebrate the goddesses of corn and manioc. The Curupira nation of the eastern Tupi group believes in spirits and genies of the woods, oftentimes seen as spirits inhabiting very important trees and bushes. They are the true lords of all animals; hence, they are at the same time also the protectors of all woods. They know and keep the secrets of the woods and punish those who destroy or attempt to destroy trees. These genies are subjects of the Moon goddess, mother of all plants; hence, she inhabits all trees. For the Caribe people, instead, each tree has its own spirit; thus, humans have to placate the spirit once the tree is removed from its habitat. Finally, in the Amazon area the Water Serpent is the dangerous anaconda, linked to the powerful and terrifying mother of waters.

Medicine men possessed a more elevated level of mystic powers; they are the true visionaries of all pre-European contact societies in the Americas. Their supernatural powers are aided by an entire slew of spirits who work for them and obey them unconditionally. In addition, their supernatural powers are at the service of the entire society in which they live and operate. Medicine men are thus highly respected, feared, and venerated; hence, they dominate the entire religious and spiritual life of the clan, tribe, and nation. A visionary and a mystic, the medicine man goes into a trance in order to save and ameliorate the spiritual and corporal sides of his people.

Transgressions, such as diseases and murder, are thus taboos, or rather, violations of the preestablished code of conduct ordained by their hero-trickster. This is particularly true for the Macro-Gê and the Tupi groups. In order to save their fellow citizens, medicine men oftentimes travel to the Afterlife and ask the spirits for guidance on how to cure mortals and restore welfare in the clan, tribe, and/or nation. Unfortunately, men dominate the scene, since, like in most parts of the world, also in present-day Brazil most tribes and nations are patriarchal. Yet there are cases of medicine women in the greater Mato Grosso area.

Totems, instead, represent the mystic link between an animal, a plant species, or an entire class of natural phenomena, and a family group or clan. In most parts of present-day Brazil totems represent a specific clan, tribe, and/or nation. For example, the Guaranis in northwestern Brazil are known as the "Screaming Monkeys," whereas two Canela tribes are known as the "Wild Goose" and the *Aguti* (rodent), respectively.

ANTHROPOPHAGISM VERSUS CANNIBALISM

The precolonial and early-colonial practice among some Native Brazilian tribes of eventually eating the flesh of their enemies, almost always prisoners of war, is known as *anthropophagism*. This neologism is necessary as to distinguish this social custom/religious practice from cannibalism, the latter a term used to describe the act of eating human flesh "out of necessity," as in the case of famine, drought, epidemics, and natural disasters. Anthropophagism was always practiced within a religious and ritualistic context, since eating the flesh of the prisoner(s) would have transferred back to the members of the clan, tribe, and/or nation the powers, knowledge, virtues, and qualities of the prisoner(s). Prior to his demise, the prisoner was treated with all the honors and privileges usually associated with a dignitary, including being able to marry a local woman.

CATHOLICISM

Since their arrival on April 22, 1500, the Portuguese imposed their customs and that included their religious beliefs. All efforts were thus made to see that the native population, the African slaves and their descendants, and the European colonists strictly followed the dogma of the Roman Catholic Church. There was no freedom of religion: Catholicism was the official religion of the then Kingdom of Portugal (July 25, 1139–October 5, 1910) and of the Portuguese Seaborne Empire then in the making (1415–1974; Macau: 1557–1999; East Timor: 1769–2002). The indigenous population had to be evangelized at all costs, so were the African slaves brought over from western and southwestern Africa, later also from the southern parts of the East African Coast. As for the European settlers, there were many Sephardic Jews and their descendants, collectively known as New-Christians. Later on there were also some Protestants, as in the case of French Huguenots. Yet non-Catholics could not freely and openly worship their faith lest, if caught, they would have been imprisoned, expelled from the colony, fined, exiled/deported (usually to Africa, e.g., the Slave Coast and present-day Angola), and/or killed.

In order to start its extraterritorial expansion overseas, Portugal had to seek and receive approval from the Holy See. Hence, the first and "official" reason for wanting to expand its physical boundaries and start conquering lands and people outside Europe had to be a religious one, that is, expand the Christian faith and increase the number of Christian nations. Obviously, the most important reasons were geo-economical, social, and political. Portugal wanted to control the major African and Oriental trade and spice routes passing through the Mediterranean/the Mediterranean Sea Basin, thus bypassing and replacing the monopoly of Venice (697–1797) and the Ottoman Empire (1299–1922), until then the sole leaders of trade in these areas. Having direct access to these commercial paths would have meant having the absolute monopoly of an immense commercial web that spanned from Europe and Africa to East Timor and Japan. The "discovery" of Brazil broadened this equation. Hence, it was necessary to maintain this economic control at all costs. Religion was the means through which Portugal could thus consolidate its geopolitical stronghold in key areas of its ever-growing empire.

Evangelization began a few years after Portuguese arrival in Brazil, though the first Portuguese archdiocese in the colony was erected in 1551. The early missionaries were the Jesuits who literally took at heart the plight of the native population: they strived to protect tribes, clans, and native nations against the constant abuses and enslavement efforts of the European settlers (Portuguese as well as colonizers/settlers from other parts of Europe). From north and the northeast to central and southern Brazil many missions were thus formed in *aldeias* (villages), whereby the native population was being converted to Catholicism and taught a trade in order to be self-sufficient. The Brothers of the Society of Jesus were soon followed by the

Ruins of São Miguel das Missões, Rio Grande do Sul. Founded in 1632, this Jesuit mission was part of a network of missions to the Guaranis that covered present-day Brazil and Argentina. In 1984, UNESCO declared these missions Guarani World Heritage Sites. (Andre Tomasi)

Franciscans who in their turn were followed by the Benedictines as well as by other religious orders. As in most parts of the New World, sub-Saharan Africa, Asia, and the Pacific area, the evangelization process also sparked an interest in the indigenous languages, oftentimes coupled with the fact that knowing these languages would have been beneficial to the Europeans who were trying to convert the natives and Europeanize them, in this case "Portuguesize" the natives. In addition, knowledge of the major Native American languages would have allowed the missionaries to print religious and inspirational literature in the autochthonous languages: for example, the Psalms, biblical stories, the New Testaments, Catechism, and prayers, as well as poetry, short stories, and plays loosely based on or inspired by European morality and mystery plays, not to mention grammars and dictionaries to and from Tupi. Given the myriad of languages and dialects spoken in present-day Brazil, the missionaries in fact chose Tupi as their lingua franca, or rather, as their language of communication, written as well as oral, with a number of tribes that spoke or understood languages and dialects that belong to the Tupian language family. Known by the native population as *nheengatu* (good language) and by the Portuguese-speaking people as *língua geral* (general language), Tupi and its variant Tupinambá were not only used as the languages of communication between the native population and the European missionaries but they were also employed among the European population and their descendants. In 1759, the Portuguese prime minister Sebastião José de Carvalho e Melo, First Marquis of Pombal (1699–1782), expelled the Jesuits from Portugal and all its colonies; hence, the use of Tupi began to fade out.

African slaves and their descendants were also converted to Catholicism, most of the time when they were still on African soil, waiting to be shipped to Brazil, or even en route to the Portuguese colony in the Americas. Just like the Amerindian population, the African slaves were forced to adopt Catholicism as their official religion. Once on Brazilian soil, Africans from many different ethnic, racial, linguistic, cultural, and religious backgrounds were thus thrust together and forced to adapt to a new, deplorable, and inhuman lifestyle—slavery—as well as adopt the Portuguese language and, perforce, espouse Christianity, that is, Catholicism, as their new faith.

NOTABLE MISSIONARIES

Jesuits Manuel da Nóbrega (1517–1570), José de Anchieta (1534–1597), and António Vieira (1608–1697) are perhaps the most outstanding missionaries of colonial/imperial Brazil inasmuch as they were instrumental in founding missions, villages (*aldeias*), and cities, as well as colleges, schools, and seminaries in most parts of present-day Brazil. Their dedication to the preservation of native life and mores, though under the tutelage of the Catholic faith, was instrumental for the survival/perpetuation of Amerindian cultures. Torn between their Christian "duty" to convert the natives and their egalitarian desire to protect them from physical/cultural destruction, these men dedicated their lives to the advancement of the Amerindian people by fostering their cultures and promoting tolerance and acceptance. Their sermons, treatises, and literary works are a living legacy of their efforts.

PROTESTANTISM

Catholicism was the official religion of colonial Brazil (1500–1808) and imperial Brazil (1808–1822). Independence from Portugal in 1822 brought tolerance of other religions; yet Catholicism remained the official religion of the young and short-lived Brazilian Empire (1822–1889). In 1889 a military coup d'état deposed the second and last emperor of Brazil, Pedro II (1831–1889), and instituted the Republic of the United States of Brazil. With the First Republican Constitution (1891), Catholicism ceased to be the only religion recognized by the republic. The separation of Church and state thus allowed the gradual yet ever-increasing formation of other religious groups and denominations, Christian as well as non-Christian, as in the case of Protestants, Mormons (as of the 1940s), Orthodox Christians, Jews (mainly Ashkenazim), Muslims, Bahá'ís, Buddhists, and Shintoists. This was an important step since with the new immigration waves that had started a few decades prior, there was indeed a need for tolerance and freedom of worship.

The abolition of slavery in 1888 opened the doors to immigration from other parts of the world. In 1818, in the state of Rio de Janeiro, 2,000 immigrants—mostly German-speaking—founded the Swiss colony of Nova Friburgo, the first non-Portuguese colony in Brazil. Four years later, in 1822, German immigrants established in the state of Rio Grande do Sul the São Leopoldo colony. In 1824, German colonization of Brazil began. Three years later, German immigrants arrived in Paraná. As of 1830 almost 7,000 German immigrants arrived in Brazil. Among these German-speaking immigrants there were Protestants from many denominations.

Before and after World War II (1939–1945), particularly starting from 1836, Brazil—mainly the states of Rio de Janeiro, São Paulo, Paraná, Santa Catarina, and Rio Grande do Sul—was the receptacle of many Italian immigrants. According to 1914 records, the states of Espírito Santo and Minas Gerais received ca. 50,000 and 90,000 Italians, respectively.

Italian, German, Japanese (as early as 1908 working on coffee plantations), and Jewish immigration to Brazil was mainly due to grants, agreements, and/or collaborations between different public and private entities, Brazilian as well as from the respective country/area of the world from where these immigrants originated. Because of this Japanese presence, Buddhism soon became the largest of all "minority" religions. By accepting or inviting Europeans and Asians (e.g., Japanese, Korean, and Chinese) to live and work in Brazil, the Brazilian government was thus trying to accomplish something common at this time in other parts of South America, that is, "whitening" their race, in this case, the Brazilian "race." From the religious point of view, these immigrants brought a wide spectrum of diversity, from Protestantism (all the major denominations), Judaism—for example, Sephardic, Ashkenazi, *Italkian* (Jewish-Italian), and *Zarphatic* (French-Jewish)—to Buddhism, Shintoism, and Daoism.

Many of the immigrants from the German-speaking world and from the future Italian area known as Süd Tirol (southern Tirol), then citizens of the Austro-Hungarian Empire (1867–1918)—who eventually settled in Colônia Dona Isabela (1875), and later renamed Bento Gonçalves, in today's state of Rio Grande do Sul—as well as settlers hailing from the Italian regions of Valle d'Aosta, Piedmont, Lombardy, Trentino-Alto Adige, and Friuli Venezia Giulia, were Protestant, thus contributing to the religious makeup of modern Brazil.

Historically though, the first Protestants who arrived in present-day Brazil were French-speaking Calvinists (Huguenots) who settled in Guanabara, Rio de Janeiro, when the French occupied the coastal area between Rio de Janeiro and Cabo Frio (France Antarctique: 1555–1567). In 1685, Louis XIV of France (1643–1715) revoked the Edict of Nantes (April 13, 1598), which guaranteed freedom of worship to all French citizens and replaced it with the Code Noir, thus prohibiting Jewish and Protestant presence on French soil. Many Jews and Huguenots sought refuge in England and the Anglo-American colonies, Canada, and the Antilles, particularly the Bahamas and Jamaica. Yet a few others chose Brazil as their refuge.

During the 30 years of Dutch presence in Brazil (1624–1654), under the aegis of the Dutch Reformed Church, European settlers residing in Brazil were granted freedom of religion: Sephardic Jews, Protestants, and Catholics were thus free to worship without fear of retaliation. In 1654, when Portugal regained full control of its colony, Jews and Protestant either left Brazil or began practicing their faith secretly. Two hundred and thirty-seven years later, with the First Republican Constitution (1891), Brazilians finally regained the right to choose and follow a religion other than Catholicism. The presence of immigrants from many parts of the world—thus adhering to many different religions and religious denominations—then necessary for an economy in transition from a slave-based system to a modern society centered on salaried work, finally opened the doors to freedom of religion for all.

Though Catholicism was and still is, at least "formally," the most numerous religion in Brazil (a little over 64% in the 2010 census), during these past four decades, the universal increase of secularism that has reached all Western societies, and the constant growth of Protestant denominations throughout the country (a little over 20% in the 2010 census), have seriously "challenged" the centuries-old, undisputed leadership of the Catholic Church. Mormonism, Jehovah's Witnesses, and a proliferation of Protestant churches have in fact seen a rise in Brazil, particularly Evangelical, Pentecostal, and Neo-Pentecostal denominations. Though boasting presence in almost all Brazil, "traditional" Protestants as the Baptists, Lutherans, Methodists, and Presbyterians (mostly present in Minas Gerais, Rio Grande do Sul, and Santa Catarina, as well as parts of Rio de Janeiro and Espírito Santo) and the Anglican Episcopal Church of Brazil are now being "challenged" by an ever-increasing presence of Evangelical, Pentecostal, and Neo-Pentecostal denominations (Minas Gerais, Paraná, Rio de Janeiro, and São Paulo), including Seventh-Day Adventists.

From a "religion" (better yet, a religious branch of Christianity) mainly brought over or at least revived by American missionaries in the second half of the 19th century, Protestantism gradually fulfilled the inner void and solved many of the unanswered questions that Catholics had for centuries, and not only in Brazil: how to reconcile Christian dogma and tradition(s) with the need to freely express a direct connection to God. Syncretism with African religions brought over by the slaves, inevitably intertwined with the socioreligious mores of the autochthonous population of Brazil, was the answer for those who opted to remain within the Catholic fold. Yet for those who wanted a clear break with what they perceived a stifling and encroaching religion, Protestantism was the answer. Though syncretic forms of Protestantism have existed and do exist today, combining the Native Brazilian and the African religious tradition with the Protestant denominations, including cases of hybrid forms of Catholicism and Protestantism, during the past 40 decades many Protestant denominations have managed to remain "unadulterated" and grow in numbers never imagined before, Pentecostals being the most numerous throughout the country, as the case of the Brazilian-born Universal Church of the Kingdom of God (*Igreja Universal do Reino de Deus*) (July 19, 1977), a Neo-Pentecostal Protestant denomination that asserts adherence to millions of people throughout the world.

JOSÉ HIPÓLITO TRIGUEIRINHO NETTO (1931–)

Trigueirinho's universal spiritual message is a natural continuum of current movements (e.g., Esotericism, New-Age syncretism, and Ufology), as well as an adaptation of older forms of mystic/theological schools (e.g., Theosophy, Anthroposophy, Post-Modern Spirituality, and Rosicrucianism), whereby only a select number of its members have access to esoteric truths, read minds, prophesize, speak different languages, and have contact with (spiritual) beings from higher hierarchies. Yet Trigueirinho proposes a new kind of spirituality, one that all Humankind, through proper training and guidance, could thus have not only an insight into Nature, the world below the terrestrial surface, the Universe, and the Spiritual World, but also, and more importantly, become One with the Supernatural/Extraterrestrial Being(s), thus enjoying everlasting Peace and Harmony, transferrable to Humanity, the Earth, and the entire Universe.

THE *IRMANDADES (LAY BROTHERHOODS)*

Perhaps the first attested presence of African slaves in Brazil dates back to 1532. By 1539, toward the end of the rule of the first governor general of Brazil, Tomé de Sousa (1549–1553), as well as his two immediate successors, Duarte da Costa (1553–1557) and Mem de Sá (1557–1572), slave ships were indeed a reality in the new Portuguese colony in the Americas, transporting African slaves to Brazil and other European colonies in the New World. Trading posts (*feitorias*) along the West African coast—for example, Arguim, in present-day Mauritania, São Jorge da Mina, in present-day Ghana, and the Fort of St. John the Baptist of Ouidah, in modern Benin—as well as from the archipelagoes of Cape Verde and São Tomé and Príncipe, were the receptacles of African slaves ready to be sent to Brazil and the rest of the New World. On African soil, aboard the slave ships, or on their arrival in Brazil, African slaves were "baptized" and nominally introduced to Catholicism.

Slavery in Brazil lasted for more than three centuries (1532–1888). Apart from escaping and forming their own *Quilombos* (hiding places in the hinterland), African slaves and their descendants in Brazil had only one option to overcome the hardships of labor, abuse, and torture, or rather, they tried to adapt to the living and working conditions in which they were placed and, at the same time, assimilate to the culture(s) and mores of their masters, including their religion. The *Irmandades* were Catholic lay brotherhoods made of slaves or former African slaves and their descendants that were formed in Brazil during the colonial (1552–1808) and imperial eras (1808–1822), as well as the Brazilian Empire era (1822–1889). The *Irmandades* thus had a religious-humanitarian goal, or rather, they assisted its members in having their own voice and space within an extremely hierarchical and racially divided/stratified society. Being a member of a lay brotherhood gave the African and the Afro-Brazilian population the necessary tools of learning how to become and feel part of a group and eventually save enough money for their own manumission or of a loved one, while at the same time it gave them the chance of practicing a syncretic form of Catholicism as well as traditional forms of religion/spiritualism tied to their ancestral homelands in Africa.

RELIGIOUS SYNCRETISM

Syncretism is following two or more religions and/or ways of life, yet officially still belonging to (almost) one religion. In a Brazilian context, it means being Christian, mostly Catholic, and at the same time believing in or practicing ceremonies tied to traditional African (mainly Yoruba) and Amerindian religions or spiritual rituals, though the African component is the most predominant one.

Candomblé—"ka'nome," Kimbundo for "drum," + ilé, Yoruba for "house," hence "house of drums"—is a religion tied to Nature worship with roots in the area covering present-day Nigeria-Benin, which was brought to Brazil by the African slaves and later resurfaced in its syncretic form among the Christian (for the most part Catholic) population of all Brazilians, mainly Afro-Brazilians. *Axé*, "magic force," is the dynamic element of Nature that gives life to African gods and humans. The word is now used in prayers and as a form of salutation among Candomblé practitioners; yet it is also tied to the inner mystery/secret of life. Hence, all human beings have an individual *axé*. The *Orixá*, instead, is the "deity" or "agent" of Nature/the Universe summoned and worshipped according to his or her specific manifestations. *Egum*—"dead," "ancestral," or a person who has finally reached the "status of belonging to the deities"—controls the morals and the religions of the *orixás* while, at the same time, it holds together the sociocultural organization of the Candomblé.

In the New World, African slaves and their descendants from West Africa overtly practiced Catholicism, yet covertly worshiped their ancestral gods and followed their spiritual cosmogony. However, in order to practice their "unique" form of Catholicism in a colony (1500–1822) and later an empire (1822–1889) that only contemplated Catholicism as the official religion, Afro-Brazilians had to camouflage certain obviously "African" aspects of their faith. Hence, the *orixás* were conveniently associated with Catholic saints and their supreme deity, Olódùmarè, lord of the Universe, was associated with the God of the Judeo-Christian tradition. The *orixás* thus controlled all aspects of Nature and human life. If properly adored, the *orixás* would thus help their followers. This perfectly fit within the Catholic frame of mind of the Portuguese and later (as of 1822) the Brazilian authorities. Worshipping saints (*orixás*) in order to reach God (Olódùmarè) was perfectly fine. Yet in Catholicism only priests can function as intermediaries between the saints/Mary/God and the believers. Thus, Africans and Afro-Brazilians had to disguise the intermediary role of priests and priestesses, the "pai-de-santo" (father-of-saint, a linguistic calque from the Yoruba *babalorishá*, "father of the spirit of the ancestors") and "mãe-de-santo" (mother-of-saint), as "folkloric" religious ways of worshipping saints, Mary, Jesus, and God. Given the absence of a sacred text, in Candomblé, the pai-de-santo and mãe-de-santo are thus responsible for transmitting to the new generations their ancestral beliefs, though modified and adapted to Catholicism. For instance, one of the many *orixás*, *Iemanjá*, was worshiped as *Nossa Senhora dos Navegantes* (Our Lady of the Seafaring), whereas *Iansã* was worshiped as Saint Barbara. Modern forms of Candomblé are Umbanda and Quimbanda. Followers of Candomblé call themselves *Povo do santo* (People of the Saint), *Povo de santo* (People of Saint), or simply *do santo* (of the Saint). Like Candomblé, Umbanda is a composite of many rituals and rites; yet there are some beliefs that are common to all followers. An example is faith in *Olorum* (Lord, the Sun, also known as *Zambi* [Lord]) who can have different representations. A southern Brazilian variant is known as *Batuque*, though it differs consistently in liturgy. The *Xangô*, instead, is the *orixá* of fire and lightening who punishes thieves, liars, and evildoers. He is usually identified as Saint Jerome or Saint Peter. *Xangô* is thus associated with death and bright colors, the latter representing *Egum*. The *Tambor de Mina* (Mina's Drum) is usually practiced in the Amazon, Maranhão, and Piauí. Candomblé and Umbanda temples are known as *terreiros* (public square, yard) and/or *casas* (houses). A table that shows the five stages of the development of the syncretic forms of the Afro-Brazilian religions follows.

TABLE 5.1

Afro-Brazilian Religions/Syncretic Forms		
African Ethnic/ Racial/Linguistic Groups	**Areas in Africa**	**First Areas/States in Brazil (And then to the rest of the country)**
Bantu	Angola Cabinda (Angola) Benguela (Angola) Congo	Rio de Janeiro Bahia Maranhão Pernambuco
Bantu	Mozambique	Rio de Janeiro
Western Sudanese	Yoruba (Nagô)—Nigeria Dahomey (Jeje)—Benin Fante/Twi (Mina)—Central/ Western Ghana Ashanti (Mina)—Ghana	Bahia Pernambuco Maranhão, etc.
Islamized Western Sudanese	Hausa—Northern Nigeria Peul (Fula) Mandinga—Mali Tapa (Nupe)—Northern Nigeria/Niger	Bahia Pernambuco Rio

Afro-Brazilian Religions/Syncretic Forms: First Stage

Candomblé	Western Sudan	Nagô—Yoruba =>	Ijexá—Bahia Keto—Bahia Oyo—Rio Grande do Sul Xangô—Maranhão
Candomblé	Western Sudan	Jeje—Benin/Nagô	Bahia
Candomblé	Western Sudan	Mina-Jeje	Maranhão
Candomblé	Western Sudan	Muçurumim (Muslims)	Bahia
Candomblé	Bantu	Angola	Bahia, Rio de Janeiro
Candomblé (19th century)	Bantu	Congo	Bahia, Rio de Janeiro
Candomblé (19th century)	Bantu	Angola-Congo	Bahia, Rio de Janeiro

Afro-Brazilian Religions/Syncretic Forms: Second Stage

Nagô—Pajelança	Candomblé de Caboclo	Bahia
Angola-Congo	Toré Catimó	North Northeast

Afro-Brazilian Religions/Syncretic Forms: Third Stage

Nagô Muçurumim (Muslims) Angola-Congo Cabinda Candomblé de Caboclo	Macumba => Feitiçaria (witchcraft)	Rio de Janeiro

Afro-Brazilian Religions/Syncretic Forms: Fourth Stage

Macumba (original) Catholicism Kardecist Spiritism Occultism	Umbanda (Esoteric)	Rio de Janeiro
Macumba (transformed)	Quimbunda (popular form)	Rio de Janeiro

Afro-Brazilian Religions/Syncretic Forms: Fifth Stage

Umbanda or, at times, Candomblé	Umbanda + Candomblé	Umbanda—Angola Ermelocô Umbanda—Nagô Umbanda—Jeje	Brazil
Umbanda or Kardecism	Umbanda de Branco or Umbanda de Caritas		Brazil

Source: Table was composed based on information found in Pietro Canova. 1985. *I culti afrobrasiliani. Una sfida alla società e alla chiesa.* Bologna: EMI, pp. 88–89; 114.

JUDAISM

The first Jews who set foot on Brazil were Sephardim, that is, Jews from the Iberian Peninsula. During Dutch occupation (1630–1654), Jews were granted freedom of worship. In 1636, *Kahal Zur Israel* was built in Recife, the first synagogue in the Americas. During the first 300 years of European presence—or, rather, expansion and (im)migration to the New World—the Jewish component in Brazil, particularly during the colonial period (1500–1822), was made up of Sephardic Jews, mainly *Conversos*, also known as New-Christians, Crypto-Jews, and, though pejoratively, *Marranos* (pigs). As of 1822, the year of Brazil's independence, and with the subsequent elevation to the Brazilian Empire (1822–1889), followed by the proclamation of the Brazilian Republic (1899–) and the 1891 Constitution Law granting freedom of religion for all, the descendants of the early Sephardic Jews finally had the opportunity of publicly returning and declaring adherence to Judaism without any fears of being persecuted religiously. In other words, Sephardic Jews and New-Christians could not any longer be persecuted for their (alleged) ethnic and religious background.

During the first decades of the 19th century, the Amazon region was the receptacle of a major wave of Sephardic immigration, mainly from the Maghreb, the Levant, including Egypt, and, though to a lesser degree, the southern Balkans. At the same time, though, Ashkenazi and Zarphatic (French-speaking) Jews from Alsace began settling in Rio de Janeiro and surrounding areas. In a little over a generation these Jewish communities showed signs of complete assimilation to Brazilian mores, eventually losing all ties to Judaism. Adapting to the needs of life in a new environment, the last "frontier," became the priority of most of these immigrant Jews and their descendants. Hence, Jewish way of life was sacrificed in exchange for daily survival and adaptation to the new environment. Ethnic and religious identity was thus transformed into collective, national identity: in other words, they were now Brazilian Jews, defending their rights as Brazilians, yet of Jewish ancestry.

Nevertheless, during the last two decades of the 19th century and the first four decades of the 20th century, Brazil was the receptacle of yet another wave of Jewish immigration, this time from Northern and Eastern Europe: first from newly born Germany (1871), up until 1886, and then followed the immigration of other Ashkenazim from Austria and Eastern Europe (present-day Poland, Ukraine, Russia, Byelorussia, Romania, and Moldavia).

These Jews eventually settled in the states and cities south of Rio de Janeiro, particularly São Paulo, Paraná, Santa Catarina, and Rio Grande do Sul, though São Paulo, Curitiba, and Porto Alegre were the cities with a high concentration of Ashkenazim given that the climate of this geographic area is very similar to the one they were used to back in Europe.

The majority of these new immigrants chose the Brazilian frontier for economic reasons as well as for the opportunity of finally being able to create Jewish communities where they could in a sense avoid the inevitable acculturation to Brazilian society. This was one of the main reasons why many Sephardim and "other" Jews of the Diasporas chose Brazil as their new home between the end of the 19th century and the beginning of the 20th century, settling at the very opposite frontiers: the Amazon and the Pará, in the north, and the São Paulo area in the south.

Just like its North American counterpart, the frontier in Brazil and Argentina offered these *Diasporic* Jews a wide and unlimited range of opportunities and freedom, whereby they could build and reconstruct their identity. It was exactly against this frontier space that the Brazilian-Jewish identity of many Sephardim, Ashkenazim, and "other" Jews of the Diaspora was born.

In 1823, a small contingent of Sephardic Jews from present-day Morocco settled in Belém, Pará. Their quick assimilation to the local customs was mainly due to the remoteness of the place and to the total absence of other Jewish communities in the area. Ethnic and racial mixing thus led to complete assimilation and loss of most of their Jewish traditions. Thirty-five years later, in the aftermath of the Spanish–Moroccan War (1859–1860), more than 150 Sephardic families joined their coreligionists in the Amazon area.

This voluntary Sephardic Diaspora was triggered by the political climate of the time. In the aftermath of the Berlin Conference (1884–1885), whereby the European countries had to physically occupy any claimed territory in Africa, the Middle East, Asia, and the Pacific area, it became increasingly hard for the Sephardic communities to continue living in Islamic lands as they hitherto had done.

The decline of Moroccan economy, counterbalanced by the economic growth of the Levant, triggered cases of intolerance against rich Sephardic traders and merchants residing in Morocco, especially those who had economic ties with the Middle East, France, and the United Kingdom. Some Sephardic Jews migrated to neighboring Algeria and Egypt, whereas others opted for emigrating to the New World.

During the last decade of the 19th century, a little over 1,000 Sephardic Jews from present-day Morocco thus emigrated to the Pará drawn by the burgeoning economy of the Amazon Rubber Boom (1850–1920). Most of these Sephardim prospered economically and obtained Brazilian citizenship, thus being fully integrated with Brazilian society; yet some of these Moroccan Jews of Sephardic ancestry opted for returning to Morocco.

In 1893 between 35,000 and 40,000 Jews, mostly German-speaking Ashkenazim, emigrated to the Americas. A few years later (1901), many reached Porto Alegre, capital of the future Brazilian state of Rio Grande do Sul, whose Republican Party was very much in favor of new Jewish immigrants in its midst, whereas the other German-speaking Ashkenazim settled in Santa Fe, Entre Ríos, and Buenos Aires in neighboring Argentina. In 1891, the Bavarian-Jewish philanthropist, Baron Maurice von Hirsch (1831–1896), created agricultural colonies sponsored by the Jewish Colonization Association (Associação de Colonização Judaica/Yidishe kolonizatsye gezelschaft). In 1904, 38 Jewish families arrived to the newly formed colony of Philippson (1902) in central Rio Grande do Sul, soon followed by other families. Franz Philippson was a Belgian president and owner of the Compagnie Auxiliaire de Chemins de Fer au Brésil (Auxiliary Company of the Railroads of Brazil). Philippson was located very close to the railroad linking Rio de Janeiro to Montevideo, Uruguay. In 1909, 92 families formed a second Jewish colony in Quatro Irmãos, in central-northwest Rio Grande do Sul, encompassing present-day Erechim and Getúlio Vargas along the railway road linking São Paulo to Porto Alegre. Despite its small number, roughly a little over 2,000 people, these Eastern European Jews never fully assimilated to the Brazilian culture, perhaps owing to the fact that they remained isolated in this rural area.

Also during this time, almost 300 Sephardic families, hailing from the Otto-man Empire, mainly present-day Turkey and Egypt, arrived at Porto Alegre, Rio Grande do Sul. Between 1912 and1938, they were joined by more than 110 Russian- and Romanian-speaking Ashkenazim from the then Russian region of Bessarabia (1812–1924), today part of Moldova (1991).

As of 1923, the Jewish centers in Rio Grande do Sul moved to Passo Fundo, Santa Maria, Cruz Alta, and Porto Alegre, cities where they established "alternative" Jew-ish communities. Within a generation, the arrival of Ashkenazim, Sephardim, and "Oriental" Jews, or rather, Jews from the Maghreb as well as Levantine Jews, led to the formation of Jewish neighborhoods divided by ethnicity and area/country of origin.

During 1939–1945 and immediately after the Holocaust, Brazilian Jews were able to overcome the obvious ethnic, folkloric, linguistic differences, reflected in the innu-merous ways of feeling and expressing Judaism, thus uniting in the struggle for equal-ity as citizens of a new country that had to give and guarantee freedom of worship. The same attitude was kept during the presidential regimes (1930–1945, 1950–1954) of Getúlio Vargas (1883–1954) and the following period of dictatorship (1964–1988).

Even though assimilation has been a constant in the loss of Jewish identity, be it Sephardic, Ashkenazi, or "Oriental"—by a simple act of volition or, as it often was the case, because of "mixed" marriages—a Jewish cultural identity is still part of many Brazilian Jews, despite their ethnic and racial background.

ANTÓNIO JOSÉ DA SILVA, O JUDEU, (1705–1739)

Son of a prominent New-Christian family, António José da Silva, "o judeu" (the Jew), was born in Rio de Janeiro where he remained until 1713, the year in which, together with his father, he followed his mother to Lisbon where she was sentenced for being a practicing Jew. While studying at Coimbra, António was imprisoned under suspicion of Judaizing. On being released, after having been tortured and punished for his "crimes," António began practicing law and writing dramas acclaimed by many as masterpieces of satirical theater. Yet his work also drew the attention of the Inquisition. He was again imprisoned, together with his mother and young wife, though the latter two were eventually released. António, instead, was found guilty of reverting to Judaism and was eventually tortured and burned at the stake.

ISLAM

With the arrival of the first caravels in the New World, Arabic and Islam started to have a role, though minimal, in the life of the European colonists on the American continent, from the Anglo-American colonies and the Caribbean to Brazil. Even though they did not have a decisive role in the colonization process of the Ameri-cas, Muslims were deeply involved in the commercial transactions and the political events that contributed to the formation of a new space, namely, the Atlantic world, a geographical area that linked all continents.

Ethnic and racial diversity of the Muslim slaves in the Americas embraced a vast geographic area of the Western Sudan, as well as central-southern Africa. Most of the slaves from the Western Sudan were Muslims or highly Islamized Africans; hence, they possessed a high degree of familiarity with or knowledge of Arabic, written and spoken, as well as of the Islamic precepts, at least its basic tenets.

Undoubtedly, the first Muslims who resided in Brazil were the African slaves who were brought over by the Portuguese during the first half of the 16th century. The majority of the slaves who were sent to Brazil and the rest of the Americas came from the Senegambia, the Upper Guinea area, the Windward Coast, the Gold Coast, the Bight of Benin, the Bight of Biafra, West Central Africa, and, later in the slave trade era, East Africa. Barring the Bight of Biafra and West Central Africa, most of these slaves were Muslim or partly Islamized Africans. It is thus understandable that a good number of the slaves who were eventually sent to Brazil managed to keep some aspects of their monotheistic faith, even if it was already intertwined with the traditional African beliefs or, once on Brazilian soil, with Catholicism.

Islam was transported to Salvador da Bahia by Hausa and other Muslims, among the latter, the Tapa, also known as Yoruba or Nupe. As Muslim or heavily Islamized slaves in Brazil, these men knew how to speak, read, and write in Arabic as well as their native languages, yet in Arabic script (*aljamia* in Portuguese).

Because of the Atlantic slave trade (16th–19th centuries), communication between Brazil and the Western Sudan was constant. Given its geopolitical and strategic position, Salvador da Bahia held a decisive role in this Diasporic nexus between the two sides of the Atlantic. Each expedition brought to Brazil African "news" and, at the same time, contacts and Afro-Brazilian presence were established along the Slave Coast in the Gulf of Guinea (present-day Togo, Benin, and western Nigeria), particularly after a few Afro-Brazilian freedmen, between 3,000 and 8,000, settled in this area (first decades of the 19th century), known locally as *Agudas*, *Tabom*, or "Brazilians." This continuous contact between the two sides of the Atlantic reinforced the ties and connections with traditional African religions and Islam.

With the opening of Brazil to immigration during the imperial age (1822–1889) and the Republican period (1889–), the first modern-era Muslims arrived from present-day Lebanon, Morocco, Palestine, and Syria, later joined by Muslims hailing from Southeast Asia. In 1956, the first mosque on Brazilian/South American soil opened its doors in São Paulo. Today there are roughly 60 Muslim institutions spread throughout Brazil, covering Islamic centers, mosques, Qur'ānic schools, *sociedades beneficentes* (charitable societies), and hospitals, serving a total of almost 90 Muslim communities, from Manaus to Porto Alegre, with the aim of preserving the Arabic language, the cultures, and civilizations, as well as the social and religious festivities of the entire Islamic world. Major Muslim communities are found today in the greater São Paulo and Santos area (and 40% of the entire Muslim population), Curitiba, and Paraná. Sunnis outnumber Shiites; the latter are the result of a more recent immigration to the same areas chosen by their coreligionists. Like most religious "minorities," Brazilian Muslims are well integrated and assimilated to the local milieu. Just like Muslims around the world, Brazilian Muslims worship in mosques (*mesquitas*) and study in Qur'ānic schools (*madrasa* in Arabic or *escola corânica* in Portuguese). Also like in most parts of the word today, there are conversions to Islam of people originally not from the Middle East or from areas where Islam is not the dominant religion, known in Brazil as *Muslimas* (Arabic for Muslims). Marriages outside the faith and assimilation to Brazilian life are the main reasons for Islam still maintaining low numbers when it comes to adherence to all Islamic precepts and regularly attending services.

THE MALÊ REVOLT: 1835

In January 1835, African and Afro-Brazilian slaves and freedmen, Muslim as well as Christian, were encouraged to revolt by Muslim teachers (*Imale* in Yoruba and *Malê* in the Portuguese spoken in Bahia). The Malês were highly respected freedmen and slaves who knew how to read and write in Arabic and Portuguese—in Latin script as well as in Arabic characters (*aljamia* in Portuguese)—were knowledgeable of the Qur'ān, and proselytized among the population. The insurgents wore amulets with pieces of paper containing verses from the Qur'ān folded and set in sewn leather pouches, wore long white robes (*abadás*), and wore various kinds of rings on their fingers. Though it was not the only freedmen/slave uprising in Brazil, and despite its failure, the Malê Revolt was significant because of its socioreligious significance.

THE RELIGIOUS SCENARIO IN BRAZIL TODAY

With the end of dictatorship (1964–1988) Brazil began a sociopolitical and economic climb that eventually led the country joining other emerging nations around the world. Interest in religion accompanied this ascent and overall economic welfare, to the detriment of the Catholic faith which ever since has seen a sharp decline in numbers. Though most people still claim to be Catholic, in reality they are nonpracticing Catholics and/or they also practice other forms of religions. Some may combine Catholic rituals, or better yet, cultural practices associated with Catholicism, with a Protestant "attitude," "outlook," or "way of life," regardless of the denomination, though Evangelicalism is the majority. This is particularly the case of Brazilians who are "in transition" between the two main branches of Christianity (Catholicism and Protestantism). Others, instead, may follow or would be attracted to Spiritism or (Afro-Brazilian/indigenous Brazilian) Spiritualism while still claiming ties to Catholicism. However, most nominal Catholics, some practicing Catholics, and, to a lesser degree, some Protestants, do mix their Christian faith with Candomblé and/or Umbanda, to name the most common syncretic practices.

With roots in the American Midwest and California in the first two decades of the 20th century, modern Pentecostal movements reached Brazil in the 1950s. In less than a decade Brazilian Pentecostals were able to open their own churches and spread their evangelization to different ethnic, racial, and socioeconomic groups of the country. The gradual modernization and socioeconomic amelioration of Brazil also saw the gradual proliferation of religions, in general, and Protestant denominations, in particular (e.g., Assemblies of God, Baptists, Evangelicals, and Pentecostals), as well as Jehovah's Witnesses and Mormons. Hence, in a little over 50 years the religious scenario in Brazil changed drastically, addressing the needs of a population that, unlike before, now wants to express its religiosity in a more direct way.

Though the 1891 Constitution separated religion from the state, thus making all religions equal in the eyes of the law, the Catholic Church in Brazil maintained its influence until the late 1970s when people gradually began to look for other alternatives to Catholicism, namely Protestantism. Today more than ever, in an effort to regain momentum, the Catholic Church in Brazil is trying to engage people in actively living their Catholicism; this includes "accepting" more effusive ways of expressing faith, like performing music, dancing, and singing, thus resembling the Afro-Brazilian way of addressing religiosity, yet devoid of any syncretic element. The challenge is enormous, since during the span of 40 years the Catholic Church in Brazil has lost almost 40 percent of its followers, mainly to Protestantism (all denominations, primarily Evangelicals) and other world religions, particularly Afro-Brazilian and Asian/Neo-Asian beliefs. The choice of a South American pontiff, Pope Francis (2013–), was perhaps a strategic attempt to halt a trend common to the rest of Latin America, namely the proliferation of Protestantism.

The Brazilian Buddhist community is the third largest in the Americas, after the United States and Canada, encompassing most of the Buddhist rituals of the world. A little less than 50 percent of Japanese-Brazilians are either Buddhists or practice some sort of Buddhist ritual. Like followers of other minority religions, Buddhists in Brazil face the problem of assimilation to Brazilian mores; yet there is an increasing interest in Buddhism by many who are seeking an alternative to the Judeo-Christian religious tradition.

With origins in Allan Kardec's 19th-century doctrine, Spiritism is the second-largest religion in Brazil, after Protestantism. Belief in God as the Supreme Intelligence, in spirits and in the existence of other life forms in the Universe, in reincarnation, and in the constant upgrading of the soul until it reaches perfection, perhaps, make this doctrine more appealable to urban people (educated adults) from a higher social class since it requires reading and understanding different complex texts as well as performing individual self-discovery and meditation, something that the lower classes cannot do because of lack of proper schooling and, most of all, because they are more concerned with more pressing matters, that is, making ends meet. Yet there are followers, particularly from the middle and lower classes, who practice a hybrid form of Spiritism, strongly imbued with Catholic elements.

As mentioned earlier, because of the low numbers of strict followers, in Brazil Islam is still somewhat considered an "ethnic" religion. However, just like what happens with other "minority" religions associated with particular ethnic groups—as in the case of Jews and Eastern Orthodox Armenians, Greeks, Lebanese, Russians, Syrians, and Ukrainians—in Brazil there is today a revival whereby the new generations are striving to reconnect to their linguistic, ethnic, and/or religious past. In addition, because of the universality of Islam and current sociopolitical and economic world events, Brazilian Muslims are oftentimes taking this revival a step further, thus wanting to reconnect not only to their own specific ethnic group (if any) but also, and more importantly, to the entire Muslim world, *ummah* in Arabic, and join a more universal cause. The same can be said of the many and diverse Jewish groups: there is a renewed interest in asserting one's Jewishness and reconnect it to a more universal sense of belonging to the Jewish world, one that encompasses Brazil into this more universal outlook to what it means being Jewish today in a globalized world.

CHICO XAVIER (1910–2002)

Nominated twice for the Nobel Peace Prize (1981, 1982), Francisco de Paula Cândido Xavier, commonly known as Chico Xavier, was a very popular Spiritist medium and philanthropist. Born in Pedro Leopoldo, Minas Gerais, Xavier wrote 468 books through *psychography* (automatic writing); that is, spirits took control of his hand, in his capacity of medium, and in a waking state. Xavier's spiritual guide was Emmanuel who, in his past lives, had been Senator Publius Lentulus—a Roman Consul during the reign of Augustus (63 BCE–14 CE)—Damian, a Spanish priest, and a professor at the Sorbonne. Like all Spiritists, Xavier emphasized the fact that mediums are only channels through which spirits operate and that mediums cannot contact the dead unless the latter want to be contacted. Spiritism became widely known in Brazil, thanks to Xavier's frequent appearance on television in the 1960s and early part of the 1970s.

Alas, as the number of followers of the different Protestant denominations increases, so do the cases of intolerance toward and prejudice against followers of Candomblé and Umbanda. Actually, not only are the *Povo do Santo* being discriminated against but also any person/believer who practices or manifests some sort of syncretic form of Christianity clearly influenced by these Afro-Brazilian belief systems. The same can be said of the Catholic Church: though in the past it "closed" an eye on some hybrid forms of Catholicism, now, perhaps under the influence of some more conservative Protestant denominations, as the Neo-Pentecostals, the Brazilian Catholic Church is outwardly condemning such fused practices.

Overall, in a country of almost 200 million people, religion is by far the most intriguing aspect of Brazil, particularly the way it is being lived and expressed by communities from all walks of life and all ethnic and racial backgrounds and all in between. The way these religions deal with their own dogmas and interact with one another obeys to the necessity to resist or at least attenuate the inevitable in all societies, particularly today: assimilation, in this case, assimilation to Brazilian mores which by their very nature are hybrid and multinational.

REFERENCES

Alden, Dauril, ed. 1973. *Colonial Roots of Modern Brazil. Papers of the Newberry Library Conference.* Berkeley: University of California Press.

Baranov, David. 2000. *The Abolition of Slavery in Brazil. The "Liberation" of Africans through the Emancipation of Capital.* Westport, CT: Greenwood Publishing.

Bascom, William Russell. 1972. *Shango in the New World.* Fourth Occasional Publication of the African and Afro-American Research Institute. Austin: University of Texas Press.

Bastide, Roger. 1978. *The African Religions of Brazil: Towards a Sociology of Their Interpretation of Civilizations.* Baltimore, MD: Johns Hopkins University Press.

Bilgé, Barbara J. 1986–1987. "Islam in the Americas." In Mircea Eliade (ed.), *The Encyclopedia of Religions*, 16 vols, pp. 425–431. New York: Macmillan, 7.

Bruneau, Thomas C. 1974. *The Political Transformation of the Brazilian Catholic Church*. New York: Cambridge University Press.

Burns, E. Bradford. 1962. "Introduction to the Brazilian Jesuit Letters." *Mid-America*, 44: 172–186.

Canova, Pietro. 1985. *I culti afrobrasiliani. Una sfida alla società e alla chiesa*. Bologna: EMI.

Cardozo, Manoel S. da Silveria. 1947. "The Lay Brotherhoods of Colonial Bahia." *Catholic Historical Review*, 33: 12–30.

Conrad, Robert. 1972. *The Destruction of Brazilian Slavery, 1850–1888*. Berkeley: University of California Press.

Crook, Larry, and Randal Johnson, eds. 1999. *Black Brazil. Culture, Identity, and Social Mobilization*. Los Angeles: UCLA Latin American Center for Publications, University of California.

Graham, Richard. 1999. "Free African Brazilians and the State in Slavery Times." In Michael Hanchard (ed.), *Racial Politics in Contemporary Brazil*, pp. 30–58. Durham, NC: Duke University Press.

Hamilton, Russell G., Jr. 1970. "The Present State of African Cults in Bahia." *Journal of Social History*, 3: 357–373.

Jacobsen, Jerome V. 1942a. "Jesuit Founders in Portugal and Brazil." *Mid-America*, 24: 3–26.

Jacobsen, Jerome V. 1942b. "Nóbrega of Brazil." *Mid-America*, 24: 151–187.

Kent, R. K. 1970. "African Revolt in Bahia: 24–25 January 1835." *Journal of Social History*, 3: 334–356.

Kent, R. K. 1996. "Palmares: An African State in Brazil." In Richard Price (ed.), *Maroon Societies. Rebel Slave Communities in the Americas*, pp. 170–190. 1979. Baltimore, MD: Johns Hopkins University Press.

Kiemen, Mathias Charles, O. F. M. 1973. *The Indian Policy of Portugal in the Amazon Regions: 1614–1693*. 1954. New York: Octagon.

Levi, Joseph Abraham. 2004. "Padre António Vieira." In Monica Rector and Fred Clark (eds.), *Dictionary of Literary Biography. Portuguese and Brazilian Literature*. 2 vols. Vol. 2, pp. 385–396. Bruccoli Clark Layman. Detroit: Gale Thomson. *Brazilian Literature*, 2.

Levi, Joseph Abraham. 2009. "Brazil, Jews, and Transatlantic Trade." In M. Avrum Ehrlich (ed.), *Encyclopedia of the Jewish Diaspora*. 3 vols, pp. 723–727. Santa Barbara: ABC-CLIO, 2.

Levi, Joseph Abraham. 2011. "José Trigueirinho. Calling Humanity (Mysticism: Theosophy, Anthroposophy, Rosicrucianism." In David M. Fahey (ed.), *Milestone Documents of World Religions. Exploring Transitions of Faith through Primary Sources*. Vol. 3, pp. 1604–2002. Dallas: Schlager Group, 1430–1449.

Lovell, Peggy A. 1994. "Race, Gender, and Development in Brazil." *Latin American Research Review*, 29: 7–35.

Mattoso, Katia M. 1986. De Queirós. *To Be a Slave in Brazil. 1550–1888*. 1979. Trans. Arthur Goldhammer. New Brunswick: Rutgers University Press.

"The Medicine Man." *Journal of Civilization*, 10 (1841): 145–148.

Mello, José Antônio Gonçalves de. 1961. "The Dutch Calvinists and Religious Toleration in Portuguese America." *Proceedings of the British Academy*, 47: 485–488.

Mooney, Carolyn J. May 22, 1998. "Notes From Academe: Brazil—Anthropologist Sheds Light on Jungle Communities Founded by Fugitive Slaves." *Chronicle of Higher Education*.

Pierson, Donald. 1967. *Negroes in Brazil: A Study of Race Contacts at Bahia.* 1942. Carbondale: Southern Illinois University Press.

Prince, Howard Melvin. 1978. Slave Rebellion in Bahia, 1807–1835. Dissertation, Columbia University, 1972. 1975. Ann Arbor: UMI.

Ramos, Arthur. 1980. *The Negro in Brazil.* 1939. Philadelphia, PA: Porcupine Press.

Rashid, Samory. 1999. "Islamic Influence in America: Struggle, Flight, Community." *Journal of Muslim Minority Affairs*, 19 (1): 7–31.

Reis, João José. 1993. *Slave Rebellion in Brazil. The Muslim Uprising of 1835 in Bahia.* 1986. Trans. Arthur Brakel. Baltimore, MD: Johns Hopkins University Press.

Ribeiro, René. 1961. "Relations of the Negro with Christianity in Portuguese America." *Proceedings of the British Academy*, 47: 454–484.

Ribeiro, René. 1970. "Brazilian Messianic Movements." In Sylvia L. Thrupp (ed.), *Millennial Dreams in Action. Studies in Revolutionary Religious Movements*, pp. 55–69. New York: Schocken Books.

Russell-Wood, A. J. R. 1982. *The Black Man in Slavery and Freedom in Colonial Brazil.* London: Macmillan, St. Anthony's College.

Santos, Juana Elbein dos, and Deoscoredes M. dos Santos. 1969. "Ancestor Worship in Bahia: The Égun-Cult." *Journal de la Société des Américanistes*, 58: 79–108.

Schwartz, Stuart B. 1970. "The Mocambo: Slave Resistance in Colonial Bahia." *Journal of Social History*, 3: 313–333.

Schwartz, Stuart B. 1985. *Sugar Plantations in the Formation of Brazilian Society: Bahia, 1550–1835.* Cambridge: Cambridge University Press.

Schwartz, Stuart B. 1996. "The Mocambo: Slave Resistance in Colonial Bahia." In Richard Price (ed.), *Maroon Societies. Rebel Slave Communities in the Americas*, pp. 202–226. 1979. Baltimore, MD: Johns Hopkins University Press.

Verger, Pierre. 1955. "Yoruba Influence in Brazil." Trans. Elizabeth Bevan. *ODU: Journal of Yoruba and Related Stu*dies, 1: 3–11.

Verger, Pierre. 1976. *Trade Relations between the Bight of Benin and Bahia from the Seventeenth to the Nineteenth Century.* Trans. Evelyn Crawford. Ibadan: Ibadan University Press.

Wagley, Charles. 1971. "Race and Social Class in Brazil." In Richard Frucht (ed.), *Black Society in the New World*, pp. 168–176. New York: Random House.

Warren, Donald, Jr. 1965. "The Negro and Religion in Brazil." *Race*, 6: 199–216.

Social Classes and Ethnicity

Simone Bohn

Popularly known as the giant of South America, Brazil has grown a reputation for having delightful warm weather, stunning beaches, exotic national feasts, and a population in a seemingly perennial state of joy, where individuals from distinct ethnicities supposedly live in harmony. On closer inspection, one will notice that Brazil in fact possesses a very complex and segmented society; a relatively newer, consolidating democracy; and a large economy, in an upward trend, but whose wealth distribution is markedly skewed across social groups, ethnicities, genders, and regions of the country.

This chapter focuses on Brazil's highly fragmented social structure, particularly the intersection between social class and ethnicity. The reader will note that Brazil has undergone important social transformations in the recent years, resulting primarily from the latest spurt in economic growth, and more encompassing social policies. Nevertheless, despite these positive social changes, the South American giant remains as one of the most unequal countries in the world. Tellingly, nowhere else are these inequities more visible than when ethnicity is on the spotlight, which calls into question Brazil's image of a "racial democracy" (Twine 1998).

In order to uncover the contours of this delicate class–race relationship, one needs to understand the peculiar features of Brazil's cultural anchorage and its pattern of economic development. This chapter, thus, will examine the key milestones in Brazil's transformation from a rural, premodern society into an upper middle-income industrialized, urban country, with a substantial service sector, an enlarged "new" middle class, and a still sizable contingent of individuals who remain below the poverty level. These socioeconomic processes have shaped Brazil's class structure rather decisively, and within it the positions occupied by blacks, browns, whites, and Brazilians of all ethnicities.

Brazil's regional disparities date back to the colonization period and are still visible nowadays. The country has 26 states and a federal district spread across five regions: the North, the Northeast, the Southeast, the South, and the Midwest.

The Southeast is by far the most populous region, with over 42 percent of the total Brazilian population, most of which living in urban areas. As Brazil's economic powerhouse, the region accounts for over 55 percent of the country's gross domestic product (GDP). The Southeast possesses high levels of literacy rate, comparatively better basic infrastructure, and high averages of per capita family income. Its economic vibrancy has attracted migrants from other regions, transforming most cities of the Southeast into essentially multiethnic and multicultural places.

As the epicenter of European immigration to Brazil, particularly of Germans and Italians, the South of Brazil has the highest regional level of human development Index in the country: less than 5 percent of its population (or less than 9% in the rural zones) lacks literacy skills, and the smallest stock of poverty.

The Northeast and the North, by comparison, are known for their high rates of poverty and extreme poverty, lack of basic infrastructure, and high levels of illiteracy, particularly in their rural areas. Three of the Northeastern states have the largest concentration of afro-descendants in Brazil: the states of Bahia, Maranhão, and Piauí. The North, where the Amazon Forest is located, concentrates the largest number of demarcated indigenous lands and of areas awaiting the end of the official land titling process.

Even though the indigenous are not by far the key ethnic group in the Midwest, this region also has some pockets of demarcated Native Brazilian reserves. Since at least the 1990s, the Midwest has been experiencing substantial economic growth, owing primarily to the intensification of agribusiness.

Before embarking on this historical overview, the next section provides the outlook of the class-ethnicity conundrum in contemporary Brazil.

THE INTERSECTION OF CLASS AND ETHNICITY IN CONTEMPORARY BRAZIL

According to the 2010 census, conducted by the Brazilian Institute of Geography and Statistics (IBGE 2011: p. 114), Brazil nowadays possesses close to 191 million inhabitants, making it the fourth most populous nation in the world (UN-DESA 2011). Reflecting its asymmetrical path of economic development (which will be thoroughly discussed later), Brazil's population is unevenly dispersed across its territories, having both densely populated regions (and cities within them) and rather sparsely inhabited areas. Its Southeast region accounts for approximately 80.3 million inhabitants, or 42 percent of the total population. Some of this region's cities rank high among the world's largest metropolises; the municipality of São Paulo, for instance, is one of the top 10 most populous cities in the globe.

In terms of ethnic background, the former slave-owning country, with a large indigenous pre-colonization population, nowadays is composed of 47.7 percent whites, 7.6 percent blacks, 1.1 percent Brazilians of Asian descent, and 0.4 percent Native Brazilians. According to the 2010 IBGE Census, the latter comprised 305 different ethnic groups who speak approximately 274 indigenous languages, totally over 896,000 individuals. The most populous indigenous groups are the Tikúna, the Guarani Kaiowá, the Kaingang, the Makuxí, the Terena, the Tenethara, the Yanomámi, the Potiguara, the Xavante, and the Sateré-Mawé (FUNAI n.d.).

GILBERTO FREYRE

Gilberto Freyre spent several years as a foreign college and graduate student in the United States in the late 1910s and early 1920s, having studied at Baylor University in Waco, Texas, and at Columbia University in New York (Chacon 2001).

This stay and, in particular, the culturalist studies at Columbia had a profound impact on Freyre, as he was stunned by the experience of racial segregation in public spaces, which was prevalent in the United States. He strove to distinguish between race and culture, and to focus on the socioeconomic determinants of human development; the latter, rather than race itself—Freyre argued—is what shapes the fate of the different racial groups (Ventura 2010).

In *Casa-Grande e Senzala*, Freyre never employed the term "racial democracy"; he used it only in marginal works, long after the expression had been attributed to him. Later in life, he emphasized numerous times (see, e.g., Freyre 1956) that his arguments about the harmonious coexistence of ethnicities in Brazil did not imply that racism does not exist in the country.

In addition, approximately 4 in every 10 Brazilians (43.1%) defined themselves as brown, or *pardos*, in Portuguese (IBGE 2011). In an impressive "spectacle of the races" (Schwarz 1993), the latter have in their heritage two or more ethnicities, whether black and white, indigenous and white, indigenous and black, and so on. Brazilian social and political thought interpreted this miscegenation process rather differently. Multiracialism was the target of profound criticisms by Brazil's pioneer "scientific" community. According to some Brazilian authors, such as Sílvio Romero, Nina Rodrigues, Oliveira Vianna, and Euclides da Cunha, miscegenation would engender a substandard race, which would ultimately sow the seeds of economic, social, and moral backwardness; these authors argued that this fate could be prevented only through the implementation of a public policy of deliberate whitening of the population, brought about by European migration to Brazil (more on this later).

Independent, post-slavery Brazil in fact never had legally sanctioned racial barriers to private and public careers; there were no racially segregated schools, churches, or buses; and neither were there legal impediments to the marriage of individuals from different ethnicities.

Despite recognizing these differences between the United States and the South American giant, some authors (e.g., Bastide and Fernandes 1959; Cardoso 1962; Fernandes 1965; Franco 1974; Ianni 1962) began, in the late 1950s, to contend that racial discrimination, indeed, exerts an important, *independent* effect on an individual's life chances in Brazil.

After these pioneer studies, several other works (Skidmore 1974; Telles 2004; Twine 1998) emphasized the view that racial democracy in Brazil is a myth, as nonwhites systematically have worse life conditions than white Brazilians. Others, such as Hanchard (1994), even suggest that this myth had the deleterious effect of preventing the Brazilian state, for several decades, from developing public policies to address racism and its effects.

It was primarily with the publication of the groundbreaking work of Gilberto Freyre in 1933, entitled *Masters and Slaves* (*Casa-Grande e Senzala*, in Portuguese), that multiracialism began to be portrayed in a positive light. According to Freyre, ethnic amalgamation was the source of genuine Brazilianness, and, as such, of a real *nation* (as opposed to just an independent state) characterized by the harmonious coexistence of a multiplicity of races, in an experience that stood in stark opposition to the dreadful segregationist experiences found in other national contexts (Freyre 1956). Over time, Freyre's *Masters and Slaves* was widely misread and profoundly criticized.

Nonetheless, despite this second image of the country as a "racial democracy," the race question has never completely left center stage in Brazil, especially due to the substantial gaps in educational and material attainment across ethnic groups. Ultimately, there is a substantial correlation—albeit not determination—between race and class belonging in Brazil. For instance, in terms of race-based wage disparity, the 2009 IBGE National Household Research revealed that, on average, Afro-descendants and *pardos* in Brazil make only 57.4 percent of what a white Brazilian earns (Table 5.2). Lamentably, a gap in earnings exists across every single region of Brazil and it is visible even when one controls for education; that is, when we compare two Brazilians of similar level of formal education, there is a high likelihood that the white Brazilian will have a higher salary than the nonwhite.

Similarly, poverty is more prevalent among nonwhites than whites. Still according to IBGE's 2009 National Household Research, whereas only 24.1 percent of the

124 | *Brazil*

TABLE 5.2 Ratio of Hourly Wage That Nonwhites Earn in Comparison to White Brazilians, by Years of Formal Schooling and Region (2009)

Regions	Total Black	Total Brown	Up to 4 Years Black	Up to 4 Years Brown	5 to 8 Years Black	5 to 8 Years Brown	9 to 11 Years Black	9 to 11 Years Brown	12 or More Black	12 or More Brown
North	69.1	63.9	72.3	79.8	82.4	88.2	76.8	72.8	87.7	71.8
Northeast	62.8	62.2	96.8	88.4	93.4	80.2	79.9	80.3	64.5	76.4
Southeast	56.9	60.0	84.1	81.1	80.3	77.3	74.0	81.0	71.1	74.7
South	61.1	62.7	71.8	78.4	80.1	78.8	77.7	78.2	56.4	63.7
Center-West	61.0	62.4	89.8	79.9	81.9	83.7	73.3	77.7	69.0	75.3
Total	57.4	57.4	78.7	72.1	78.4	73.0	72.6	75.8	69.8	73.8

Source: IBGE (2009).

whites had a *per capita* monthly income of less than half a minimum wage (which was R$510 at the time of that study, roughly US$284 in mid-2009), the proportions were much higher for nonwhite Brazilians. Approximately one-third (33.8%) of those with Asian descent belonged to this income category, as well as 38.9 percent of the indigenous, 44.4 percent of blacks, and 47.3 percent of browns. Among the 1 percent richest in Brazil, in the same year, 82.5 percent were white, 1.8 percent were black, and 14.2 percent were *pardos*.

A similar racial-based gap is found in other areas. Whereas 7.2 percent of whites were illiterate in 2010, 10.0 percent of Brazilians of Asian descent did not know how to read and write. The illiteracy numbers for other Brazilians were even higher: 14.2 percent for *pardos*, 15.0 percent for blacks, and a shocking 26.3 percent for Native Brazilians (IBGE 2011). Similarly, in 2009, among Brazilians 25 years and older, 15.0 percent of whites had a university degree, in contrast to 5.3 percent of browns and 4.7 percent of blacks. This gap in access to higher education will tend to continue in the near future, as in 2009, 20.6 percent of the Brazilian whites were enrolled in universities (as undergraduate or graduate students), as opposed to 8.4 percent of browns and 7.4 percent of blacks (IBGE 2009).

Apart from its racial component, social inequality remains a powerful feature of the Brazilian society. When measured by the GINI Index, which assesses the degree of income concentration in a society (it varies from 0 = perfect equality to 100 = perfect inequality), Brazil's level was at around 54.7 in 2009, which makes the South American giant rank amid the top 10 most unequal nations in the world. As one can see in Table 5.3, as opposed to the overall trend of the 1980s "lost decade," since 2002 Brazil's degree of wealth concentration has been declining progressively, albeit slowly. Later we will see that more encompassing, universalist-like social policies and the economic growth of the mid-2000s are thought to have played a key role in this decrease of income concentration.

Nevertheless, despite this current downward trend in social inequity, income remains highly concentrated in Brazil. In 1981, for instance, whereas the 10 percent wealthiest Brazilians controlled 45.9 percent of the country's GDP, the 10 percent poorest held only 0.8 percent. In 2009, the same numbers were 42.9 percent and

As is common in a number of Brazilian metropolitan cities, a favela stands in stark contrast to modern high-rise buildings in Belo Horizonte, capital of the state of Minas Gerais, Brazil. (Rogério Medeiros Pinho)

TABLE 5.3   GINI Index in Brazil

	GINI Index	Country	Country Code	Year	Year Code	GINI Index (World Bank Estimate) [SI. POV.GINI]
		Brazil	BRA	1981	YR1981	–
		Brazil	BRA	1982	YR1982	–
		Brazil	BRA	1983	YR1983	–
		Brazil	BRA	1984	YR1984	58.38
		Brazil	BRA	1985	YR1985	55.59
		Brazil	BRA	1986	YR1986	58.46
1984	58.38	Brazil	BRA	1987	YR1987	59.69
1985	55.59	Brazil	BRA	1988	YR1988	61.43
1986	58.46	Brazil	BRA	1989	YR1989	63.3
1987	59.69	Brazil	BRA	1990	YR1990	60.49
1988	61.43	Brazil	BRA	1991	YR1991	–
1989	63.3	Brazil	BRA	1992	YR1992	53.17
1990	60.49	Brazil	BRA	1993	YR1993	60.12
1992	53.17	Brazil	BRA	1994	YR1994	–
1993	60.12	Brazil	BRA	1995	YR1995	59.57

(Continued)

TABLE 5.3 (continued)

	GINI Index	Country	Country Code	Year	Year Code	GINI Index (World Bank Estimate) [SI.POV.GINI]
1995	59.57	Brazil	BRA	1996	YR1996	59.89
1996	59.89	Brazil	BRA	1997	YR1997	59.8
1997	59.8	Brazil	BRA	1998	YR1998	59.61
1998	59.61	Brazil	BRA	1999	YR1999	58.99
1999	58.99	Brazil	BRA	2000	YR2000	–
2001	59.33	Brazil	BRA	2001	YR2001	59.33
2002	58.62	Brazil	BRA	2002	YR2002	58.62
2003	58.01	Brazil	BRA	2003	YR2003	58.01
2004	56.88	Brazil	BRA	2004	YR2004	56.88
2005	56.65	Brazil	BRA	2005	YR2005	56.65
2006	55.93	Brazil	BRA	2006	YR2006	55.93
2007	55.23	Brazil	BRA	2007	YR2007	55.23
2008	54.37	Brazil	BRA	2008	YR2008	54.37
2009	53.87	Brazil	BRA	2009	YR2009	53.87
2011	53.09	Brazil	BRA	2010	YR2010	–
2012	52.67	Brazil	BRA	2011	YR2011	53.09
		Brazil	BRA	2012	YR2012	52.67
		Brazil	BRA	2013	YR2013	–
		Brazil	BRA	2014	YR2014	–

0.8 percent, respectively (World Bank 2012). In other words, very little has changed. According to the 2010 IBGE Census, 8.5 percent of Brazilians (or 16.3 million) face extreme poverty, which the Brazilian institute considered to be a per capita monthly income of less than R$77 (or US$39 as of June 30, 2010). Among this impoverished group, 11.4 million have a per capita monthly income between R$1 and R$77, and 4.3 million extremely poor individuals are devoid of any income (IBGE 2011).

Brazil's class structure, thus, is highly fragmented. Besides a large contingent of poor, the country also possesses a considerable group of extremely poor individuals. This substantial degree of class segmentation also applies to the middle-income groups: when defined solely by income, Brazil possesses several middle classes. As shown in Table 5.4, the Brazilian social class pyramid has many layers, each of which encompasses a different proportion of families, with similar unit sizes but rather distinct purchasing power. Needless to say, the adults and children from the 2.2 million families (roughly 3.8 percent of the total) at the top of the pyramid have much better life chances than the 12.5 million families (21.6 percent) at the bottom.

How did this pattern of social inequality, and especially its ethnic undertones, come about? The next sections will briefly examine the historical roots of Brazil's intricate social configuration.

TABLE 5.4   Percentage of Families Across Income Brackets

Monthly Family Income Brackets	Percentage of Brazilian Families	(Their Average Size)
Up to $516	21.63	3.07
$516.1–$774	17.42	3.18
$774.1–$1,547	29.36	3.38
$1,547.1–$2,579	15.38	3.42
$2,579.1–$3,869	7.23	3.48
$3,869.1–$6,558	5.18	3.47
Over $6448.1	3.81	3.3

Income in US$

Source: IBGE (2008). Pesquise de Orçamentos Families, Sidra Table No. 1595, http://www .sidra.ibge.gov.br.

COLONIZATION

The South American giant began its independent life only on September 7, 1822 (it was still ruled by a Portuguese monarch until 1889, when it became a republic). Prior to 1822, the Portuguese America, later Brazil, endured more than three centuries of Portuguese colonization. The latter was a complex enterprise with significant, long-lasting political, economic, social, and cultural implications.

In terms of the social class-ethnicity ramifications, there are three main aspects to be emphasized: the particular pattern of occupation of the Brazilian territory, the nature of the organization of the productive process in the colony, and the roots of the miscegenation culture.

Similar to the remaining countries of Latin America and the Caribbean (besides Africa and parts of Asia), Brazil was a colony of exploitation and not of settlement. Settler colonies, by and large, were characterized by the organization of the productive process to supply the domestic market, free trade with overseas countries, the prevalence of small- or mid-sized rural properties devoted to the cultivation of several crops, and the presence of free men and women leading the colonization enterprise. The settlers left their mother countries to fix residency permanently in the new territories; enjoyed substantial political, religious, and cultural freedoms; and, given this context of self-government (Hall 2002), were on the whole able to reproduce their old societal structure in the new land. This pattern of colonization was typically the case of the British upshots, such as the United States, Canada, Australia, and New Zealand.

In contrast, in Brazil, the Portuguese colonizers had a different attitude and mentality altogether. Informed by the mercantilist ideas then in vogue (Maxwell 1973), the colonists initially aimed at the fast accumulation of wealth, particularly bullion. When the objective changed to a more permanent colonization (albeit still quintessentially exploitative in nature), private entrepreneurs, whom the Portuguese Crown had already ceded partial colonization rights, were also granted very sizeable parcels of land, the so-called hereditary captaincies (Holanda 2012). Despite this mixed private–public nature of the colonization process in Brazil, the roles that one would usually attribute to the state—such as external defense, domestic control of violence, and administration of justice—were, in practice, exercised by the private power of individual actors, which outstripped the public power in strength and presence in the national territory (Faoro 2001). Needless to say, this pattern of occupation of the colonial territory had important ramifications, as it gave rise to substantial regional inequalities in terms of economic development (some of them are still visible in today's Brazil). Moreover, given that land plays a crucial role in structuring both economic and *social* relations (Furtado 1976), the concentration of land in few hands created the conditions for the production (and, later, reproduction) of an economic elite, with an incredible ability to metamorphose itself into a powerful political elite as well (Martins 1976; Pereira 2003). This pattern of land concentration remains a bone of contention in Brazil until today, as it structures the conflict among important social actors: agribusiness entrepreneurs permanently interested in expanding their cultivation area, Native Brazilians committed to the protection of their ancestral lands, and landless groups claiming the democratization of the access to land (more on this later).

Second, the nature of the organization of production in colonial Brazil also had significant and enduring consequences for the structuring of Brazilian society and the country's position in the international division of labor. The Portuguese colonizers implemented the so-called metropolitan exclusiveness (Novais 1979); that is, the economic agents in the colony could only sell to and buy from entrepreneurs approved by the Portuguese Crown, who, with rare exceptions, were Portuguese merchants. The latter group accrued large profits by reselling exotic colonial products in the international market (Arruda 1980). The Portuguese America, thus, had an externally oriented economy (Bulmer-Thomas 1996), which focused primarily on the production of specific crops (e.g., sugar) that catered to the overseas market, bringing about Brazil's seemingly perennial Achilles's heel: the dependence on the international price of specific commodities, given the metropolitan prohibition of the development of manufacture in the colony (Novais 1979) and Brazil's position as an international provider of cheap produce and raw materials (Prado Junior 1963), which transformed it (and all the colonies of exploitation) into the foundation from which industrial capitalism and imperial powers drew their strength.

Three important components of this externally oriented economic model were the *latifundio*, which, as mentioned before, were enormous swaths of land, used in the production process; the monoculture, that is, the cultivation of a major, single crop, to supply foreign markets; and, more importantly, the widespread use of enslaved labor (Fausto 2006). In fact, unfree labor was the backbone of the Portuguese America's economy. Since the beginning of the colonization not only of Brazil but also of Latin America and the Caribbean in general (i.e., Spanish America and others), there were intense debates about the practical suitability and moral appropriateness of employing indigenous labor in the productive process (Casas 2003). Given the sparse population in Portugal and the unsuccessful initial experiences with indigenous forced labor, the end result of this controversy in Brazil was the introduction of a massive number of slaves of African origin into the soil of the Portuguese America (Prado Junior 1963). Most estimates indicate that between 3.5 and 4.0 million

African slaves forcefully toiled in this colony, which corresponded to approximately 38 percent of the overall Atlantic slave trade (Bethell 1970). These Africans forced migrants faced long days of intense, manual work, and led incredibly arduous and short lives (Karasch 1987).

Interestingly, despite the essentially exploitative nature of their integration into the Portuguese America's economy, African slaves left an indelible mark in Brazil's society, culture, and religious traditions. The Portuguese colonizers were already fond of interracial marriages in their own home country; the African presence in Portuguese America only intensified in the colony the trend toward multiracialism already visible in Portugal. If Brazil nowadays is the most "African" country outside of Africa, given its large contingent of African descendants, it also has, as indicated previously, a substantial proportion of *pardos*, some of whom also share African heritage.

Besides the African contribution to Brazil's ethnic mix, the forced laborers' cultural traditions influenced decisively Brazilian music, arts, dance, sports (Ribeiro 1995), and cuisine. Moreover, the African religions transplanted to Portuguese America changed substantially the previously existing religious landscape. Despite the ill-informed prejudice against Afro-Brazilian faiths (Mariano 2007) still in existence, Candomblé, a religious practice of African origin, remains active in Brazil (Bastide 1978; Goldman 2005; Prandi 1991; Verger 1981), as well as even more religiously syncretic creeds, such as the Afro-Brazilian religion known as Umbanda (Ortiz 1988).

One final important aspect of the colonization process—though with more distant consequences regarding ethnicity in Brazil—was the introduction of a specific type of bureaucratic tradition in Brazil's soil, especially with the arrival of the Portuguese royal family in 1808. Fleeing the Napoleonic threat in Europe, the Portuguese Crown brought to Brazil its entire court (Maxwell 1973), and, with it, a patrimonial bureaucratic tradition of an enlarged and encompassing state in charge of the economy, with personalistic rules of recruitment and promotion of public workers, and a propensity toward rent-seeking behavior and corruption (Faoro 2001). Needless to say, this type of bureaucratic structure has had long-term impact on the pattern of state intervention into the economy and society, as it created incentives toward the development of a type of state capitalism, in which the business class has been rather dependent on state protection, its direct participation in the economy, and the state regulation of social conflict—particularly of labor unrest (Mello e Souza 1999).

THE POST-INDEPENDENCE PERIOD

Both Brazil's independence in 1822 and the so-called wave of independent movements in Hispanic America from the 1810s to the 1830s coincided with an economic period of great international demand for cheap commodities abundant in the region. As a consequence, despite well-conceived and rather prescient attempts at implementing models of inward-looking economic development, such as the one championed by the Banco de Avío in Mexico (Bulmer-Thomas 1996), the economic climate of the post-independence period prompted Brazil and most of the Latin American countries to further deepen the external orientation of their economic structures.

In the decades after its independence, Brazil, in particular, intensified the externally oriented trajectory of its economic model of development. Prompted by its high international price in the first two decades of the 19th century, coffee became the key commodity of Brazil's export sector. Besides the high rates of return, the abundance of cultivable land and availability of slave labor also played a role in the expansion of coffee, whose cultivation area spread particularly in the Southeast region of Brazil—initially in Rio de Janeiro and later in the Paraíba Valley and other parts of the state of São Paulo (Fausto 1995).

Under intense international pressure, particularly from Great Britain, Brazil ended slave trade in 1850 (Bethell 1970), but abolished slavery only more than 30 years later: in 1888. Needless to say, the ruling elite justified this very gradual process on the grounds of the need to replace slave labor in Brazil, which, as mentioned before, represented the backbone of the economically active population. Policy makers resorted massively to European immigration for two reasons: The first objective was to supply the rapidly growing coffee production in the state of São Paulo. The second objective—alluded to previously—was to accelerate the process of whitening of the Brazilian population, which, in practice, amounted to an attempt at attenuating the African slave legacy in Brazil.

As a consequence of this shift in immigration pattern, approximately 1 million immigrants entered Brazil between 1884 and 1903, most of who came from Italy (Beiguelman 1968). Just to put this number in perspective, in 1872, the year of the first Brazilian census, the overall Brazilian population was estimated to be close to 10 million (Diretoria 1872). By 1890, primarily with the acceleration of free-labor immigration, the population climbed to around 14 million. Evidently, this process did impact the ethnic makeup of the Brazilian population, given that it gave rise to spatially circumscribed clusters of Brazilians of European descent. At the same time, the implementation of this new immigration policy—which changed the focus from slaves of African descent to free white labor from Europe—paralleled the lack of public policies devoted to absorbing former slaves as free economic actors and citizens of the country. In fact, the conditions of the former slaves did not improve significantly, even with the end of the monarchy in Brazil and the adoption of a republican system in 1889, which had a strong rhetoric of attempting to widen citizenship rights in Brazil (Carvalho 1987, 1990).

Furthermore, the most dynamic sector of the Brazilian economy in the period—the coffee cultivation in São Paulo—was then practically devoid of the contribution of former slaves. As the expansion of agriculture in São Paulo associates with the rise of modern capitalism in Brazil—as it led to the incipient and later intense mechanization of the production, development of railway system, and the widespread use of free-labor contracts—former slaves did not occupy a prominent position in the new capitalist order in formation (Hirano 2005). In addition, even though the emerging process of capitalist development in Brazil did change significantly the class composition of the society—with the development of an autochthone domestic capitalist class and a budding industrial proletariat—it did not alter substantially the pattern of relationship between upper classes and the state. On the contrary, as in the colonial period, the state remained as a close defender of the interests of particular economic sectors. The Brazilian state's large purchases of coffee (and subsequent burning of it), particularly when this commodity reached rock-bottom prices in the international market, attest to this highly protectionist type of state-market interaction (Fausto 1997).

THE I930s–I960s

As with the rest of Latin America, the 1929 economic crisis prompted in Brazil the rethinking of its externally oriented model of economic development (Thorp 1998). The Great Depression made unequivocally clear that the heavy dependency on commodities for export—even if they were at one time immensely profitable—made the country's economy extremely vulnerable to external shocks. The pressing need to find an alternative path amid the economic and social devastation brought about by the 1929 crisis (Fausto 1969) encouraged the resumption of inward-looking trajectories of development.

Central to these alternatives was the idea that the promotion of domestic industrialization was the only way to overcome the Achilles's heel of the Brazilian economy, namely its profound external vulnerability, and the ensuing stop-and-go trajectory of economic growth (Prebisch 1950). Also consensual was the view that the domestic bourgeoisie—the different class groupings (Weber 2004) of the autochthone entrepreneurial sector—was unable by itself to lead the industrialization project (Furtado 2007). Instead, the public sector, that is, the state, should take over the role of spurring on the domestic industrialization through several measures: the expansion of the lines of loans and subsidies to the entrepreneurial segment, partnerships with the private sector, and, interestingly, the direct state involvement in production particularly to solve the so-called structural bottlenecks in production and distribution (Baer 2008). This inward-looking model of development materialized in the now renowned "import-substitution industrialization" (Weaver 2000), which initially focused on fostering the domestic consumer goods industries, and later—with the goal of promoting the vertical integration of the economy—the capital goods industries (Baer 1972). Evidently, as mentioned before, state capital played a key role in this process, and the public sector grew into an economic agent of vital importance. State-controlled companies became huge leviathans in strategic areas such as the electrical, petrochemical, steel, and metallurgic sectors (Baer 1974; Furtado 2007). In addition, the state invested massively in the modernization of the transportation system, by building roads, railways, and harbors.

The social corollary of this process of accelerated industrialization was the expansion of the already incipient industrial working class (Fausto 1977; Pinheiro 1977), particularly in the Southeast region of Brazil. This expansion was particularly intense during the so-called Vargas's era. The latter extends from the period after the 1930 coup, but consolidated in 1937–1945, years after the promulgation of the 1937 "authoritarian" Constitution, which marked the beginning of the "Estado Novo" (the New State). Throughout these years, Getúlio Vargas—a politician from Rio Grande do Sul—ruled Brazil. In close resemblance to projects of conservative modernization that took place elsewhere (Moore, Jr., 1993), Vargas put in motion a series of public policies of economic modernization amid a conservative, authoritarian, and centralizing political framework. Despite its inherently "fascist" features, Vargas's Estado Novo is credited with a significant expansion of labor rights, such as the official establishment of the eight-hour working day, the improvement of pension entitlements of urban workers, and the regulation of evening work and of the work activities of minors and women (Skidmore 1967). Paradoxically, despite this increase in labor rights, unions and workers' associations were not independent actors; on the contrary, the state created tripartite labor committees, where public sector actors and business and labor representatives decided on workers' demands and disputes (Campello de Souza 1976; Levine 1998). Moreover, it is important to note that the incorporation of workers into the social safety nets brought about by labor rights was essentially based on class lines, and not ethnicity. In other words, during this period there were no significant social policies to improve the conditions of racially marginalized segments (e.g., the descendants of former slaves) in Brazil.

QUILOMBOLAS IN BRAZIL

Quilombos (or "refuge" in an African—Bantu—language) were formed by slaves trying to escape the horrors of servitude before the end of slavery in 1888, or, in smaller proportion, by freed slaves right after the abolition. Their members are known as *Quilombolas* (Moura 2001).

There are no solid data on their numbers. According to the Palmares Cultural Foundation, nowadays there are 743 Quilombola communities. The Ministry of Agrarian Development, on the other hand, recognizes 1,300 communities. Finally, SEPPIR (the federal Special Secretariat of Policies for the Promotion of Racial Equality) identifies approximately 2,450 Quilombola communities, or 2 million individuals (MDS 2007).

SEPPIR is presently working with the identified communities to grant them land titles, improve local infrastructure, foster sustainable and environment-friendly local economic activities, and empower the *Quilombolas* to be able to attain their citizenship rights.

By the mid-1940s profound social transformations had taken place. Approximately one-third of the population was already living in cities (Valle Silva and Barbosa 2006: p. 48). Furthermore, by 1945, the industrial sector was already responsible for 20 percent of Brazil's GDP (Thorp 1994: p. 55). The transition to democracy in 1946 initiated a period of great political effervescence, in terms of the expansion of political parties—particularly, left-leaning parties—and of attempts at the creation of a unionism devoid of its state sponsorship. Propelled by the social and economic transformations following the Cuban Revolution, the heterogeneous left in Brazil (and elsewhere in Latin America) helped foster the radicalization of several social sectors. The progressive sectors of the Brazilian society divided into several different groups, primarily between those who espoused nationalistic ideas (and emphasized national development through a reformist path) and those with unequivocally socialist aspirations (who preconized an egalitarian social order as the ultimate goal, to be pursued by revolutionary means if necessary) (Cardoso 1964; Marini 1969; Prebisch 1949; Sunkel 1969; Toledo 1977). This period of intense and open clash of ideas—capitalism versus socialism, modernization theory's development versus dependency, the role of the national bourgeoisie (or whether the latter is just an illusion)—came to a halt with the 1964 coup in Brazil, which was followed by a wave of dictatorships throughout South America (Huntington 1991; Smith 2005).

The year 1964 was the first of several of the darkest years of Brazil's history (Skidmore 1988). Through violence, terror, extra-judicial killings, and disappearances, the military dictatorship disarticulated organized civil society, destroyed the previously existing system of political parties (Fleischer 1980, 1981; Kinzo 1988), and attempted to physically annihilate the different segments of the revolutionary left and even of the more reformist leftist groupings. Economically, there was a considerable expansion of the state's intervention into the economy, particularly in the basic industries and the development of new technologies under the Second National Development Plan (Bacha 1977; Coutinho 1981; Lessa 1977). Although this intensification of the state's presence in the economy did contribute, to a certain extent, to the so-called economic miracle (Singer 1972)—years in which Brazil's rate of economy growth was over 9 percent a year (1968–1973)—the fragility of this model of economy growth became evident later, after the second oil shock, and especially in the aftermath of the 1982 debt crisis. Under the military dictatorship (1964–1985), one key element sustaining Brazil's relatively high rates of economic growth was the accrual of an enormous foreign debt: it jumped around US\$3–4 billion in the mid-1960s to over US\$101 billion by the mid-1980s (Bresser-Pereira 1990: p. 26; Cruz 1984: p. 12). Needless to say, after the completion of the transition to democracy in 1985, with the transfer of power to a civilian president (the first civilian president was elected only in 1989), the governments of "New Republic" inherited a difficult "fiscal crisis of the state" (Sallum, Jr., and Kugelmas 1993), marked not only by the weight of a large foreign (and domestic) public debt but also by hyperinflation.

BRAZIL AFTER RE-DEMOCRATIZATION

If 1985 marks the resumption of a democratic regime, the 1980s also witnessed a different kind of transition in Brazil: the change from a large, highly interventionist state, which for a long time had operated as a direct producer in several industries, to a pattern of state-society-market relationship, marked by the dictates of the so-called Washington Consensus (Williamson 1990). The social corollary of the imposed fiscal discipline—which led to a decade of net negative economic growth in the 1980s—was the weakening of the labor unions (Cook 2002), the retrenchment of the (already feeble) social rights (Sposati 2002), and the increase of socioeconomic inequality (Hira 2007).

Even though Brazil's hyperinflationary problems essentially ended in 1994, successive governments, particularly that of President Cardoso (1994–1998 and 1998–2002), have since striven to maintain tight measures of fiscal discipline. Cardoso's portfolio of social policies was widely expanded with the arrival of President Lula da Silva, Workers' Party, to the central government in 2002, and after his reelection in 2006. These social policies—coupled with the growth effect stemming from a commodities boom (related in part to China's rise)—have contributed significantly to the reduction of socioeconomic inequality, as mentioned earlier. Brazil has gone from a country where the less affluent were the most numerous contingent to a nation where the so-called new middle class is the largest social segment (Neri 2014; Neri, Melo, and Monte 2012).

Besides the Bolsa Família (or Family Grant), which is a conditional cash transfer program providing income to poor and extremely poor families, a large portfolio of social policies has been created to help foster better living conditions for poor adults and/or to improve the life chances of the children of impoverished families. Most of these programs fall under the umbrella of the Zero Hunger initiative. Examples of its extensive list of public policies are social benefits for the elderly; special lines of credit to encourage home ownership, and entrepreneurship, and to allow family farmers to invest in technological improvements; the "water for all" and "electricity for all" programs; free distribution of medicine to less affluent individuals; programs to promote eye and dental health in poor communities; and programs to foster adult literacy and to aid and prepare women for stable and environment-friendly jobs, particularly in poverty-stricken areas (Melo 2008).

Since the 2000s, Brazil has seen initiatives, particularly from the federal government, which went well beyond the attempt at redressing centuries-old, deep-seated pattern of class inequality. There have been also public policies aiming at attenuating the socioeconomic and political marginalization of specific ethnic or racial groups in Brazil. One important victory of great symbolism for Afro-Brazilians was a 2011 Law officiating November 20th as Zumbi's National Day of Black Consciousness—to pay homage to a 17th-century leader of a community of runaway slaves located in the Northeast of Brazil. As of November 2014, approximately 20 percent of the Brazilian cities transformed the date into an official city holiday (Portal Brasil 2014). When it comes to substantive projects, one of the most important programs has been the introduction of a quota system to increase the access of Afro-descendants, *pardos* and the indigenous to tertiary education, particularly federal universities. Even though there has been intense debate around the issue (Maggie and Fry 2004; Moehlecke 2002; Munanga 2001), over 60 percent of the Brazilian population in 2013 indicated that they support this affirmative action initiative and only 16 percent manifested their opposition to it (IBOPE 2013).

Whereas improved access to university education is an important, first step toward the still distant goal of leveling the playing field for historically marginalized groups, for the indigenous in Brazil another struggle is equally important. It is the fight for the demarcation of their lands. Although the 1988 Constitution (in its Article 231) guarantees to the indigenous in Brazil the "original" right to their ancestral lands as well as the "exclusive" right of possessing and using these specific lands, the process of demarcation of indigenous lands has been slow-moving. The National Foundation for the Indigenous (FUNAI) indicates that, as of mid-2015, there were over 200 indigenous lands still awaiting their final regularization, and, unless there are substantial procedural changes, it may take a few years for these processes to be completed. In the past, some demarcation cases have lasted around 20 years. Furthermore, there is still work to be done regarding the life conditions in the indigenous lands themselves, as 32.3 percent of the indigenous living in indigenous lands are illiterate; less than 31 percent of their homes have running water; and their life expectancy is still shockingly low: 45 years (in contrast to 73 years for the general population). On a more positive note, Brazil is moving to protect the isolated indigenous groups (i.e., those without any contact with the rest of the population). There have been reports of 77 of such communities, 30 of which have been confirmed by FUNAI. The federal government has been issuing the so-called decrees of interdiction, with the goal of protecting these isolated indigenous groups, as well as the recently contacted groups.

NATIVE BRAZILIANS

The Native Brazilians are overwhelmingly young (47% of them are under 19 years old), have a much lower life expectation (45.6 years) than the average Brazilian (73.4 years), and are disproportionately affected by a higher-than-average rate of infant mortality rate. Around 58 percent of them live in officially recognized indigenous lands. Some groups remain isolated; in fact, the Brazilian Census Bureau (IBGE) estimates that approximately 29 percent of the Native Brazilians who live in indigenous lands do not speak Portuguese, only their mother language. One in every three Native Brazilians living in indigenous lands is illiterate.

Native Brazilians are spread throughout the Brazilian territory, with the largest concentrations in the states of Amazonas, Mato Grosso do Sul, Pernambuco, and Bahia. The expansion of agricultural activities (particularly the cultivation of soy and rice, among other staples) and of cattle-ranching in the North and Center-West has put agribusiness at odds with indigenous interests, and the invasion and occupation of indigenous lands—and the ensuing violence from both sides—has grown more common.

In sum, several factors in Brazil's history have engendered a society with a highly fragmented class structure with clear racial undertones. In this sense, Brazil has been for a long time a nation of sharp contrasts. Whereas its 1 percent wealthiest citizens have access to the same goods and services available to their counterparts from the Global North, its less affluent citizens still struggle daily with food security issues. Nevertheless, more inclusive and socially targeted public policies of the past years are slowly, but steadily, producing important changes in the social fabric of the South American giant.

REFERENCES

Arruda, José Jobson de Andrade. 1980. *O Brasil no comércio colonial*. São Paulo: Ática.

Bacha, Edmar. 1977. "Issues and Evidence on the Recent Brazilian Economic Growth." *World Development*, 5 (1): 47–67.

Baer, Werner. 1972. "Import Substitution and Industrialization in Latin America: Experiences and Interpretations." *Latin American Research Review*, 7 (Spring): 95–122.

Baer, Werner. 1974. "The Role of Government Enterprises in Latin America's Industrialization." In David Geithman (ed.), *Fiscal Policy for Industrialization and Development in Latin America*, pp. 263–292. Gainesville: University Press of Florida.

Baer, Werner. 2008. *The Brazilian Economic. Growth and Development*. Boulder, CO: Lynne Rienner Publishers.

Bastide, Roger. 1978. *The African Religions of Brazil*. Baltimore, MD: The John Hopkins University Press.

Bastide, Roger, and Florestan Fernandes. 1959. *Brancos e Negros em São Paulo: Ensaio Sociológico sobre Aspectos da Formação, Manifestações Atuais e Efeitos do Preconceito de Cor na Sociedade Paulistana*. São Paulo: Editora Nacional.

Beiguelman, Paula. 1968. *A Formação do Povo no Complexo Cafeeiro: Aspectos Políticos*. São Paulo: Livraria Pioneria Editora.

Bethell, Leslie. 1970. *The Abolition of the Brazilian Slave Trade*. Cambridge: Cambridge University Press.

Bresser-Pereira, Luiz Carlos. 1990. "Da Crise Fiscal à Redução da Dívida." In João Paulo dos Reis Velloso (ed.), *Dívida Externa e Desenvolvimento*. Rio de Janeiro: José Olympio.

Bulmer-Thomas, Victor. 1996. "The Struggle for National Identity from Independence to Mid-Century." *The Economic History of Latin America since Independence*, New York: Cambridge University Press.

Campello de Souza, Maria do Carmo. 1976. *Estado e partidos políticos no Brasil (1930 a 1964)*. São Paulo: Alfa-Omega.

Cardoso, Fernando Henrique. 1962. *Capitalismo e Escravidão no Brasil Meridional: O Negro na Sociedade Escravocata do Rio Grande do Sul*. São Paulo: Difusão Européia do Livro.

Cardoso, Fernando Henrique. 1964. *Empresário industrial e desenvolvimento econômico*. São Paulo: Difusão Européia do Livro.

Carvalho, José Murilo de. 1987. *Os bestializados. O Rio de Janeiro e a República que não foi*. São Paulo: Companhia das Letras.

Carvalho, José Murilo de. 1990. *A formação das almas. O imaginário da República*. São Paulo: Companhia das Letras.

Casas, Bartolomé de las. 2003. *An Account, Much Abbreviated, of the Destruction of the Indies*. Indianapolis, IN: Hackett Publishing.

Chacon, Vamireh. 2001. *A Construção da Brasilidade: Gilberto Freyre e sua geração*. Brasília and São Paulo: Paralelo 15/Marco Zero.

Cook, María Lorena. 2002. "Labor Reform and Dual Transitions in Brazil and the Southern Cone." *Latin American Politics & Society*, 44 (1): 1–34.

Coutinho, Luciano. 1981. "Inflexões e crise da política econômica: 1974–1980." *Revista de Economia Política*, 1 (1): 77–100.

Cruz, Paulo Roberto Davidoff Chagas. 1984. *Dívida Externa e Política Econômica*. São Paulo: Brasiliense.

Diretoria Geral de Estatística. 1872. *Recenseamento do Brazil 1872*. Rio de Janeiro: Diretoria Geral de Estatística.

Faoro, Raymundo. 2001. *Os Donos do Poder: a formação do patronato politico brasileiro*. Porto Alegre: Globo.

Fausto, Boris. 1969. *A Revolução de 1930. História e Historiografia*. São Paulo: Brasiliense.

Fausto, Boris. 1977. *Trabalho Urbano e Conflito Social (180–1920)*. São Paulo: DIFEL.

Fausto, Boris. 1995. "Imigração e participação política na Primeira República: o caso de São Paulo." In Boris Fausto, Oswaldo Truzzi, Roberto Grün, and Célia Sakurai (eds.), *Imigração e política em São Paulo*. São Paulo: Editora Sumaré.

Fausto, Boris. 1997. *A Revolução de 1930*. São Paulo: Companhia das Letras.

Fausto, Boris. 2006. *História do Brasil*. São Paulo: EDUSP.

Fernandes, Florestan. 1965. *A Integração do Negro na Sociedade de Classes*. São Paulo: Dominus, EDUSP.

Fleischer, David. 1980. "O Bipartidarismo (1966–1979)." In David Fleischer (ed.), *Os Partidos Políticos no Brasil*. Brasília: Editora da UnB.

Fleischer, David. 1981. "A evolução do bi-partidarismo brasileiro, 1966–1979." *Revista Brasileira de Estudos Políticos*, 51: 155–185.

Franco, Maria Sylvia de Carvalho. 1974. *Homens Livres na Ordem Escravocrata*. São Paulo: Ática.

Freyre, Gilberto. 1956. "Prefácio." In René Ribeiro (ed.), *Religião e Relações Raciais*. Rio de Janeiro: Ministério da Educação e Cultura.

FUNAI. n.d. "O Brasil Indígena." http://www.funai.gov.br/index.php/indios-no-brasil/o-brasil-indigena-ibge.

Furtado, Celso. 1976. "Characteristics of the Agrarian Structures." *Economic Development of Latin America: Historical Background and Contemporary Problems*. New York: Cambridge University Press.

Furtado, Celso. 2007. *Formação econômica do Brasil*. São Paulo: Companhia das Letras.

Goldman, Marcio. 2005. "Formas do Saber e Modos do Ser: multiplicidade e ontologia no Candomblé." *Religião e Sociedade*, 25 (2): 102–120.

Hall, Catherine. 2002. *Civilising Subjects: Metropole and Colony in the English Imagination, 1830–1867*. Chicago, IL: University of Chicago Press; Cambridge, UK: Polity Press.

Hanchard, Michael. 1994. *Orpheus and Power: The Movimento Negro of Rio de Janeiro and São Paulo, Brazil, 1945–1988*. Princeton, NJ: Princeton University Press.

Hira, Anil. 2007. "Did the ISI Fail and Is Neoliberalism the Answer for Latin America? Re-Assessing Common Wisdom Regarding Economic Policies in the Region." *Brazilian Journal of Political Economy*, 27 (3): 345–356.

Hirano, Sedi. 2005. *Formação da Sociedade moderna no Brasil. Capitalismo Moderno: relações Brasil/Estados Unidos, mercado e migrações internacionais*. São Paulo: Humanitas.

Holanda, Sérgio Buarque de. 2012. *Roots of Brazil*. Notre Dame, IN: University of Notre Dame Press.

Huntington, Samuel. 1991. *The Third Wave. Democratization in the Late Twentieth Century*. Norman: University of Oklahoma Press.

Ianni, Octávio. 1962. *As Metamorfoses do Escravo: Apogeu e Crise da Escravatura no Brasil Meridional*. São Paulo: Difel.

IBGE. 2009. *Pesquisa Nacional por Amostra de Domicílios*. Rio de Janeiro: IBGE.

IBGE. 2011. *Censo Demográfico de 2010. Resultados do Universo*. Rio de Janeiro: IBGE.

IBOPE. 2013. "62% dos Brasileiros são Favoráveis às Cotas em Universidades Públicas." Accessed on April 12, 2013. http://www.ibope.com.br/pt-br/noticias/Paginas/62-dos-brasileiros-sao-favoraveis-as-cotas-em-universidades-publicas.aspx.

Karasch, Mary. 1987. *Slave Life in Rio de Janeiro*. Princeton, NJ: Princeton University Press.

Kinzo, Maria D'Alva Gil. 1988. *Oposição e autoritarismo. Gênese e trajetória do MDB*, São Paulo: IDESP/Vértice.

Lessa, Carlos. 1977. "Visão crítica sobre o II Plano Nacional de Desenvolvimento." *Tibiriçá*, 2 (6).

Levine, Robert. 1998. *Father of the Poor? Vargas and His Era*. New York: Cambridge University Press.

Maggie, Yvonne, and Peter Fry. 2004. "A reserva de vagas para negros nas universidades brasileiras." *Estudos Avançados*, 18 (50): 67–80.

Mariano, Ricardo. 2007. "Pentecostais em ação: demonização dos cultos afro-brasileiros." In Vagner Gonçalves da Silva (ed.), *Intolerância religiosa. O Impacto do Neopentecostalismo no Campo Religioso Afro-brasileiro*. São Paulo: EDUSP.

Marini, Ruy Mauro. 1969. *Subdesarrollo y revolución*. Mexico City: Siglo XXI.

Martins, Luciano. 1976. *Pouvoir et développement économique formation et évolution des structures politiques au Brésil*. Paris: Anthropos.

Maxwell, Kenneth. 1973. *Conflicts and Conspiracies: Brazil and Portugal, 1750–1808*. Cambridge, UK: Cambridge University Press.

MDS. 2007. *Relatório GT População Quilombola*. Brasília: MDS-CNAS.

Mello e Souza, Laura. 1999. "Raymundo Faoro: Os Donos do Poder." In Lourenço Dantas Mota (ed.), *Introdução ao Brasil. Um banquete no Trópico*. São Paulo: Editora Senac.

Melo, Marcus André. 2008. "Unexpected Success, Unanticipated Failures: Social Policies from Cardoso to Lula." In Peter R. Kingstone and Timothy Power (eds.), *Democratic Brazil Revisited*. Pittsburgh, PA: University of Pittsburgh Press.

Moehlecke, Sabrina. 2002. "Ação Afirmativa: História e Debates no Brasil." *Cadernos de Pesquisa*, 117 (November): 197–217.

Moore, Jr., Barrington. 1993. *Social Origins of Dictatorship and Democracy: Lord and Peasant in the Making of the Modern World*. New York: Beacon Press.

Moura, Clóvis. 2001. "A quilombagem como expressão de protesto radical." In Clóvis Moura (ed.), *Os Quilombos na Dinâmica Social do Brasil*. Maceió: EDUFAL.

Munanga, Kabele. 2001. "Políticas de Ação Afirmativa em Benefício da População Negra no Brasil: Um Ponto de Vista em Defesa de Cotas." *Sociedade e Cultura*, 4 (2): 31–43.

Neri, Marcelo. 2014. "The New Brazilian Middle Class and the Bright Side of the Poor." *Poverty in Focus*, v. 26, UNDP, pp. 17–19.

Neri, Marcelo, Luisa Melo, and Samanta Monte. 2012. *Superação da pobreza e a nova classe média no campo*. Rio de Janeiro: Editora FGV.

Novais, Fernando. 1979. *Portugal e Brasil na crise do antigo sistema colonial (1777–1808)*. São Paulo: Hucitec.

Ortiz, Renato da Silva. 1988. *A morte branca do feiticeiro negro. Umbanda e sociedade brasileira*. São Paulo: Brasiliense.

Pereira, Anthony. 2003. "Brazil's Agrarian Reform: Democratic Innovation or Oligarchic Exclusion Redux?" *Latin American Politics and Society*, 45 (2): 41–65.

Pinheiro, Paulo Sérgio. 1977. *Política e Trabalho no Brasil (dos anos vinte aos 1930)*. Rio de Janeiro: Paz e Terra.

Portal Brasil. 2014. "Mais de mil cidades tem feriado no Dia da Consciência Negra." Accessed November 22, 2014. http://www.brasil.gov.br/governo/2014/11/mais-de-mil-cidades-tem-feriado-no-dia-da-consciencia-negra.

Prado Junior, Caio. 1963. *Formação do Brasil Contemporâneo: Colônia*. São Paulo: Brasiliense.

Prandi, J. Reginaldo. 1991. *Os Candomblés de São Paulo*. São Paulo: EDUSP-Hucitec.

Prebisch, Raúl. 1949. "O desenvolvimento econômico da América Latina e seus principais problemas." *Revista Brasileira de Economia* 3 (3): 47–111.

Prebisch, Raúl. 1950. *The Economic Development of Latin America and Its Principal Problems.* New York: United Nations.

Ribeiro, Darcy. 1995. *O Povo Brasileiro.* São Paulo: Companhia das Letras.

Sallum, Brasilio, Jr., and Eduardo Kugelmas. 1993. "O Leviatã Acorrentado: A Crise Brasileira dos Anos 80." In Lourdes Sola (ed.), *Estado, Mercado e Democracia: Política e Economia Comparadas.* Rio de Janeiro: Paz e Terra.

Schwarz, Lilia. 1993. *O Espetaculo das Raças. Cientistas, Instituições e Pensamento Racial No Brasil: 1870–1930.* São Paulo: Companhia das Letras.

Singer, Paul. 1972. *O Milagre Brasileiro—Causas e Conseqüências.* São Paulo: Caderno Cebrap, N. 6.

Skidmore, Thomas. 1967. *The Politics of Military Rule, 1964–1985.* New York: Oxford University Press.

Skidmore, Thomas. 1974. *Black into White. Race and Nationality in Brazilian Thought.* New York: Oxford University Press.

Skidmore, Thomas. 1988. *Politics in Brazil 1930–1964: An Experiment in Democracy.* New York: Oxford University Press.

Smith, Peter H. 2005. "Cycles of Electoral Democracy." *Democracy in Latin America. Political Change in Comparative Perspective.* New York: Oxford University Press.

Sposati, Aldaiza. 2002. "Regulação social tardia: característica das políticas sociais latino-americanas na passagem entre o segundo e terceiro milênio." VII Congreso Internacional del CLAD sobre la Reforma del Estado y de la Administración Pública, Lisboa, Portugal, October 8–11, 2002.

Sunkel, Oswaldo. 1969. "Política nacional de desarrollo y dependencia externa." In Andrés Bianchi et al. (eds.), *América Latina: Ensayos de interpretación económica.* Santiago de Chile: Editorial Universitária.

Telles, Edward. 2004. *Race in Another America. The Significance of Skin Color in Brazil.* Princeton, NJ: Princeton University Press.

Thorp, Rosemary. 1994. "The Latin American Economies in the 1940s." In David Rock (ed.), *Latin America in the 1940s. War and Postwar Transitions.* Berkeley: University of California Press.

Thorp, Rosemary. 1998. "Industrialization and the Growing Role of the State: 1945–1973." *Progress, Poverty and Exclusion. An Economic History of Latin American in the 20th Century.* Washington, DC: IADB.

Toledo, Caio Navarro de. 1977. *ISEB: Fábrica de ideologias.* São Paulo: Editora Ática.

Twine, France W. 1998. *Racism in a Racial Democracy: The Maintenance of White Supremacy in Brazil.* New Brunswick, NJ: Rutgers University Press.

UN-DESA. 2011. "World Population Prospects, the 2010 Revision." United Nations, Department of Economic and Social Affairs. Accessed March 12, 2012. http://esa.un.org/wpp/Analytical-Figures/htm/fig_11.htm.

Valle Silva, Nelson, and Maria Lígia de O. Barbosa. 2006. "População e estatísticas vitais." In *Estatísticas do Século XX.* Rio de Janeiro: IBGE.

Ventura, Roberto. 2010. *Casa-Grande e Senzala.* São Paulo: Publifolha.

Verger, Pierre. 1981. *Orixás: Deuses Iorubás na África e no Novo Mundo.* Salvador: Corrupio.

Weaver, Frederick. 2000. "Import-Substitution and Semi-Industrialization in Latin America, 1930–1970s." *Latin America in the World Economy: Mercantile Colonialism to Global Capitalism.* Boulder, CO: Westview Press.

Weber, Max. 2004. "The Distribution of Power in Society: Classes, Status Groups and Parties." In Sam Whimster (ed.), *The Essential Weber: A Reader*. London: Routledge.

Williamson, John. 1990. *Latin American Adjustment: How Much Has Happened*. Washington, DC: Institute for International Economics.

World Bank. 2012. Accessed March 12, 2012. http://databank.worldbank.org/data/home.aspx.

Gender, Marriage, and Sexuality

Maria Lúcia Rocha-Coutinho

This is a brief overview of the status of Brazilian women relative to male counterparts, gender role, parenting and household tasks, education, and representation in the workforce and politics from colonial times up to the present moment.

THE PATRIARCHAL FAMILY

During colonial times, family—based on a legalized union—was practically nonexistent in Brazil. In part, this was due to the fact that most Portuguese men did not move to the new land with the intention to get established there, but, rather, to get rich and go back to Portugal. Thus, women usually remained in Portugal and men frequently had relationships in Brazil with the Amerindian women and the African slaves. It was only with the concession of land grant that family started to gain an increasing importance in Brazil. There emerged, then, the patriarchal family, composed of a central nucleus—the white couple and their legitimate children—and a peripheral one, not always well delimited, which included slaves, aggregates, Amerindians, African slaves, and mestizos, as well as the patriarch concubines and their illegitimate children.

The main authority in the patriarchal family was the man, head of the family during colonial times *(pater familias)* who held the power over slaves, aggregates, his sons and daughters, and even over his wife. The marriages were arranged by the parents and, usually, they aimed social or political interests. Consanguineous marriages were very common, as well as marriages in which the husband was much older. Sex with the spouse aimed at perpetuating the family name, although men were allowed to have sexual intercourse with other women, especially slaves. The same was not accepted in the case of women, who had to remain chaste and pure until their wedding and, later on, give birth to healthy children, especially males. That is, a double sexual morality invigorated at the time.

These first families constituted autonomous groups that took care of production, administration, justice, and self-defense. Thus, the patriarchal organization in Brazil did not restrict itself to family life, but, rather, the landowners of the large sugar

The family of a civil servant during a stroll in Rio de Janeiro in colonial Brazil, from Voyage Pittoresque et Historique au Bresil *by Jean-Baptiste Debret. At that time, settler women could not leave home without a male escort. (Getty Images)*

plantations and, later on, of the coffee plantations, were also responsible for the political, social, and economic matters.

Women's behavior in Brazil during colonial times varied according to their social class: the lower-class women worked hard, although they could enjoy more personal freedom. As for the ones from the elite, not all of them were excluded from the public life and confined to the private sphere of family life. Some of these women, on account of their husband's death or illness, managed properties, slaves, and even politics, exerting the patriarchal command with the same vigor and, sometimes, with even more energy than men, something that make us believe that women's oppression in Brazil was related not only to their sex but also to economic, social, and political factors. This did not change, though, the role of women in patriarchal society.

However, we may say that the strength of the *pater familias* never eliminated completely the power and influence of women who, although indirectly, frequently intervened, whenever necessary, in the public or administrative affairs in Brazil, especially to protect their children, in-laws and/or god sons. It was due to their interference and effort that many of them got a position or job—some of them of great importance—not only during colonial times but also later on, during imperial times.

The strength of the mother figure in Brazil might be observed in the sentimental and mystic cult of the mother in the country, identified, as pointed out by Gilberto Freyre (1987a), with sanctified and protected figures and institutions such as the Virgin Mary and the Church, images of devotion and sacrifice. Thus, the mother in Brazil was regarded as the symbol of honor and moral solidarity of the group, occupying a central position in Brazilian family.

Contrary to what one could imagine, during colonial times, women in Brazil had a very important role. They were the ones who brought the knowledge about food preparation—adapting their Portuguese recipes to the fruits and herbs they encountered in Brazil—notions of hygiene, and the care of children and of sick people, a knowledge that was amplified and adapted to use the products they encountered in the new land. We owe them much of the dishes and sweets, made with the fruits, vegetables, and roots of the land, such as the banana, mango, and guava sweets and pastes and the manioc and corn products, such as the manioc flower, used to make the Brazilian rolled manioc flower pancake and the manioc and corn cakes and breads so appreciated in Brazil.

Besides, during these first years of colonial times, women were responsible for weaving the cotton used in bath towels, sheet, linen, and even in the clothes used inside the property. Also, due to the scarcity of specialized hand labor, they served as a doctor, nurse, and teacher, taking care of sick people—making use of the herbs that they encountered in the new land—and teaching their children how to read and write.

Despite the importance of women inside the home, providing their family the material and emotional support they needed, the social and juridical authority of the patriarch over his family members—including his wife—his property, and his domains remained unquestionable during colonial times and, in part, even during the empire and after the republic was proclaimed. Thus, during those days, fathers and husbands were allowed to place their daughters and wives in convents, with the connivance, or even the support, of justice, whenever they were dissatisfied with their behavior, as to obstruct marriages that were against their will.

THE 19TH CENTURY AND THE SENTIMENTALIZING OF THE FAMILY

The arrival of the Portuguese court to Brazil in 1808 changed the behavior of men and women in Brazil, especially that of the ones from the upper class. Women and men were more frequently seen in operas, theaters, parties, and balls, as the social life was encouraged by the court members.

In the same way, the social order changed rapidly, not only with the increasing number of workers and European immigrants in the cities, especially in the capital, the city of Rio de Janeiro, but also with the increasing number of professionals, such as doctors, lawyers, military men, and business men. There emerged, then, a middleclass in a country which was composed of basically two social classes, the landowners and the others.

Many of those professionals were landowners' sons who were sent to Europe to study and who brought on their way back the modern European liberal ideas of the Illuminist philosophers, which fit the process of modernization that the Portuguese court and, later on, the Brazilian emperors were trying to establish in Brazil.

These changes, which strengthen the power of the state in Brazil, brought on the decadence of the old patriarchal family, the most important institution for the formation of Brazilian society and that, during the first years of the colony, had an important organizing and disciplinary role, since it represented the only stable and organized group during this period, for it had the main source of power.

The despotism of the *pater familias* started to decline, although the father maintained his position of power in the family. His power was limited, though, since he had to share his authority in the public life with other institutions of social control, such as the doctors, the lawyers, the merchants, the military men, and the small industrialists. Marriages at the time changed a bit, since weddings based on love between the partners started to be accepted. Sex based on love was permitted among married couples, although they were still under control, that is, it aimed at procreation. The double sexual standard morality remained, as well as the family's hierarchy in which men were supposed to be the providers and women were supposed to take care of the house and children. The care of children was the main objective of a conjugal union, since it was necessary to strengthen the new national state and for that it was necessary to have strong men and women to defend their sovereignty.

As far as the labor division inside the home was concerned, women, especially those from the upper classes, continued to occupy a secondary position—that of taking care of the house and the family—distinct from that of men—who should dedicate themselves to the commercial, political, and cultural affairs of public life—something that was reinforced by the different natures attributed to the two of them, women and men. This secondary position occupied by women was reinforced by the fact that they had very little formal education, something that started to change over time, as they began to question this limitation imposed on them and to invest in a better education. Marriage, thus, was the only possibility opened to women, since to remain single, besides being, for most of them, financially impractical, implied a lack of prestige.

Marriages, though, changed. Different from colonial times, when they were arranged aiming at economic and political interests, now they started to be based on love. Sex between husband and wife, although highly controlled, began to be seen as a means to perpetuate the species and, thus, as a means to develop the new nation. The double sexual morality for men and women, though, continued: while men could have their affairs outside the family, women were expected to be pure, sober, and dedicated to their household, their husband, and to the care and education of their children.

Up to the end of the 19th century, women remained almost exclusively inside their homes. The only jobs opened to women—limited to the ones who had no money or a husband—were those of elementary school teacher, nurse, or housemaid. With the increasing industrialization of the country in the 19th century, the demand for people to work in the factories increased, and single women from the lower class were allowed to work in the factories to support themselves and to help support their families. In São Paulo, for example, the 1872 census showed that of the 10,256 workers in the cotton industry, 9,514 were women.

Early in the 20th century, besides their activity in industry, women began to work in commerce (as sellers) and in business offices (as secretaries), besides increasing their participation in the industries and as elementary school teachers. The qualities attributed to women workers, pointed out as a positive factor for their incorporation in the job market, were the fact that they were more obedient, more patient, and more dedicated to their work, and had no or very low aspirations.

ORGANIZED WOMEN'S MOVEMENTS

The first organized women's movements occurred in the North and Northeast regions by the end of the 19th century and aimed at abolishing slavery in Brazil. In 1882, it was founded in the province of Ceará, called the Liberating Women's Society of Ceará, presided by Maria Tomásia Figueira, together with Maria Correia do Amaral and Elvira Pinto. They acted in favor of liberty, founding associations in the capital and in the interior of the province, and greatly contributing to the end of slavery in the province of Ceará in 1884 by the Legislative Assembly of the province four years before the end of slavery in Brazil. In the same year, in the city of Manaus, the capital of the province of Amazon, the Liberating Amazon Women Association was founded by Elisa de Faria Souto, Olímpia Fonseca, Filomena Amorim, among other women who belonged to the local white elite. They fought for the emancipation of all slaves in the Amazon province, something that happened on March 30, 1887, a year before the signing of the Aurea Law that abolished slavery in Brazil.

The greater enrollment of women in the job market, though, did not reduce their participation at home as the person responsible for the household duties and the care and education of children. On the contrary, feminine work was only accepted by the Brazilian society during this time insofar as it complemented the family's income and it attended the interests of the increasing demand of labor workers by the growing industrialization of the country. By the end of the 19th century in Brazil, more precisely in the North and Northeast regions, there emerged the first women's organizations, mostly involved with the abolitionist cause, but also fighting for the right of women to vote and be voted. It is worth mentioning Nísia Floresta, an important Brazilian feminist, here. She was probably the first woman to break through the limits of private and public spheres in Brazil, publishing articles in newspapers in defense of the rights of women, Indians, and slaves.

After the proclamation of republic, elections were convoked to form a Constituent Assembly by the provisory government. At that time, a woman obtained the right to enlist, invoking an imperial law published in 1881 that gave the right to vote to all citizens who had a minimum wage of 2,000 réis (Brazilian currency at the time). In 1910 Leolinda Daltro and other feminists founded the Feminist Political Party in Rio de Janeiro (the country's capital) with the main objective of fighting for the right of women to vote and be voted, organizing several public manifestations to vindicate the right of women to full citizenship. In 1928, a few months before England, the governor of Mossoró, in the state of Rio Grande do Norte, authorized women to vote, something that is part of the records of the world's feminist movements. Celina Guimarães Viana, from Rio Grande do Norte, was the first woman to vote, after sending a petition to the judge of her state requiring the right to vote based on the electoral law of the state that allowed all citizens "with no sex distinction" to vote and be voted. After that a great national movement led women from several states to do the same. Other conquests followed, such as the first woman to be elected federal deputy in Brazil. Women's right to vote was nationally approved during Getúlio Vargas's dictatorship. Nowadays, although women constitute 51.3 percent of Brazilian population, feminine participation in politics is still very low. According to data from IBGE (2014), among the 513 deputies, only 47 are women, and among the 81 senators only 10 are women, that is, only 9.5 percent of the Congress members are women, although since 2011 Brazil has a woman president, Dilma Rousseff, who has just been elected for a second term.

The labor workers' strikes, the anarchist movement, the end of World War I, and the higher educational level of Brazilian women from the urban elite by the end of 1910 led to the emergence of a new generation of feminists. In the 1920s, several groups gathered under the name of "Brazilian Federation for Feminine Progress." They constituted the embryo to the foundation, in 1922, of the Brazilian Federation for Feminine Progress, founded and directed by Bertha Lutz, and which had a fundamental role in the conquest of feminine suffrage and, by extension, in the fight for equal political and social rights for women. The federation had great visibility and stood out as the feminist organization with the greater insertion in the public spheres of political power of its time. Since many of its members were journalists, they published articles in newspapers, organized Congresses, articulated political alliances, registered candidacies, distributed pamphlets inside planes, and became Brazilian representatives in meetings, seminars, and Congresses outside the country, exerting pressure on politicians to approve women's right to vote and be voted in the text of the Constitution of 1934, being finally successful.

Bertha Lutz, one of the major figures in Brazil's women's rights movement, at the beginning of the 20th century. Having studied biology at the Sorbonne in Paris and holding a law degree from Rio de Janeiro's National School of Law, Lutz was the first woman to direct Rio de Janeiro's National Museum. She was also the creator of the "League for the Intellectual Emancipation of Women" in 1919 and of the "Brazilian Federation for Feminine Progress" in 1922. (Library of Congress)

It was only in 1932, during Getúlio Vargas's dictatorship, with the approval of the Electoral Code of 1932 that women were allowed to vote, although only two years later, in 1934, when the new democratic state elaborated the second Republican Constitution, political rights granted to women became a constitutional right. This Constitution stated that, although all 21-year-old men were obliged to vote, for those older than 60 years and for all women the vote was optional. This has changed only in 1965, with the edition of the Electoral Code that invigorates until today and that obliges all men and women from 18 (recently changed to 16) to 80 years to vote.

POST–WORLD WAR II WOMEN: THE FULL-TIME MOTHER

The post–World War II period unleashed a profound transformation in European and North American societies that affected the behavior of women and their roles in society. Women began to get married and have their children earlier, and it was expected then that their lives would be totally devoted to their home and family. Brazilian women were very much influenced by the ideology transmitted by the media (books, magazines, and films) in those countries and reinforced by the work of professionals such as psychologists and psychoanalysts on the importance of the mother for the health and well-being of their children. Thus, during the 1950s and 1960s, Brazilian society reinforced the idea that marriage was the ultimate goal of women and that the birth of a child was the natural and desirable consequence of it. The role of spouse and mother should be regarded by women as something sublime that would fulfill their lives, and, thus, a professional career was considered practically inconceivable. Although it was common, especially among the middle- and upper-class women to enter the university, a professional career was discouraged, since it would interfere in their main roles, those of a spouse and a mother.

Women and men, fathers and mothers, had different and well-established roles both inside the home and in society. Men should be the family's financial provider, while women should care for the house, husband, and children. Men should be the authority figure inside the home and, thus, they were expected to be rational, while women were regarded as emotional and delicate, and, for that reason, they demanded attention on the part of their husbands, who sometimes treated them as a bibelot.

Regarding sexuality, the double standard of morality continued to invigorate at the time, that is, women should remain pure to their husbands, while men, due to their different "nature," were allowed to be unfaithful, although they should be discrete, and the sexual involvements should not interfere in their family lives. Marriages, though, were now based on the love between the partners and the conjugal vows were indissoluble. There was much smaller control over pleasure for both partners during sexual intercourse, although the double sexual morality continued to invigorate. Family relations were based on love and on a greater intimacy among its members. The relationships between parents and children, while opened some more space to dialogue, still had a strong hierarchical base. There still remained the old roles of the father as the main provider and of women as the person responsible for household duties and child care. The woman was at that time considered the "Queen of the House."

DIVIDED AND MULTIPLIED: NOWADAYS
WONDER-WOMEN

During the 1960s a series of social movements, developed in different parts of the world, questioned old dichotomies and put into check the socially institutionalized power, including the power of men over women. Old political ideologies, among them the one that segregated public and private spheres, giving supremacy to men in the public world and confining women to the private world, were contested.

As a part of these movements, there emerged, in different parts of the world, a series of feminist movements. Reinforced by the appearance of more effective contraceptives, women were allowed to set apart sexuality and procreation. The distinction between men and women, thus, started to be regarded as a political and not a biological one, actually hiding the dependency and exploitation of women by men. Shut down in their homes, women, along the years, produced and reproduced the workforce that society needed, and this function, until then related to feminine biological characteristics, proved to be actually a social-economical category. The feminism movements of the 1960s helped, thus, to say "no" to the old patriarchal order that gave women a secondary place in the world, and we may say that, in general, the movement was successful, changing considerably men and women relations and the role of women in society.

Despite the fact that Brazil was under Getúlio Vargas's dictatorship (1930–1945) and that there was a great political repression to all sorts of movements, under the pressure of the feminism movement of the early 20th century, the Brazilian Congress finally approved, in 1962, changes in the Civil Code, which enabled married women to work without having to be authorized by their husbands. The guardianship of husbands over their wives finally came to an end. After this important conquest, Romy Medeiros da Fonseca and other women founded the National Council of Brazilian Women. The main fights during that time were, among others, those for equality between women and men both in society and inside the house and for the approval of the divorce law.

During military dictatorship (1964–1985), women organized themselves, independently of political party, age, and social class, and engaged themselves in the fight against the military government. With the institution of the International Year of Women by the United Nations in 1975, women's problems became a theme of discussion within the universities and among the liberal professional associations. In the same year, the International Congress of Women was initiated in Mexico, where Brazil was represented by Berta Lutz, and in Brazil the Research Week on the Role and Behavior of Brazilian Women was organized.

BERTHA LUTZ

Bertha Lutz, born in São Paulo, Brazil, in 1894, to a pioneer scientist in the study of tropical diseases and an English nurse, was sent to Europe to complete education; she got in contact with the Suffragist Movement in England. She graduated in natural sciences from Sorbonne University in 1918 and went back to Brazil; she became a biologist at the National Museum, where she worked all her life, and was an energetic advocate of Brazilian women's rights. In 1918, she published an article in *Revista da Semana*, claiming women to found an association to fight for their rights, which constituted the basis for the feminist movement in Brazil. In 1919, Bertha represented Brazil in the International Feminine Council, which defended equal salaries for women and men and the inclusion of women in the service of protection for workers. In 1936, she assumed a mandate as a deputy and claimed for changes in labor legislation with respect to women's right to work, the prohibition of children's work, the right of women to maternity license, and equal rights and salaries for men and women. In 1944, she represented Brazil on the committee of Feminine Affairs at the International Conference of Work in Philadelphia (the United States) and was a member of the Brazilian delegation at the San Francisco Conference in 1945, convoked to write the final text of the Letter of the United Nations Organization. There Bertha's main contribution was her political support for the delegate from South Africa to include a compromise with equality, both between women and men and among the different nations. In 1975, she was invited to integrate the Brazilian delegation in the Mexican conference, in which the United Nations established that year as the "International Year of Women." She died in Rio de Janeiro in 1976.

In September 1975 the Center for Brazilian Women, an organization responsible for intermediating and articulating the feminists' main objectives under the form of a collective action, was created. Many women who had lived in exile came back and gave several important contributions to the Center for Brazilian Women; they organized a center for women's studies that promoted seminars, discussions, and research on the conditions of women in Brazil. Several publications in specialized journals, books, and the media emerged from these studies.

In 1977 a political commission was established in the Congress to investigate the situation of women in the job market and other activities that brought to light facts that were known only to small groups of people. Some of the issues that shocked society were the knowledge that, in rural areas, women received only one-fifth of the salary received by men to do the same job; that, despite the fact that legislation obliged companies with more than 30 women workers to keep day care centers, this law was not being respected; that state companies obstructed the access of women to some sectors and that these impediments were against the law; that pregnant women were fired with no actual reason; that agencies maintained with foreign allocations of money were promoting the indiscriminate sterilization of women, among other accusations.

Still during the 1970s, study and research groups devoted to the discussion of women's issues were organized in almost all scientific societies' meetings—such as those of the National Association of Research and Graduate Studies in Social Science, the National Association of Research and Graduate Studies in Psychology, and the National Association of Research and Graduate Studies in Language, Literature and Linguistics.

WOMEN TODAY

The profile of the new generation of Brazilian women has changed a lot during the past years. Today, despite the appearance of more effective contraceptives, like the pill, that separated sexuality from procreation, and of the fact that an increasing number of women in Brazil are on the job market—some of them occupying important positions—things did not seem to have changed as much as it appears to have. Although social discourse usually describes the ideal marriage as one in which husbands and wives share the responsibility for family care and support, studies have shown that most men and women still believe that women should be the person responsible for the house and children and men should provide family's financial needs. They also still seem to believe that the unity of mother–child is basic, universal, and psychologically more appropriate both for the child's healthy development and for the fulfillment of the mother ("motherhood mystique"). Thus, despite many changes, old patterns of gender behavior seem to coexist in Brazilian society with more modern ideas that proclaim the equality of rights and duties inside and outside the home. It is worth mentioning here that this is so despite the fact that some women nowadays are even better prepared than men and some of them occupy positions of power and prestige in society, attaining financial independence and greater control over their lives.

As one can observe in the last IBGE census (2010), as well as in data of Catho Group (an RH Consulting Company), Brazilian women surpassed men in terms of educational level if we consider not only undergraduate studies but also MBAs and graduate studies. As far as executive women are concerned, 24 percent of them have completed undergraduate studies, 40 percent have MBA degrees, 28 percent have MA degree, and 8 percent have PhD degrees, greater percentages than those of men. Besides, they have a better knowledge of the English language—so important for today's global market.

Regarding their professional ambitions, as pointed out by a study developed by Hewlett and Rashid (2011), 80 percent of Brazilian women aim to reach top command positions against only 52 percent of American women in the same condition. Data from Catho Group on the increase in the number of women in different leadership positions point out to the fact that during 1996 and 2008, the number of women in top leadership positions (those of CEO, director, and manager) more than doubled, while in the intermediate levels the increase was smaller. From 2008 to 2015 the number of women in top leadership position grew even higher. In 2008, 20.56 percent of the CEOs, 25.88 percent of the directors, and 32.03 percent of the managers in big corporations were women (see Rocha-Coutinho and Coutinho 2011). In 2015, according to data from Catho Group, these numbers were even higher: 67.96 percent of the CEOs, 47.94 of the directors, and 82.17 percent of the managers in national and international corporations in Brazil were women. This might confirm not only that women aim at reaching these positions of command and leadership but also that they are slowly getting there. The number of women has been growing not only in business but also in other areas. The Federal Council of Medicine pointed out that the profile of MDs has been undergoing a historical transformation, with the number of female doctors gradually growing (from 25.62% of women against 74.38% of men in the age group of 65 to 69 to 53% of women against 46.69% of men in the age group equal or lower than

28 years). Thus, for the first time, the majority of doctors in Brazil are women, in the age group of 29 years old or younger (see Leite 2013, p. 69).

A series of factors seem to have contributed to these changes, among which is the appearance of more effective contraceptives, such as the pill, that allowed women to be the owners of their own sexuality and be able to decide if and when they want to have children, which we will be addressing later. Also, the greater investment of women in a better education, engaging in university areas traditionally considered masculine, such as business, economy, medicine, and law school, increased their possibilities to get better jobs and reach higher positions, including those of power and control, occupied before mostly by men. Hence, their salaries became higher, sometimes surpassing those of men. In a research developed with women from the BRICS (which are Brazil, Russia, India, China and South Africa) countries, Hewlett and Rashid (2011) pointed out that approximately 20 percent—or even more—of women who work full-time in Brazil earn more than their mates. Thus, as stressed by the authors, it is possible that "the power of the purse, even more than educational opportunities and career aspirations, is helping women break through the social traditions that held back many in their mothers' generation" (Hewlett and Rashid 2011, p. 23).

In part, as a consequence of the possibility to provide or to contribute to the providing of their families' needs, Brazilian women nowadays do not suffer the same pressure their grandmothers and mothers did some years ago to leave their jobs after marriage or even after the birth of their first child. Besides, as pointed out by Bruschini (2000), since these women are more qualified, they generally engage in more qualified jobs and, thus, earn more money, enough to pay for the domestic infrastructure that is necessary to substitute them when they are out working. Moreover, the activities that they develop at work are more gratifying than the ones developed by women with a lower level of instruction. Thus, the expectations relative to middle-class women and men regarding their performance both on the job market and at home have been changing in the country, although these changes might be more related to the sphere of work than to that of home.

Regarding family composition in Brazil, by the end of the 20th century and the beginning of the 21st century a diversity of different family arrangements arose, as pointed out by several researchers (see Rocha-Coutinho 2015) in opposition to the traditional modern conjugal family, composed of the male provider and the housekeeping mother, which for decades constituted the dominant model in Brazil, especially in the urban middle classes.

The data relative to the Brazilian family released by the last IBGE census (2010) seems to confirm this emergence of a diversity of nontraditional family arrangements. According to the data, there was a growth in the number of consensual unions, of reconstituted families, of families headed by women, of childless families, among others. The number of childless families, for example, grew from 14.9 percent in 2002 to 20.2 percent in 2010, not to mention the fact that the percentage of childless couples in 2010 (22.1%) was close to the number of couples with children (23.5%). Besides, according to the IBGE census data, in 2010, one-third of women who have a graduate degree become mothers around the ages of 30 and 34 and the mean age of children is 1.14. From 2000 to 2010, according to IBGE data, the number of women who have their first child at 30 years or more grew from 27.6 percent to 31.3 percent.

During these 10 years, there was a growth in the number of women head of the family (from 22.2% to 37%) and a decrease in the number of the men head of the family (from 77.8% to 62.7%). It is also worth mentioning the fact that the number of families headed by women without a mate had a significant increase (from 19.5% to 46.4%). The 2010 census also pointed out that 16.3 percent of the families with children might be considered reconstituted families, that is, families in which the children are from both members, only from the household head, only from one of the members, or even a combination of these situations.

Regarding the level of instruction, women surpassed men, according to the 2010 census (9.9% of male population has at least an undergraduate degree against 12.5% of the female population). Among the ones who have an undergraduate, an MA, and a PhD degree, most of them have a degree in social sciences, business, and law (37.8%), and for each level separately, 38.4 percent have an undergraduate degree, 31.3 percent have an MA degree, and 25 percent have a PhD degree. The increasing educational level of women, including those with an undergraduate degree, was observed in all sectors of the economic activity. It is worth mentioning here that the largest difference between women and men with an undergraduate degree occurred in the case of civil engineering, a traditionally male sector, where the increase of women with an undergraduate degree was 8.3 percent, while the increase of males with an undergraduate degree was 0.6 percent (in 2011, 28.6% of women had an undergraduate degree in this sector against 4.7% of men). The sectors of public administration and services to corporations were the ones with the higher proportion of women, including those with an undergraduate degree. Despite that, the average salary of women in 2011 was R$1.343,81 a month (the equivalent, in December 2011, to US$731.15), 72 percent of the average salary of men in the same position. In 2003 the medium salary of women was 70.8 percent of those of men. When we compare women's salaries to those of men's, from 2003 to 2011, while women's salary increase was 24.9 percent, men's salary increase was 22.3 percent.

One might say that these changes were made possible thanks to the increase in the number of middle-class women with undergraduate degrees, especially in the more valued areas, such as business and law, together with the possibility to reduce, or even opting for not having children, opened up by the appearance of more effective contraceptives. These factors opened new options to them and seem to be changing the values with respect to the role and position of women in society. Brazilian women nowadays are reaching positions of more power and prestige on the job market, and their salaries are getting higher, sometimes even higher than their male counterparts. This massive entrance of highly qualified middle-class women in top positions of corporations has been affecting not only gender and family relations but also the functioning of the corporations themselves in Brazil.

Despite this situation, the discrimination by gender has not been eliminated completely in Brazil, although the behaviors of both women and men have become more complex. The changes in family structure and on the job market, as well as the increasing number of middle-class women occupying important positions on the job market, among other facts, have increased the demand for domestic workers. The difficulties involved in conciliating professional and domestic life have led Brazilian women to count on the help of maids to care for their children and elderly parents, which has been increasing the difference between working and middle-class women.

The proportion of women who have paid jobs in the country (66%) is still low, and the number of domestic informal workers (8.1%) is high, according to IBGE (2010).

Among the younger women (from 16 to 24 years), 69.2 percent are engaged in informal types of work, and among the older women (60 years or more), this proportion is 82.2 percent. The higher participation of these two groups of women in informal types of work might be related to several aspects, among them, the low schooling and the difficulty to engage on the formal job market, in the case of young women, and the need to complement their low salaries after retirement, in the case of older women. This data might reflect not only the precariousness of this type of work as the exploitation of cheap hand labor.

Domestic work is Brazil is a feminine occupation par excellence, since 93 percent of the workers are women (IBGE 2010). In 2009, 55 percent of them belonged to the age group of 25 to 44 years. An expressive number of them did not have any of the rights that regular paid jobs in Brazil do and they usually work in very precarious conditions. Most of them had around six years or less of schooling, and the great government challenge was to implement public policies to formalize this type of work. By the end of 2014, a law was approved in Brazil, giving all domestic workers the same rights and obligations of all other formal workers. The consequences are yet to come, but it is already provoking some changes, and both employers and employees will have to make arrangements to adjust to the new situation, since many middle-class women who now count on the help of a maid will not be able to afford the increasing cost of having a full-time maid to help them in housework and child care and some of them are opting to have a part-time maid. As for the maids themselves, they are trying to make arrangements to cover all weekdays, such as getting jobs in different houses along the week so as to provide for theirs and their families' needs.

The fertility rate of women in all age groups and social classes has decreased during the last decade, according to data from IBGE (2010), although this decrease was greater for those who have completed undergraduate courses (1.14 children in 2010) than for the lesser-educated women (3.00 children in 2010 for women with incomplete elementary school education). Nevertheless, it is worth pointing out here that, even among the less educated women, the fertility rate has decreased in Brazil (from 3.43 children in 2000 to 3.00 in 2010), especially among younger women (between the ages of 15 and 24).

More than half of the Brazilian women (54.2%) in 2011, according to IBGE, had 11 or more years of schooling, against only 40.6 percent of the men. In 2011, of the 39.5 million women working, 17.6 percent were engaged in commerce; 16.8 percent were in education, health care, and social work; and 15.6 percent were in domestic labor. The percentage of women who receive smaller salaries than those of men decreases as the salaries increase, although women's salaries are lower than those of men for all groups. According to IBGE data (2010), in 2009, women with 12 years of study or more received 58 percent of men's salaries with the same level of education, and this difference was a little higher (61%) for all other educational-level groups. Besides, this difference in salary has reduced very little from 1999 to 2009. It is worth mentioning here, though, that men have a medium weekly journey of work longer than that of women in all kinds of work (while men work 43.9 hours, women work 36.5 hours).

Less educated women (those with up to 8 years of schooling) work less (32.9 hours) than more educated women (38.3 hours for those with 9 to 11 years of schooling, and 36.3 percent for those with 12 or more years of schooling). The opposite is true for men—that is, while those with up to 8 years of schooling work 43 hours, those with 12 years or more of schooling work 41.2 hours—except for the ones with 9 to 11 years, who work 43.5 hours a week. One possible explanation is that the less educated women are engaged in more precarious types of work, sometimes due to their need to complement their families' income, while the others are engaged in formal types of work that demand more strictly and fixed working hours of their workers.

In Brazil, women spend more than double the hours that men do in household tasks and the care of children. In 2009, while women spent around 22 hours per week in domestic tasks, men spent only around 9.5 hours per week (IBGE 2010). Women with 12 or more years of schooling dedicate themselves less to household duties (17 hours/week) than those with 8 or less years of schooling (25.3 hours/week). This might happen because the first group of women have better-paid jobs and can, thus, pay someone to help them with these tasks. This is an import data, since to conciliate productive work, family and personal life has been an important stressing factor in women's lives. Besides, it points out to the massive entrance of women in the job market and to the fact that women continue to have a double day's journey. It is necessary, then, to develop more public actions to promote a greater quality of life to women population, in general, and to poor women, in particular, since they are the ones who dedicate more time of their day to domestic affairs.

Despite that, in general, the situation of women in Brazil seems to be changing, although at a slower pace for lower-class women than for middle-class women. The recent transformations that occurred in Brazil in gender social relations, family composition and relationships, and female work, as pointed out by Araújo and Scalon (2005), pervade all social segments and strata, and point out to a weakening of more hierarchical relations to more horizontal and symmetrical types of relationships both inside the family and in society, in general. Nowadays a multiplicity of different forms of family and conjugal relationships co-occur in Brazilian society. According to statistical data (IBGE 2010), families are getting smaller in Brazil in all social classes. There is an increase in the number of consensual unions, of families headed by women, of childless families, of families composed of same-sex couples, and of reconstituted families, among others. The average age Brazilian women, in general, are getting married is increasing, especially in the middle class, something that might indicate that many of them are giving priority to their professional lives. There was an increase in the number of divorces, something that might indicate that nowadays women and men in the country are searching for personal happiness and fulfillment.

However, women face now the challenge of overcoming the burden of being more autonomous in their choices, breaking emotional ties that are no longer satisfactory to them, having more freedom to exert their sexuality, and being more competitive and efficient on the job market and, at the same time, continuing to fulfill their old responsibilities at home, of taking care of household duties and child care and of fulfilling the emotional charges that fall upon them of being the main provider of care and affection to their family members.

REFERENCES

Alencastro, L. F. (org). 1997. *História da vida privada no Brasil—império: a corte e a moderni-dade naciona*. Vol. II. São Paulo: Companhia das Letras.

Arantes, A. A., B. Feldman-Bianco, C. R. Brandão, M. Corrêa, R. Slenes, S. Kofes, V. Stol-cke. (orgs.). 1994. *Colcha de retalhos: estudos sobre a família no Brasil*. Campinas, SP: Editora da UNICAMP.

Araújo, C. and C. Scalon (orgs.). 2005. *Gênero, família e trabalho no Brasil*. Rio de Janeiro: Editora Fundação Getúlio Vargas.

Besse, S. K. 1996. *Restructuring Patriarchy: The Modernization of Gender Inequality in Brazil, 1914–1940*. Chapel Hill, NC: The University of North Carolina Press.

Bruschini, C. 2000. "Gênero e trabalho no Brasil: novas conquistas ou persistência da dis-criminação?. In M. I. Baltar (org.), *Trabalho e gênero: mudanças, permanências e desafios*. São Paulo: Editora 34.

Bruschini, C., and M. R. Lombardi. 2003. "Mulheres e homens no mercado de trabalho: um retrato dos anos 1990." In M. Maruani and H. Hirata (orgs.) *As novas fronteiras da desigualdade: homens e mulheres no mercado de trabalho*. São Paulo: SENAC.

Callcott, M., J. Hayward, and M. S. Caballero. 2011. *Maria Graham's Journal of a Voyage to Brazil*. Anderson, SC: Parlor Press.

Cândido, A. 1951."The Brazilian Family." In T. L. Smith and A. Marchant (eds.), *Brazil: Portrait of Half a Continent*. New York: The Dryden Press.

Costa, J. F. 1989. *Ordem médica e norma familiar*, 3rd ed. Rio de Janeiro: Edições Graal.

Cowling, C. 2013. *Conceiving Freedom: Women of Color, Gender and the Abolition of Slavery in Havana and Rio de Janeiro*. Chapel Hill, NC: The University of North Carolina Press.

Damatta, R. A. 1995. *The Brazilian Puzzle. Culture on the Borderlands of the Western World*. David J. Hess and Roberto DaMatta (eds.). New York: Columbia University Press.

Dell Priore, M. (org.). 2004. *História das mulheres no Brasil*. São Paulo: Editora Contexto.

Dias, M. O. S. 1995. *Power and Everyday Life: The Lives of Working Women in Nineteenth-Century*. New Jersey: Rutgers University Press.

Dias, M. O. S. 2006. *História do amor no Brasil*, 2nd ed. São Paulo: Contexto.

Féres-Carneiro, T. (org.). 2003. *Família e casal: arranjos e demandas contemporâneas*. Rio de Janeiro: Ed. PUC-Rio/São Paulo: Loyola.

Freyre, G. 1987a. *The Masters and the Slaves (Casa-Grande e Senzala): A Study in the De-velopment of Brazilian Civilization*. Transl. Samuel Putman. Oakland, CA: University of California Press.

Freyre, G. 1987b. *Mansions and the Shanties: The Making of Modern Brazil*. Transl. Harriet de Onis. Oakland, CA: University of California Press.

Freyre, G. 1987c. *Order and Progress: Brazil from Monarchy to Republic*. Transl. Rod. W. Horton. Oakland, CA: University of California Press.

Freyre, G. 2007. *Brazil: An Interpretation*. New York: Alfred A. Knopf.

Graham, S. L. 1992. *House and Street: The Domestic World of Servants and Masters in Nineteenth-Century Rio de Janeiro*. Austin, Texas: University of Texas Press.

Guedes, M. C. 2004. As mulheres de formação universitária: a reversão da desigualdade de gênero e seus reflexos no mercado de trabalho no Brasil: 1970–2000, Dissertação de Mestrado. Programa de Mestrado em Estudos Populacionais e Pesquisas Sociais (ENCE/IBGE).

Hewlett, S. A. 2007. *Off-Ramps and On-Ramps. Keeping Talented Women on the Road to Success.* Boston, MA: Harvard Business School Press.

Hewlett, S. A., and R. Rashid. 2011. *Winning the War for Talent in Emerging Markets: Why Women Are the Solution.* Boston, MA: Harvard Business Review Press.

Hirata, H., and M. Maruani (orgs.). 2003. *As novas fronteiras da desigualdade: homens e mulheres no mercado de trabalho.* São Paulo: Editora SENAC.

Holanda, S. B. 1971. *Roots of Brasil—ND Kellogg Inst Int'l Studies.* Transl. Harvey Summ. Notre Dame, IN: University of Notre Dame Press.

IBGE. 2009. *Pesquisa Nacional por Amostra de Domicílio.* Rio de Janeiro. IBGE.

IBGE. 2011. Censo Demográfico de 2010. Resultados do Universo. Rio de Janeiro. IBGE.

IBGE. 2014. Pesquisa Nacional. Rio de Janeiro. IBGE.

Leite, M. I. M. G. (2013). Transpondo barreiras, vencendo limites: escolhas, satisfações e desafios de mulheres médicas cariocas, Tese de Doutorado. Programa EICOS (UFRJ) [PhD dissertation on Interdisciplinary Studies on Community and Social Ecology – Federal Univ. of Rio de Janeiro].

Matthews, C. H. 2012. *Gender, Race and Patriotism in the Works of Nisia Floresta.* Woodbridge, Suffolk: Boydell & Brewer.

Meade, T. 2005. *"Civilizing" Rio: Reform and Resistance in a Brazilian City, 1889–1930.* State College, Pennsylvania: Penn State University Press.

O'Connor, F., and B. S. Drury. 1998. *The Female Face in Patriarchy: Oppression as Culture.* East Lansing, MI: Michigan State University Press.

Rocha, M. I. B. (org.). 2000. *Trabalho e gênero: mudanças, permanências e desafios.* São Paulo: Editora 34.

Rocha-Coutinho, M. L. 1994. *Tecendo por trás dos panos. A mulher brasileira nas relações familiares.* Rio de Janeiro: Ed. Rocco.

Rocha-Coutinho, M. L. 1999. "Behind Curtains and Closed Doors: Brazilian Women in Family Relations." *Feminism & Psychology*, 9: 373–380.

Rocha-Coutinho, M. L. 2003. "New Options, Old Dilemmas: Close Relationships and Marriage in Brazil." In Anna Laura Comunian and Uwe P. Gielen (eds.), *It's All about Relationships*, pp. 111–119. Lengerich, Alemanha: Pabst Science Publishers.

Rocha-Coutinho, M. L. 2007. "Família e emprego: conflitos e expectativas de mulheres executivas e de mulheres com um trabalho." In T. Feres-Carneiro (org.), *Família e casal: saúde, trabalho e modos de vinculação*, pp. 157–180. São Paulo: Casa do Psicólogo.

Rocha-Coutinho, M. L. 2011. "De volta ao lar: mulheres que abandonaram uma carreira profissional bem-sucedida com o nascimento dos filhos." In T. Feres-Carneiro (org.), *Casal e família: conjugalidade, parentalidade e psicoterapia*, pp. 133–148. São Paulo: Casa do Psicólogo.

Rocha-Coutinho, M. L. 2012. " 'Mulher moderna é assim, dá conta de tudo . . . ': a difícil arte de equilibrar trabalho e família." In M. N. Strey, A. Botton, E. Cadoná, and Y. A. Palma (orgs.), *Gênero e ciclos vitais: desafios, problematizações e perspectivas*, pp. 193–220. Porto Alegre: Edipucrs.

Rocha-Coutinho, M. L. 2015. "Investimento da mulher no mercado de trabalho: repercussões na família e nas relações de gênero." In T. Féres-Carneiro (org.). *Família e casal: parentalidade e filiação em diferentes contextos*, pp. 103–117. Rio de Janeiro: Ed. PUC-Rio/Ed. Perspectiva.

Rocha-Coutinho, M. L., and R. R. Coutinho. 2011."Mulheres brasileiras em posições de liderança: novas perspectivas para antigos desafios." *Revista Economia Global e Gestão—INDEG*, XVI (1): 61–80.

Schwarz, R. 1992. *Misplaced Ideas: ao vencedor as batatas. Essays on Brazilian Cultures. Critical Studies in Latin American and Iberian Cultures.* John Gledson (ed.). London/New York: Verso/New Left Books.

Skidmore, T. E. 2009. *Brazil: Five Centuries of Change.* Oxford, England: Oxford University Press.

Véras, E. Z. 2014. *Women in Management: Old and New Challenges in Brazil and China.* Germany: LAP LAMBERT Academic Publishing.

Wolfe, J. 1993. *Working Women, Working Men: São Paulo and the Rise of Brazil's Industrial Working Class, 1900–1955.* Durham, NC: Duke University Press.

Education

Antonio Lima da Silva and Volnei M. Carvalho

INTRODUCTION

Before the Portuguese arrived in Brazil in the 16th century, there used to be a peculiar way of educating that had been developed by the Amerindians. It worked without the presence of a teacher, of the institution, without methods or systematic planning of the learning process. Unquestionably, the children and the young men learned to become members of those communities through the teaching from the elders; for example, they learned the rules of the tribes, the rituals and the whole oral tradition, and the male/female social roles.

From a historic point of view, formal education in Brazil began with the arrival of the Society of Jesus in 1549. That religious mission played an important role in the development of Brazilian education: the mission created several schools in different regions of the colony; it brought experienced priests to work with education in Brazil. These priests applied teaching methods of their own—*Ratio Studiorum* (Plan of Studies)—and taught generations of settlers and Amerindians in schools, seminaries, and settlements in villages and missions. At the end of the first period of Jesuitical teaching, the student knew how to read, write, count, and pray resourcefully. The Jesuits led formal education in Brazil on an exclusive basis for a little longer than two centuries, until their expulsion by the Marquis of Pombal in 1759. That was the beginning of a five-decade period during which Brazilian education experienced great disorder in the systematization of the learning process under the so-called royal classes. It lasted until the arrival of the royal family from Portugal in 1808. This event contributed to the rise of cultural life in the colony, allowing the outbreak of military academies, medical and law schools, and the launch of the Royal Press (Bello 2001).

There are several key moments in the history of education in Brazil. Paiva Bello (2001) points out 10 periods in the Brazilian formal education process: (1) the Jesuitic Period (1549–1759), (2) the Marquis of Pombal Period (1760–1808), (3) the D. João VI Period (1808–1821), (4) the Imperial Period (1822–1888), (5) Former Republic Period (1889–1929), (6) Second Republic Period (1930–1936), (7) New State Period (1937–1945), (8) Later Republic Period (1946–1963), (9) the Military Regime Period (1964–1985), and (10) the Political Opening Period (1986–present).

In that sense, we should understand that an education model, with its own characteristics, was instituted at each historical period. Depending on the sociopolitical conditions of that time, the new breakdown caused by the new model brought new requests, interests, and/or regressions along with it. This means that many actions that took place in a previous historical moment could not be accomplished properly or simply disappeared from a period to the other.

Even after the proclamation of the very wanted republic in 1889, education, in fact, had never been a priority of the governments that queued along this period. During the 20th century, the high growth of the population favored the rural exodus and the overpopulation of the cities, bringing consequences that reach our present days. Two of them are the amount of illiterate people that overpasses the rate of 10 percent of the population—not to mention the "functional" illiterates ones—, and a low and unequal level of schooling, which several government surveys show.

In this chapter about education in Brazil, we intend to focus on the organization, structure and functioning of the educational system, the main educational policies of the present, some aspects of the management of the systems and school units, the funding in education, and recent statistics data about this theme that are available from official sources such as the Ministry of Education (MEC) and its council of educational research, the INEP (Anísio Teixeira's National Institute of Pedagogical Studies).

ORGANIZATION OF BRAZILIAN EDUCATION: STRUCTURE AND FUNCTIONING

The first National Education Guiding Law—LDBEN (1961)—ruled all the educational activity in the country. Between 1968 and 1971, the military governments repealed a great deal of the LDBEN 4024/61 passing two federal laws, one for the university level (1968) and the other for primary and secondary schools (1971).

The second "Master Law" in education was passed in 1996, as part of the 1988 Constitution. It is known as Law 9394/1996, and has been in place for 17 years. Nowadays, all formal education in Brazil has this law as a reference. In that sense, when the congressmen made such a law they expressed an agreement on what education should be. The first article states: "Education is meant to be the formative process that takes place in the family environment, in the social relations at work, but also at the teaching and researching institutions, the social and cultural demonstrations, and the civil society organization" (*Brazilian Constitution*). This article reveals, therefore, the law's plural and varied understanding of the meaning of education. Moreover, the current LDBEN specifies, among others, the structure and the functioning of the Brazilian educational system, as we shall see next.

The LDBEN points out in its eighth article that the Union, the states and the counties will organize, on a collaborating basis, their specific teaching systems. In that perspective, in the section that deals with the organization of the Brazilian education, the law establishes the following teaching systems for Brazil:

Federal Educational System: Corresponds to the educational institutions supported by the state, the higher education institutions maintained by private initiative, and the country's normative and executive organizations (National Education Council and the MEC).

The States and the Federal District Educational Systems: Correspond to the educational institutions supported by the states' and counties' public authorities; the primary and secondary educational institutions created and supported by private initiative, the higher education institutions supported by the municipalities or districts, and the normative and executive organizations (National Education Council and the State Educational Bureau).

Educational Systems of the Municipalities: Correspond to the institutions that carry out the primary, secondary, and childhood education funded by the cities' public authorities or by private initiative, and by the normative and executive organizations (Educational City Council and the Municipal Educational Bureaus). Nevertheless, the integration between the District and States Educational Systems is optional. That results in a unified system of primary and secondary education.

In Brazil, private institutions are allowed to engage in educational activity, as stated by the Constitution and the LDBEN, as long as some requirements are met. They include following the general education regulations, and granting the public authorities the power to authorize their operation and evaluate their educational quality. The results achieved by private primary and secondary schools are usually more efficient than the one obtained by the public institutions. The same is not true for higher education. Public universities have a very good reputation. Monthly tuition fees at the private institutions vary from US$100 to US$3,000. The most expensive undergraduate programs, such as medical studies, are the ones that bring social prestige.

The LDBEN also establishes in its Article 21 that school education is formed by basic education and higher education. The first one comprehends three stages: childhood education, primary, and secondary. It is a subjective public right. That means that there may be legal consequences to the authorities, if it is not offered regularly. The second stage is composed of courses and programs such as undergraduate, graduate, and extension courses. Next, we will see some organizational and pedagogical aspects of the formal education.

Childhood Education

Childhood education is one of the changes from the 1988 Constitution, in the sense that the Brazilian state finally recognizes children's right to attend public school. It should be maintained by the counties, and is aimed at the complete development of children aged zero to five years, regarding their physical, psychological, intellectual, and social aspects.

Although the law dates from the 1980s, it was only two decades after that acknowledgment that the public authorities explicitly undertook the responsibility for its operation. That was because it was only in 2007 that the financial resources were guaranteed by FUNDEB (Basic Education Funds), as we shall see ahead in this text. From that moment on, new day care centers and preschool establishments were created and the number of enrollments at the public system for children at that age range grew considerably.

The first part of the basic education is formed by day care centers for children from zero to three years old, and preschools for children from four to five years old. Unlike what happens in the elementary and high school regimes, the institutions are under no obligation to fulfill the 800 hours or 200 school days. Students cannot fail at this level. Their assessment is done through the monitoring and recording of their development. Minimum qualifications for teachers are a *licenciatura* (undergraduate teaching degree) or a high school teaching degree. Data from 2012 published by INEP show that children education totals 7,295,512 children registered in day care centers and preschools in the whole country.

Elementary Education

Elementary education corresponds to the longest duration of the basic education period. It lasts nine years, and encompasses students ranging from 6 to 14 years old. Its objective is the basic formation of the individual, and there are different ways to divide its functioning: by cycles, years, series, semesters, non-serial groups, age, competence, or by any other forms that the learning process requires, as suggested by the LDBEN. In this case, the school shift is four hours long of full schoolwork.

The guiding principles of this intermediate level of basic education are autonomy, responsibility, and solidarity; respect for public welfare, rights, and duties; and critical consciousness, esthetic principles, sensitivity, creativity, diversity, and cultural and artistic expression. Data from 2012 published by INEP show that there are 29,702,498 children and adolescents enrolled in elementary schools in the country.

Secondary Education or High School

The final period of basic education is the high school level. It enrolls 15- to 17-year-old students, and it has an introductory focus toward either university education or preparation at the technical level for the workforce.

There is also a humanistic view that aims to prepare the youth to exercise their citizenship, and guidance to make sure that the pedagogical work will establish relationships between theory and practice. Besides, this level is composed of a common base that is formed by different areas of knowledge: languages, codes and their technologies; natural sciences; mathematics and its technologies; and human sciences and their technologies. The pedagogical principles are identity, diversity, autonomy, interdisciplinarity, and contextualization. Just like the primary school, the high school requires the mandatory number of 200 days of true pedagogical work with a four-hour-a-day workload.

Students in a Brazilian classroom. (Pcphotography69/Dreamstime. com)

There are specific guidelines for each stage of basic education. They control public policies and the elaboration, planning, execution, and evaluation of pedagogic and political proposals for each educational level. These guidelines always take into consideration the Curricular Guidelines for Basic Education elaborated by the Basic Education Chamber (CEB) of the National Education Council (CNE). Data from 2012 published by INEP show that there are 8,376,852 young people enrolled in high schools throughout the country at this moment.

University Education

In Brazil, the history of higher education started in the 19th century with the creation, by the Portuguese Empire, of isolated colleges in important cities of the time. Undergraduate medical and law schools, for example, were constructed in cities like Salvador, Olinda, São Paulo, and Rio de Janeiro. However, it was only in the following century that Brazilian universities appeared. The University of Rio de Janeiro was created under decree in 1920 to pay homage to King Albert of Belgium with the title of *Doctor Honoris Causa* during his stay in Brazil for the celebration of the centennial of Brazil's independence. The real very first Brazilian university, the University of São Paulo (USP), was founded and started its operations in 1934. Therefore, if compared to other Latin American nations such as Mexico, the

Dominican Republic, and Peru, the Brazilian university tradition is very recent and late, since universities were founded in these other Latin American countries in the 16th century.

According to the LDBEN, the goal of higher education is to form professionals of different areas of knowledge, and to promote the dissemination of cultural, scientific, and technical knowledge. Its aim is to enhance cultural creation, and to encourage teaching, scientific investigation, and extension through the development of scientific spirit and reflexive thought. Higher education in Brazil embraces undergraduate and graduate programs (master's and doctoral degrees), brush-up, and sequential and extension courses. According to the LDBEN, the sequential courses are courses of several scope levels, organized by knowledge fields, open to candidates who fit the requirements established by the institutions, as long as they have already finished secondary education or its equivalent. On the other hand, the extension courses are courses open to candidates who fit the requirements established in each case by the higher education institution, and reflect the university's commitment to social responsibility, playing its role in making the knowledge developed from its teaching and researching activities available for the community at large.

At the higher education level, the fulfillment of 200 days of school studies is mandatory as well. This period is usually divided into two semesters, excluding the final exams period. Every undergraduate program is ruled by a resolution and bylaws from the National Education Council. There are many different higher education institutions in Brazil, such as universities, colleges, educational centers, integrated colleges, and higher education institutes or schools. It is important to emphasize that at the Brazilian public universities the service is public and free of charge.

Currently, the selection of candidates to a place at a Brazilian higher education institution is achieved through two main ways: the *vestibular*, the more traditional entrance exam, and the performance on the ENEM (National Secondary Education Examination). ENEM is a yearly evaluation carried out by MEC (Ministry of Education and Culture) that measures the quality of secondary education. The results may be a "passport" to the entrance at a Brazilian public university through SISU. SISU is an online platform created by MEC in 2009 in which students who have got through ENEM run for a place at a public university, right after the publication of the exam results. In 2014, the SISU offered 171,401 vacancies in 4,723 undergraduate programs in 115 public institutions.

According to Article 52 of the LDBEN, the universities are multidisciplinary institutions whose aim is to form higher-level professionals in teaching, research, and extension programs. They are sites of excellence that foster human knowledge. These institutions are characterized by:

1. institutionalized intellectual production through the systematic study of the most relevant themes and problems, both from scientific and cultural, as well as regional and national points of view;
2. a third at least of the faculty composed of holders of master's and doctoral degrees;
3. a third of the academic staff must be full-time faculty.

Furthermore, the universities have autonomy to create, organize, and extinguish programs; to issue diplomas and confer degrees; to create curricula, programs, as well as research and extension projects; and, last but not least, to determine the number of openings.

At the Brazilian public universities, those interested in a career in higher education teaching by law have to be selected through a public competition in which they take exams and their academic titles are taken into consideration. Generally, the candidates take written tests in their fields of specialization, and their academic titles, experience, and techno-scientific production are evaluated. A committee of three professors who have doctoral degrees usually carries out the evaluation, in accordance with the public notice published by the institution. Nowadays a professor with a PhD degree in the beginning of his or her career at a federal university earns approximately US$4,000 a month.

One important contribution from the 2004–2014 decade concerning higher education is the approval of the law 12.711/2012, known as *Lei de Cotas* (Quotas Act). It grants a 50 percent reserve of enrollment at all public universities and educational or science and technology institutes for students who have completed their studies up to the secondary education level at regular public schools or at night shift schools. The other 50 percent of the vacancies are available competitively for all applying. It should be emphasized that in the 1993–2003 decade many public universities adopted affirmative action policies and guaranteed a percentage of their vacancies for students from public schools and/or Afro-Brazilians or Amerindians. Bahia's State University (Uneb), Brasília's State University (UnB), Rio de Janeiro's State University (Uerj), and Bahia's Federal University (Ufba) were the first institutions to implement those policies in Brazil. It is important to point out that before the implementation of those affirmative action policies, the great majority of the openings for the most prestigious undergraduate programs were filled by students from the most respected—and expensive—private schools.

Data from 2012 published by INEP show that there are 6,739,689 people enrolled in higher education in Brazil. Among these, approximately 85 percent attend presential courses and 15 percent prefer the non-presential method. It is important to point out that a little more than 14 percent of the population between 18 and 24 years of age are enrolled in an undergraduate program.

Education Programs

Besides the traditional education according to the age range of the students, the Brazilian educational system established "educational programs" that take into consideration the peculiarities and diversity of the public. Next, we will show the characteristics of the main official programs existing in Brazil now.

Education for the youth and adults is an educational mode supported by the LDBEN. It caters specially to those individuals who did not finish their primary or secondary studies at proper age. In order to apply for a place at the primary education, the minimum age is 14; 18 is the minimum age for the secondary or high school program. There is no maximum age. City councils and states have tried to develop educational policies to favor this social group with financial support from the government. Data from 2012 published by INEP show that there are 3,906,877 young adults and adults enrolled in this program in all the country.

Professional training is supported by the LDBEN and by several decrees that establish the principles, guidelines, and operationalization of professional qualification at the basic, technical, and technological levels—Decree 2.208 from 1997; Law 11.741/2008. Since 2005, there has been considerable expansion of technological education in the whole country, in benefit of, especially, the medium-sized towns of the country side. In the last few years, Brazil has experienced a considerable economic and social development. That demands professionals that are more qualified. With that goal, besides the higher education institutions, many non-university institutions, such as federal technology institutes, have produced a remarkable number of people for the job market. Data from 2012 published by INEP show that there are 1,063,635 students enrolled in this program at schools around the country.

Special education is an educational mode that is supported by the Constitution and by the LDBEN whose goal is to provide education for people with disabilities in regular groups or classrooms at schools designed for this kind of public. Currently, one of the neuralgic points in this matter is the inclusion of such individuals in the regular educational network, especially due to the lack of physical and operational structure of the schools and teachers' poor qualification. Even at some of the highly qualified private schools of the country, there is record of problems in the proper service for this public. Data from 2012 published by INEP show that there are 820,344 enrollments in special education programs in schools throughout the country. Among these, 199,655 are enrolled in special classes, and 620,777 are members of regular or inclusive classes.

Supported by the LDBEN, the *remote education or remote learning* is a mode that has grown rapidly and consistently in the past few years. Enrollment in undergraduate courses, in particular, now correspond to about 15 percent of the number of students registered in higher education in the country. *Remote learning* is mediated by different kinds of communication and information technology in which students and teachers are separated in time and/or places, whose curriculum contents are specially designed to meet the features of the group. It involves, above all, self-learning, dedication, and discipline.

Educational Policies in Brazil

Educational policies are created by the state, just like any other public policy. They are the result of choices and decisions that involve individuals, groups, institutions, and interests. In the past few years, a series of educational policies aimed at the basic and higher education levels have been implemented with the cooperation of federal, state, and city governments, as we shall see next. It is important to emphasize that some of these policies are implemented because of the pressure of social movements on the public authorities.

According to the Constitution of 1988, states and cities must spend at least 25 percent of their budget on education, and the federal government at least 18 percent.

Important Policies

Educational Development Plan—PDE

This is a medium to long-term collective plan designed to improve the quality of education in the country. In fact, it embraces several programs related to:

Basic Education: teacher education and national minimum wages; education salary and FUNDEB, IDEB (Basic Education Development Evaluation), and *Gestão Educacional* (Education Management).

Higher Education: REUNI (Federal Universities Restructuring and Expansion Plan), PNAES (National Plan for Student Assistance), PROUNI (University for All Program), FIES (Student Financing Fund), and SINAES (National System for the Evaluation of Higher Education).

Technological and Professional Education: Federal Institutes of Technological Education and Professional Education for Young Adults and Adults.

Plano de Metas Compromisso Todos pela Educação ("Everyone Committed to Education" Plan of Goals)

This is one of PDE's main projects. Its goal is to mobilize society to defend the quality of basic education in Brazil. Its long-term perspective is to improve education indicators. Its foundation is the collaborative regimen among the Union, states, Federal District, and the municipalities.

Articulated Actions Plan (PAR)

This is a multiannual plan of each municipality that adhered to the "Everyone Committed to Education" Plan of Goals. This plan anticipates agreements between the municipalities and the MEC, particularly in relation to technical and economic support from MEC/FNDE. It aims at enforcing the 28 directives of the Decree number 6094 from April 24, 2007, Article 2, and at gradually achieving the goals proposed by IDEB. Its foundation is the collaborative regimen between the municipalities and the states, counting with the local population social control (Local Committee for Commitment), Article 2, Directive XXVIII, of Decree 6094, from April 24, 2007. PAR elaborates the evaluation of the city's education on a participative basis, predicting educational actions in detail, the results expected and ways to accomplish them, a calendar of activities, and the people responsible for the actions.

BASIC EDUCATION-SPECIFIC PROGRAMS SUPPORTED BY THE MINISTRY OF EDUCATION

Pro-Infancy— National Program for Restructuring and Providing for the Childhood Education Public School Network: According to MEC, this is an economic assistance program for the Federal District and the municipalities to build, renovate, and acquire equipment and furniture for childhood education public day care centers and preschools. Its aim is to ensure that children have access to public day care centers and childhood education schools, especially in metropolitan areas, where more children live.

"More Education"—Full Time School: This program's main goal is to increase the school shift up to a minimum of 7 hours daily or 35 hours weekly. It aims at restructuring the curriculum toward integrative education. In this case, public city, state, and federal schools adhere to this program, respecting its specific requirements.

PDE School: It is a program based on participative planning that addresses school management. Its objective is to help public schools to improve their efficiency in management. In this case, MEC usually gives priority to low IDEB marks schools, funding them in order to support and carry out their planning.

"Way to School" Program: Supported by the National Educational Development Fund (FNDE), its goals are to renew and expand basic education students' school transportation services. Plans include the standardization of the vehicles for better security for the students, and the reduction of the costs of acquisition of the buses.

Educational Technologies Guide: This guide brings information on prequalified technologies that pair off with MEC's developed technologies that help school principals and teachers to learn about and identify techniques that can contribute to improve education in their educational network.

"Accessible School" Program: This initiative gathers resources toward improving access to the schools, teaching and pedagogical resources, and communication and information in regular public schools. The funds comes from the Money Direct to School Program (PDDE) from the MEC, which finances the construction of ramps, access aisles, adapted restrooms, and the acquisition of wheelchairs, assistive technology resources, water fountains, and accessible furniture.

NATIONAL EDUCATION PLAN (PNE)

Foreseen by Article 214 of the Federal Constitution, the PNE will last for 10 years. Its aim is to articulate the education national system to collaborate with the federalized units in order to define directives and objectives in order to implement strategies to assure the maintenance and development of instruction in its several levels, steps, and modalities. Through integrated actions of the public authorities from different federal levels, this plan includes these goals:

(1) illiteracy eradication

(2) universalization of school availability

(3) improvement of teaching quality

(4) professional training

(5) humanistic, scientific, and technological development in the country

(6) creating regulations for the use of public funds in education tied to a proportion of the GDP (gross domestic product).

The MEC is responsible for the elaboration of the bill about the National Education Plan. They forward it to the National Congress for discussion and approval. Then, the bill is sent for sanction by the president. After more than three years at the Congress, the PNE has recently been sanctioned by the president. One of the points considered very delicate is the one which allocates 10 percent of the GDP for public education. The federal government has indicated its approval of this percentage, and intends to use the profits from the exploration of the *Pré-sal* (Pre-salt) oil fields on the Brazilian coast to finance it. There is great expectation related to the president's sanction of the PNE because it establishes important goals. There are 12 articles and 20 aims in a whole for the Brazilian educational field in the next decade in the present PNE structure.

DEMOCRATIC SCHOOL MANAGEMENT IN BRAZILIAN EDUCATION

The idea of a democratic school management started to receive support in Brazil in the 1980s. The 1988 Constitution made this aspiration of important defenders real. It was ratified by the LDBEN in 1996.

Although it has been recurrent in the school system to claim the adoption of a participative administration, the implementation of this model is far from being a generalized or even a dominant practice. The essential elements for a participative school management are the Political Pedagogic Project and the introduction of councils to stimulate school units' autonomy.

These are the ways through which school principals are selected at the Brazilian public school units:

Designation: the person named is usually related to some political group in power.

Public Competition: the person named is approved by a public competition that is specific for that position of school principal. This is not a very common way of selecting principals yet.

Direct Elections: supported by the present legislation, this is the modality recommended by the democratic school management model. The person named is elected with the votes of the majority of the stakeholders in that school's educational process: teachers, workers, servants, students, and parents. States' educational systems have their own guidelines for the hiring process of principals.

Compound System of Direct Elections Plus Public Competition: in this case, only the teachers who passed the exams where they demonstrated their technical competence can run for the position of principal. Apparently, that would be the most suitable model of selection.

Principles of Participative School Management and Organization

Among the principles of participative school management, we can highlight five: the autonomy of the school and the school community; the existence of an organic relation between the administration and the participation of school members; the community involvement in the educational process; task planning; and, finally, shared assessment.

BASIC EDUCATION FINANCING: FNDE AND FUNDEB

National Educational Development Fund (FNDE)

The FNDE is a federal autarky affiliated with the MEC whose main task is to provide economic and technical support, and perform actions that will contribute to a high-quality education for all.

The amount of money managed by the fund comes in its major part from the "education wage"—a social contribution of 2.5 percent that is charged from the companies. The FNDE's funds go to those programs that involve the purchase of school meal, textbooks, books in Braille, school transportation, and investment in libraries and in the infrastructure of public schools in the whole country.

Basic Education Development Fund (FUNDEB)

FUNDEB is a special accounting fund destined for the states (26 states and the capital, Brasília) consisting, in its majority, of funds from taxes and bank transfers from the states, the Federal District, and municipalities, that are allocated to education by Article 212 of the Constitution. The amount reserved for each state/municipality depends on the tax collection and the number of enrollments in its network. MEC specifies a minimum amount per student each year by. In 2012, that amount was R\$2,096.68 (approximately US\$800) for each student attending basic education. The federal government subsidizes a part of the sum to be paid for those states unable to collect the minimum amount per student.

The legal basis for the FUNDEB is the Law 11.494/2007, which is valid until 2020. In each state, the FUNDEB is composed by percentages from several funds and taxes. The states have invested the funds that they have received mainly in the elementary and secondary education, whereas the municipalities have given priority to basic and childhood education. It is important to emphasize that FUNDEB has guaranteed resources for the three steps of basic education, as well as for young adults and adults' education (EJA). That differs from FUNDEF, a government initiative from the LDBEN/1996 that guaranteed funds only for basic education.

The law is very clear regarding the application of the resources: at least 60 percent of these annual resources from the funds will be destined to the payment of the salaries of basic education teachers working at public schools and the remaining 40 percent go to educational development and maintenance. According to INEP, the share for education on the GDP in 2011 reached 5.3 percent.

ASSESSMENT OF BASIC EDUCATION ACCORDING TO MEC

The IDEB (Basic Education Development Index) is Brazil's basic education grade. It has contributed a lot to measure the quality of education in the country. It was created in 2007 by INEP, as an integrating part of the PDE (Educational Development Plan). IDEB's calculations take into consideration the school attainment rate (passes and dropouts) and students' performance on SAEB (National System of Basic Education Evaluation) and on *Prova Brasil* (Brazil Exam), another assessment tool for the quality of education in the country.

SAEB: It is an evaluation that collects samples from students enrolled at the fifth and ninth grades of basic education, as well as from those at the third grade of high school (the last grade) from the country's public and private schools located in rural and urban areas. The results are shared with each state, region, and the country as a whole.

Prova Brasil (Brazil Exam): It is an evaluation that is applied to students at the fifth and ninth grades of public basic education, in the state, municipal, and federal networks, in rural and urban areas, at schools where at least 20 students are enrolled in those grades. This exam offers results that are used for IDEB's calculations. INEP has recently released data about public schools' performance at IDEB in 2012. These numbers show that more than half of the Brazilian states have reached the goals established for the basic education initial and last grades. On the other hand, less than half of these states have reached the goal established for secondary school.

Higher Education Expansion and Evaluation

As an acknowledgment of the strategic role of higher education institutions, and of the need to expand the access to this level of education, the growth in the number of student offerings both in the public and private spheres was remarkable in the 2004–2014 decade. The federal government surpassed the goals it had set for the public universities it administers for 2010. For example, they created 16 new universities, opened 100 new university campi, and the federal universities group has reached 188 towns, according to information from MEC. In 2013, the creation of four new federal universities was authorized—two of them in Bahia, one in Ceará, and one in Pará.

REUNI

The Support for Federal Universities Restructuring and Expansion Plan (REUNI) is a program whose objective is to increase the number of student offerings, to enforce students' continuity at the institution until their graduation, and to promote a better use of the facilities and the human resources (personnel). All the 53 existing universities participated in the program when it was launched.

The main goals REUNI sets for the universities to achieve are:

- to increase the graduation rate up to 90 percent;
- to increase the professor-student rate to 18 at the undergraduate courses;
- to improve the number of enrollments to 20 percent in undergraduate programs.

SINAES—Higher Education National Evaluation System

Created by the Law 10.861, from April 14, 2004, the SINAES has three main tasks: evaluation of the institutions, the programs of study, and the students' performance. SINAES evaluates all aspects related to these three missions: the teaching, research, social responsibility, students' performance, institutional management, the faculty, facilities, and several other aspects. This assessment program's goals are to improve the quality of higher education, to administer the expansion of student offerings, and to increase, permanently, its institutional efficiency and academic and social effectiveness.

MEC's assessment teams often visit the higher education institutions to check in loco all information they receive from them. If irregularities are found, these institutions may suffer a series of penalties, such as reduction of student offerings for certain programs of study, extinction of programs, and cancellation of the entrance exam. It is important to highlight that students are evaluated through the ENADE (Students' Performance National Exam), whose objective is to assess students' performance in their undergraduate courses, in relation to the curriculum content, their abilities, and competences. Nevertheless, students often boycott the exam, which prevents a fair final evaluation of some higher education institutions. According to the present legislation, only the public federal universities and the private ones must administer the exam. State and municipal universities are not under such obligation.

The branch of MEC that is responsible for advising and evaluating *stricto sensu* graduate programs (MAs and PhDs) is CAPES (Coordination for the Improvement of Higher Education Personnel), whose responsibilities include promoting the dissemination of qualified scientific production, investing in the formation of high-level resources in Brazil and abroad, and, last but not least, fostering international scientific cooperation.

At present, there are 5,082 graduate programs in the country, in different areas of knowledge. Among these, 2,893 offer master's degrees, 1,792 are doctoral, and 397 are professional master's programs. Every year CAPES distributes thousands of scholarships for graduate students whose GPA in their programs is equal to or above four. CAPES also encourages doctoral students to carry out part of their formation abroad, providing scholarships for a program that they call "sandwich doctoral program." Another government agency that promotes and funds research is CNPQ (National Council of Technological and Scientific Development), which is affiliated with the Ministry of Science, Technology and Innovation (MCTI). Its mission is to develop and play a role in the formulation of policies related to science, technology, and innovation, and, by doing so, contribute to foster knowledge, sustainable development, and national sovereignty.

CAPES and CNPQ manage at the moment a great institutional program named Ciência sem Fronteiras (Science Mobility Program). As stated on its official website, the program "seeks to promote the consolidation, expansion and internationalization of science and technology, and to encourage Brazilian innovation and competitiveness through cultural exchange and international mobility." The project intends to offer 101,000 scholarships within four years to promote student exchange, so that undergraduate and graduate students will be able to get training abroad in order to interact with competitive educational systems in regards to technology and innovation.

It is not a difficult task to draw up a picture of the Brazilian education. It is hard, though, to realize that Brazilians still have so many challenges to overcome in the beginning of the 21st century. They include illiteracy eradication; making basic and secondary education fully available; increasing the number of student offerings for higher education institutions (especially for students between 18 and 24 years of age, since at the moment only about 15 percent of that group are enrolled in an undergraduate program); and raising the government expenditure with education, with stronger efforts to avoid frauds and corruption with public money. It is also necessary to improve the quality of the education by implementing frequent professional training for teachers and school administrators; to revitalize the schools' infrastructure and to improve working conditions; and to promote a general revision of the educational legislation in order to define clearly what in fact is quality in public education, and what the necessary penalties or sanctions for negligent administrators will be.

Civil society and public authorities must combine forces and search for solutions for the chronic problems addressed here in the decades to come. If not, Brazilians may witness disastrous consequences—feelings of anti-citizenship and violence may emanate from their extremely unequal society.

REFERENCES

Brasil.1988. *Constituição Federal.*

Brasil. 1996. *Lei de Diretrizes e Bases da Educação Nacional*, no. 9.394.

Bello, José Luiz de Paiva. 2001a. Educação no Brasil: a História das rupturas. http://hid0141. blogspot.com.br/2011/07/educacao-no-brasil-historia-das.html Pedagogia em Foco. Rio de Janeiro. Disponível em: http://www.pedagogiaemfoco.pro.br/heb14.htm. Acesso em: 27/01/2012.

Bello, José Luiz de Paiva. 2001b. Educação no Brasil: a História das rupturas. Pedagogia em Foco. Rio de Janeiro. ATUALIZADA, pp. 1–13.

Libâneo, José C. 2004.*Organização e Gestão da Escola: Teoria e Prática.* Goiânia: Alternativa.

Libaneo, J. C, João Ferreira Oliveira, Mirza Seabra Toschi. 2009. *Educação Escolar: políticas, estrutura e organização.* São Paulo: Cortez Editora.

Meneses, J. G., and Anita Favaro. 2002. *Estrutura e funcionamento da Educação Básica.* Pioneira Editora.

Web Sites

http://download.inep.gov.br/educacao_basica/censo_escolar/resumos_tecnicos/resumo_tecnico_censo_educacao_basica_2011.pdf.

http://download.inep.gov.br/educacao_superior/censo_superior/resumo_tecnico/resumo_tecnico_censo_educacao_superior_2011.pdf.

http://download.inep.gov.br/download/superior/censo/2011/resumo_tecnico_censo_educacao_superior_2011.pdf (ACESSADO EM 30/08/2015), http://www.capes.gov.br/.

http://www.cdes.gov.br/observatoriodaequidade/relatorio2.htm.

http://www.cienciasemfronteiras.gov.br/.

http://www.cnpq.br/.

http://www1.folha.uol.com.br/mercado/2013/06/1303780-aprendendo-a-gastar.shtml.

www.ibge.gov.br.

www.inep.gov.br.

www.portal.mec.gov.br.

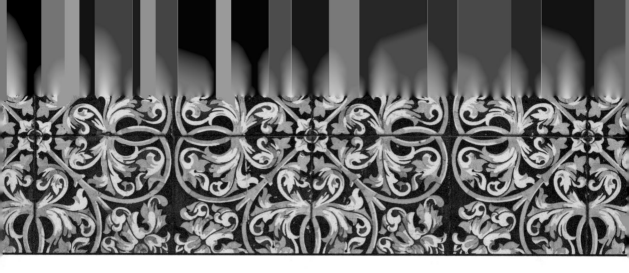

Culture

Language

Helade Scutti Santos

BRAZIL: A MULTILINGUAL COUNTRY

Brazil is the largest Portuguese-speaking country in the world, and Portuguese is a language spoken by almost the totality of its population. This created a general belief that Brazil is a monolingual country, which is not true. Portuguese is not the only language spoken in the country, and it is not the mother tongue of hundreds of thousands of Brazilians. In fact, Brazil is one of the most multilingual countries of the world, and it is the South American country with the greatest linguistic density and diversity.

INDIGENOUS LANGUAGES

In the Brazilian territory, in addition to Portuguese, there are 274 indigenous languages, as of the 2010 Brazilian Census. This may sound like a lot of languages, and it could make people believe that a great number of Brazilians are multilingual or speak at least one indigenous language. Unfortunately, the majority of the Brazilian population ignores the diversity of indigenous groups and indigenous languages present in Brazil and most of these languages are endangered because of their small number of speakers (around 200/250 speakers per language). Outside of the indigenous lands, which are areas of the Brazilian territory inhabited and exclusively possessed by indigenous people, the situation is even more dramatic: 47 percent of the indigenous languages have no more than 10 speakers. Only 24 indigenous languages have more than 1,000 speakers. In addition, many of these languages do not have a writing system.

There is great diversity among the indigenous languages of Brazil. They belong to two big different language families: Tupi and Macro-Jê. Each one of them is subdivided into branches with their respective languages, as shown in Figures 6.1 and 6.2. There are also isolated languages or groups of languages. Within the Tupi family, we will find the branch with the bigger number of languages: Tupi-Guaraní, which includes Guaraní. This language is not only spoken in Brazil but also spoken in neighboring countries, including parts of northeastern Argentina and southeastern Bolivia, and is one of the official languages of Paraguay. Guaraní is also one of the official languages of Mercosur, along with Portuguese and Spanish.

Although most of the nonindigenous population of Brazil is not multilingual, indigenous individuals are frequently bilinguals or multilinguals. Within the same group or village it is possible to hear more than one language—including Portuguese. Multilingualism is not new for indigenous populations. The cultural and social contact among different ethnic and linguistic groups has been quite frequent. Despite

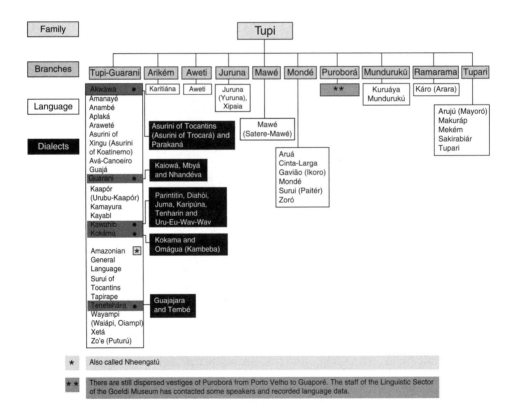

FIGURE 6.1   *Languages and branches that belong to the Tupi family*

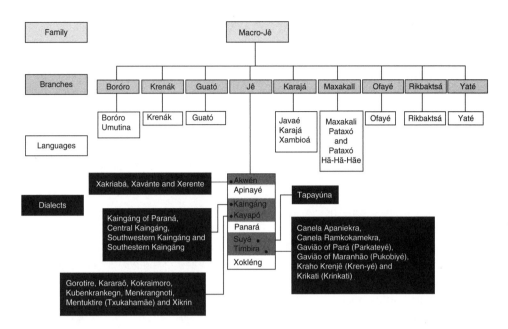

FIGURE 6.2   *Languages and branches that belong to the Macro-Jê family*

the linguistic differences among the 22 ethnic groups that inhabit the Rio Negro "Black River" area, there is intense cultural and social exchange among them. As part of their tradition, the man has to speak his father's language but must marry a woman who speaks a different language. Therefore, individuals who belong to these groups are essentially multilinguals, and the men usually speak between three and five languages. Although many of the indigenous individuals are polyglots, the number of indigenous groups (305 ethnic groups) is bigger than the number of languages. Some indigenous groups have lost their original language and speak Portuguese or have adopted the language of their neighbors. The 2010 Brazilian Census shows that only 37 percent of the indigenous individuals speak an indigenous language.

Of all indigenous languages found in Brazil, only 15 are satisfactorily described and documented, whereas 35 are completely ignored. The rest of them are only partially studied. Due to the situation of these endangered languages, linguists have a crucial role to play. Because languages are considered important repositories of cultural heritage and collective memories, when a language disappears, it is not only part of the tradition and the identity of a specific group that disappears with it. Humanity loses part of its intellectual patrimony as well. In addition, when a language becomes extinct, science loses important evidence that can help scientists to build a better understanding of the nature of human language as well as our past and history. This is why to describe, document, and preserve the indigenous languages present in the Brazilian territory is an essential and urgent endeavor. Some initiatives that point to this direction are the Documentation of Brazilian Indigenous Languages and Cultures (http://prodoclin.museudoindio.gov.br/), supported by UNESCO, and the use of indigenous languages in indigenous schools, either as languages of instruction or as part of the curricular program.

IMMIGRANT LANGUAGES

In the second half of the 19th century, Brazil started to receive substantial immigration of non-Portuguese-speaking people from Europe and Asia. This massive immigration was due to a combination of factors: Europe was going through a huge economic crisis at that time and Brazilian landowners needed workers for the expanding Brazilian agriculture after slavery was abolished. Millions of Europeans, most of them Italians, migrated to Brazil. They went particularly to the Southern and Southeastern regions of Brazil. Many of these immigrants lived in rural communities where they have kept their home language long enough to develop distinctive dialects from their original European sources. Two examples are *Riograndenser Hunsrückisch*, which is the main German dialect spoken in the south of Brazil, and *Talian* or *Vêneto* (*brasileiro*), based on the Venetian language and spoken by descendants of Italian immigrants. In some communities in the south of Brazil, such as Legado Antunes in the state of Santa Catarina, *Riograndenser Hunsrückisch* is still the preferred language of communication with family members and neighbors. On the other hand, most Brazilians with Italian ancestry speak Portuguese as their native language.

Brazil is home to the largest Japanese population outside of Japan, concentrated especially in the states of São Paulo and Paraná. Japanese immigrants began arriving in 1908, as a result of the decrease in the Italian immigration to Brazil and a new labor shortage on the coffee plantations of the Southeastern region. A great number of the Japanese immigrants went to live in *Liberdade* (the Portuguese word for "freedom"), São Paulo's own equivalent of Japantown in the United States. Although Japanese and other Asian descendants are all over São Paulo today, *Liberdade* is still a meeting spot for many groups, especially among young people who are interested in Japanese culture. It also became a popular tourist destination in the city for national and international visitors. *Liberdade* is one of the places in Brazil where you can easily hear people talking in Japanese. Japanese spoken in Brazil is usually a mix of different Japanese dialects influenced by Portuguese.

For a long time these immigrant languages were stigmatized and prohibited by the Brazilian government, especially during World War I and World War II. Today, however, immigrant languages such as German and Italian are being reintroduced into the curriculum in states of the South region like Rio Grande do Sul, where they originally thrived. Furthermore, because a high number of second- and third-generation Japanese Brazilians who went to work in Japan in the 1980s and 1990s are returning, the number of Japanese speakers in Brazil may increase.

HISTORY OF BRAZILIAN PORTUGUESE

When the Europeans arrived on the Brazilian coast in 1500, there was between 1 and 6 million people, speaking around 1,000 languages, in the current Brazilian territory. Within the past centuries 85 percent of these languages disappeared and Portuguese was little by little becoming the dominant language. In the first centuries of the colonial system, the Portuguese colonizers were minority when compared to the indigenous population and they started to learn the language of the *Tupinambá* (general name for the groups that inhabit the Brazilian coast). This language was later known as *língua geral*, "General Language," and it was informally and domestically spoken by a great part of the population of the Brazilian colonial system: Portuguese settlers, Amerindians from different groups, and African slaves. Because colonizers were mostly males, they started to have children with indigenous females, and *língua geral* was the mother tongue of this new generation. The Jesuits, in addition to composing religious hymns and plays in *língua geral*, studied, normatively described, and taught this language with the goal to convert the non-Europeans to Christianity. This is how *língua geral*, along with Portuguese, became part of the colonial system. Nevertheless, toward the end of the 18th century, Portuguese started to become dominant over *língua geral*. In 1758, the Portuguese Crown, under the administration of Marquis of Pombal, prohibited the use of the *língua geral* or any other indigenous language. In 1808 the Portuguese royal family came to Brazil and brought 16,000 Portuguese with them. With the increase in Portuguese immigration and after *língua geral* was prohibited, Portuguese started to be imposed as the national language, until it became the language spoken by the overwhelming majority of the population as it is today.

In the 19th century, the *língua geral* spoken in the north of Brazil started to be called Nheengatu. Nheengatu is still spoken in the Amazônia as the native language of the rural population of the Upper Rio Negro (Black River) region of Amazonas state. It is also used as a common language of communication between indigenous and nonindigenous individuals, as well as between indigenous individuals from different ethnic groups. Nheengatu is also a way for some indigenous groups that have lost their original language, such as the Baré and the Arapaço, to affirm their ethnic identity.

The contact with indigenous languages as well as with African languages brought by the slaves influenced the evolution of Brazilian Portuguese. This influence is more clearly detected—and less controversial among linguists—on the vocabulary of Brazilian Portuguese. Words derived from the Tupi-Guaraní family are particularly prevalent in place names (*Maracanã, Itú, Butantã, Caruaru*), animals (*colibrí* "hummingbird," *arara* "macaw," *jacaré* "alligator," *tatu* "armadillo"), vegetables (*mandioca* "manioc"), and fruits (*maracujá* "passion fruit," *abacaxi* "pineapple," *pitanga* "Brazilian berry"). Brazil, which has more people of African descent than any other country outside of Africa itself, received a large number of African slaves originated from Bantu communities. Linguists estimate that approximately 300 words from Bantu languages were incorporated into the vocabulary of Brazilian Portuguese. Some examples can be found in Table 6.1. The Bantu languages may have also influenced some features of the Brazilian Portuguese pronunciation, but this needs to be further investigated.

Although Brazilian Portuguese borrowed many words from indigenous and African languages, the influence of these languages on the phonology and grammar of Brazilian Portuguese has been very minor. Most linguists who study the Linguistic History of Brazilian Portuguese agree that many of the Brazilian Portuguese grammatical features can be traced back to the European Portuguese of the 15th century.

Why European and Brazilian Portuguese took different paths if both varieties derive from the European Portuguese spoken in the 15th century is still an unanswered

TABLE 6.1   Words in Portuguese from Bantu Languages and Their English Translations

Words in Portuguese from Bantu Languages	English Translation
bagunça	Mess, disorder
bunda	Butt, buttocks
cachaça	Sugarcane brandy
caçula	Youngest child of a family
quiabo	Okra
moleque	Young boy
orixá	Deity, god, or goddess
samba	Brazilian musical genre and dance style
xingar	Curse, insult

question. Some linguists claim that the European Portuguese was the one that changed while the Brazilian Portuguese stayed very similar to the Portuguese that was brought to South America by the colonizers in the 15th century. Some other researchers find that the separation between European Portuguese and Brazilian Portuguese happened mainly in the 19th century, when many grammatical features not attested in European Portuguese start to become part of the Brazilian Portuguese grammar.

In summary, the contemporary version of the Portuguese spoken in Brazil is a descendent of European Portuguese, but its actual shape and features derive from a complex interaction between the language of the colonizer and:

- the numerous indigenous languages that could be heard in the Brazilian territory before the Portuguese arrived;
- the also numerous African languages brought during the period of the slave trade (officially from 1549 to 1830);
- the languages of European and Asian immigrants who arrived in Brazil in the second half of the 19th century.

To this complex linguistic interaction it is also necessary to add some demographic and social information that help us understand how the contact between Portuguese and these languages took place and how Portuguese was transmitted to the non-lusophone population. Not only was Brazil widely multilingual until the second half of the 18th century, but until the first half of the 19th century only 30 percent of the population had Portuguese ancestry, whereas the other 70 percent were mostly afro descendants and Indians with non-Portuguese family languages. This means that the majority of the population learned Portuguese as a nonnative language and possibly imperfectly. Furthermore, until 1890, Brazil had 85 percent illiterates, which means that the Portuguese spoken in Brazil was not standardized and was almost never learned through schooling. This sociolinguistic situation is responsible not only for the distance between the European and the Brazilian Portuguese but also for the heterogeneity of Brazilian Portuguese.

PORTUGUESE AROUND THE WORLD

Portuguese is the sixth most spoken language in the world, with approximately 240 million speakers, including native and nonnative speakers. The majority of them, 190,732,694 (see www.ibge.gov.br), were born in Brazil. Portuguese is the official language of Brazil as well as of seven other countries: Portugal, Mozambique, Angola, Cape Verde, Guinea-Bissau, São Tomé and Príncipe, and East Timor. Portuguese is also spoken in Macau and Equatorial Guinea, and in the cities of Goa and Daman and Diu in India. There are also many immigrant Portuguese-speaking communities in different countries of the world, including the United States. According to the 2007 American Community Survey conducted by the U.S. Census Bureau, Portuguese is the 11th most spoken non-English language in the United States. The regions of the country that host the majority of the Portuguese-speaking immigrants are Massachusetts, New Jersey, Florida, and California. Massachusetts concentrates by far the higher number of Portuguese speakers of the country, with 31 percent of all the U.S. Portuguese speakers.

Brazil is the largest Portuguese-speaking country in the world, and it also contributes to the majority of the Portuguese-speaking communities in the United States. Although the first Brazilian migratory wave to the United States took place in the mid-1980s, the Brazilian government claims that there are approximately 1 million Brazilians living in the United States.

THE DIFFERENT PORTUGUESE LANGUAGES

Although the same language is spoken in Brazil and Portugal, Brazilians have difficulty in understanding the language spoken in Portugal when they first hear a European Portuguese speaker. This is mainly due to differences in pronunciation between these two varieties of Portuguese, which creates very strange combination of sounds to the ears of a Brazilian Portuguese speaker. Furthermore, Brazilians in general have had very little exposure to European speech and are quite unaware of the differences in pronunciation and vocabulary. Thus, it is necessary for a Brazilian who talks to a Portuguese to "train his or her ears" before he or she starts to understand his or her interlocutor and to clarify possible misunderstandings that may be caused by different words or different meanings of shared words. European Portuguese can be almost unintelligible to Brazilian Portuguese speakers in some contexts, and this is why it is not unusual to find subtitles in Portuguese movies when they are shown in Brazilian theaters.

The written formal language, the one that is taught in Portuguese and Brazilian schools, is more similar because prescriptive grammar and style manuals usually follow the standard of Portugal. Yet there are differences in vocabulary, grammar, and the spelling of certain words. Differences in spelling of Portuguese words are minor and similar to how the spelling of certain words in the United States differs from the spelling used in other English-speaking countries. However, when it comes to the spoken or colloquial varieties of Portuguese, the number of differences multiplies. Because the constraints that apply to written language do not affect the spoken language in the same way, some linguists claim that the grammar of the spoken variety of

Brazilian Portuguese has enough specific features, no longer shared with European Portuguese, to allow a description of these two varieties of the language as two different linguistic systems. Some grammatical features of Brazilian Portuguese that cannot be found in Portugal are:

- replacing the verb *haver* with the verb *ter* "have" when it means "there is/are" (*Tem um gato embaixo da mesa.* "There is a cat under the table");
- beginning sentences with object pronouns (*Me faz um favor.* "Do me a favor");
- ellipsis of the object pronoun (*Faz tempo que este filme está em cartaz mas ainda não fui ver ø.* "This movie has been in theaters for a while but I haven't seen it yet").

Differences are also present in the vocabulary and a few examples can be found in Table 6.2.

When comparing Brazilian Portuguese and the varieties of Portuguese spoken in Africa, there are significant differences as well. Mozambican and Angolan Portuguese are emerging varieties of the language, and the linguistic norm is still based on that of Portugal. However, there are similarities in the pronunciation of some sounds between Brazilian Portuguese and Angolan Portuguese and between Brazilian Portuguese and Mozambican Portuguese. Due to the influence of local languages, there are differences in vocabulary and expressions used in each of the Portuguese-speaking countries.

BRAZILIAN PORTUGUESE DIVERSITY

Portuguese, as any other natural language, presents variation depending on different factors. As mentioned in the previous section, Portuguese and Brazilians do not speak in the same way. This means that the place of origin of the speaker determines his or her accent, the vocabulary and also some of the constructions he or she uses. However, geographical variation is not only noticeable when different Portuguese-speaking countries are considered. Brazilians from different regions have different accents. These varieties of the language are called dialects. For a long time it was believed that Brazilians from the south to the north speak alike, with very few differences in pronunciation. More recent research has shown that there is more variation in the Brazilian dialects of Portuguese than previously thought, but the variation is more due to socioeconomic factors and level of education than to geographical distance. Therefore, it is possible to find more variation between two speakers from the

TABLE 6.2 Sample of lexical differences between European and Brazilian Portuguese

European Portuguese	Brazilian Portuguese	Translation to English
casa de banho	*banheiro*	Bathroom
pequeno almoço	*café da manhã*	Breakfast
bairro de lata	*favela*	Slum
bicha	*fila*	Line
autocarro	*ônibus*	Bus

same region, but one of them being illiterate and the other one having a college degree than between two educated speakers who come from different regions of the country.

According to linguists, there are at least two quite different varieties of Portuguese in Brazil: Standard Brazilian Portuguese, which is the variety used in written or formal oral contexts by educated people and learned through schooling, and Vernacular Brazilian Portuguese, which is quite heterogeneous and is most frequently used in informal contexts by educated as well as uneducated people. This means that most of the Brazilian population must learn to read and write in a dialect that they neither speak nor fully understand.

Although educated Brazilians end up becoming "bilinguals" in their own language and move back and forth between the standard and the vernacular varieties of Brazilian Portuguese, they believe that Brazilian Vernacular—or at least many rules typical to this variety—is nothing else but a corrupted form of the language, a bad Portuguese. Pointing to a different direction, many linguists and educators have been striving to describe the grammar of Brazilian Vernacular and to demonstrate that this is not a simplified or poorer version of European Portuguese or Standard Brazilian Portuguese. This is rather a different dialect—or set of dialects—with its own grammatical properties and complexities.

The work of linguistics in comparing Standard and Vernacular Brazilian Portuguese, as well as conducting more studies on the origin and historic development of the features that characterize the Brazilian Vernacular, is crucial to inform pedagogical practices of teaching Portuguese as a native language. Schoolteachers must have a good knowledge of the standard and the vernacular dialects and have a prejudice-free attitude toward both of them. On the one hand, Standard Brazilian Portuguese is learned mainly through schooling and is important for every Brazilian to be successful in different communicative contexts. On the other hand, many of the students arrive at school speaking Vernacular Brazilian Portuguese that they learned at home, and teachers must not only respect their dialect but also develop appropriate strategies to help them learn the standard dialect and become "true bilinguals" in their own language.

THE ROLE OF BRAZIL IN PROMOTING PORTUGUESE

As previously mentioned, Portuguese is among the 10 most spoken languages of the world, and the presence of Portuguese language and lusophone culture beyond the national borders of Portuguese-speaking countries has become more evident in the past few years. Not only is the importance of Brazil in the spread of Portuguese related to its demography and its emergence as economic power, but it is also related to proactive language policies promoting the Brazilian culture and the Brazilian variety of Portuguese. The creation of Mercosul/Mercosur, an economic and political agreement among Argentina, Brazil, Paraguay, Uruguay, and Venezuela, was responsible for crucial changes in official policies for foreign-language teaching in the countries involved. Among the actions taken by Brazilian authorities, it is worth mentioning the development of CELPE-Bras, an official certificate of proficiency in Brazilian Portuguese as a nonnative language. This is the only certificate of language knowledge officially recognized in Brazil, and it is a requirement for nonnative speakers of Portuguese who want to pursue a bachelor's degree in a Brazilian university.

Inside the Museum of Portuguese Language in São Paulo, Brazil. The museum opened on March 20, 2006. (Luisrftc/Dreamstime.com)

Other actions related to the production of linguistic instruments and the spread of the Portuguese spoken in Brazil are the publication of textbooks for the teaching of Portuguese as a foreign language as well as the creation of courses and programs for pedagogic training of Portuguese instructors.

Brazil has not only taken action on advertising Brazilian culture and the variety of Portuguese spoken in Brazil, but it has also been participating on initiatives to foster friendship and cooperation among Portuguese-speaking countries as well as to boost and strengthen Portuguese as a global language. Brazil's international actions include the signing of the constitutional declaration of the Community of the Portuguese-Speaking Countries (Comunidade dos Países de Lingua Portuguesa) in July 1996 and the active participation in creating the International Institute of Portuguese Language (Instituto Internacional da Língua Portuguesa) in 2002. A last meaningful effort at disseminating the Portuguese language and its culture that is worth mentioning is the Museu de Língua Portuguesa (Museum of the Portuguese Language), the only museum in the world especially dedicated to a language.

TEACHING OF PORTUGUESE LANGUAGE AS A NONNATIVE LANGUAGE

In the contemporary world, the national borders have been slowly vanishing, and countries have been choosing to be part of trade blocs. This along with other social and economic changes has created the need of learning other languages besides our native language. This is why multilingualism is the norm and monolingualism is an exception in the contemporary world. Inside these trade blocs, policies related to language teaching and learning, language rights, translation issues, and establishment of common technical and scientific terminology have been widely discussed. As mentioned earlier, Mercosul/Mercosur and the economic growth of Brazil in the past decade have led to an increased interest in Portuguese language and cultural aspects of Brazil.

The interest to study Portuguese has been increasing in the United States as well. The Modern Language Association publishes a report on enrollment in languages other than English in American institutions of higher education. The 2009 report ranks Portuguese as the 14th most commonly studied language in college-level courses nationwide. Portuguese is taught in 226 universities in the United States. Enrollment in Portuguese courses has been increasing in the past 20 years. In 2009 the growth in enrollment reached a double-digit figure (10%), staying only behind Arabic, Korean, Chinese, and American Sign Language.

The reason why Portuguese has become a very important language is that political, economic, and cultural aspects of the lusophone countries, especially Brazil, have received a lot of attention lately. With respect to economic activities more specifically, a study of the Brazilian Agency for the Promotion of Exports and Investments (Apex) shows that the number of business affairs made in Portuguese grew 534 percent between 2004 and 2008. With an increasing participation of Portuguese in the business world, universities and language schools, not only in the United States but also in different countries, have reported an increase in demand for Portuguese courses.

FOREIGN-LANGUAGE EDUCATION IN BRAZIL

Since 1996 Brazilian schools have been required to offer a foreign language to all students starting in the fifth grade of elementary school. The choice of which language to teach is made by the community, and the majority of Brazilian schools offer English classes starting in the fifth grade and going up to the last year of high school. More recently, due to political and economical relations with other countries of Latin America and Spain, the popularity of Spanish started to grow in Brazil. Not only did the number of language academies and the interest of people in learning Spanish started to increase, but a bill that required Spanish language to be offered as an elective subject in all secondary public and private school system was approved in 2005.

Although English, Spanish, or another foreign language is part of the curriculum in all Brazilian schools, it doesn't mean that most Brazilians are highly proficient in at least one foreign language. The Brazilian educational system is very diverse in quality and resources. Teachers with the best qualifications and training are often employed by expensive private schools, where they are usually paid better. The average teacher salary in public schools is often quite low, which makes it hard to recruit highly qualified teachers for the public system. Furthermore, not all the states of the country count on universities and courses that can adequately qualify and train foreign-language teachers.

More recently, with the increase in use of foreign languages in work and social activities, Brazilians have become more aware of the need to learn a foreign language, especially English and Spanish. In addition, Brazil hosted the World Cup in 2014 and will host the Olympic Games in 2016, and these events are providing a sense of urgency about learning languages that can allow communication with athletes, tourists, and visitors. Because most of the Brazilian school system does not fulfill the need of becoming fluent in English, Spanish, or any other foreign language, many parents from middle- and upper-class families often rely on private-language academies and after-school tutoring.

PORTUGUESE ON THE INTERNET AND SOCIAL MEDIA

According to the Internet World Stats, in 2010 Portuguese was the fifth most used language on the Internet, with 82.5 million users. The Camões Institute also has found that Portuguese became the ninth idiom in production of content on the Internet in 2011. It means that the Portuguese-speaking community is very active when it comes to the use of the Internet because only 3.9 percent of the Internet users use Portuguese on the web. The picture is not different on the social networks. Socialbakers tracked the growth of users' primary language from May 2010 through November 2012 to find that Portuguese is already the third most popular language on Facebook, only behind English and Spanish. Their numbers show that Brazil has experienced major growth in the number of new users, going from Facebook's fourth biggest country to number two. Interestingly, Facebook audience in Brazil has surpassed that of Orkut, which was the most popular social network in Brazil until 2011. Orkut was fully managed and operated in Brazil by Google Brazil from 2008 to 2014, when Google decided to close its social network.

ENGLISH AND BRAZILIAN PORTUGUESE: COGNATES AND FALSE COGNATES

Because Brazil has been culturally, economically, and politically influenced by the United States, English is everywhere in Brazil: on television, in product names, and in advertising. Most movies shown in theaters and cable television are subtitled and American music is everywhere in Brazil. This means that Brazilians listen to English all the time and use a lot of English words and expressions in their everyday language. These words and expressions are especially connected to technology, modern science, and finance, as well as social and cultural aspects. Most of these words are adapted to the Brazilian Portuguese pronunciation (self-service, hot dog, delivery, overdose, mouse, etc.) and may have also gone through changes in their orthography (*piquenique* "picnic," *futebol* "football").

Nevertheless, English and Portuguese do not only share words that are borrowed from English. Because Portuguese is a Romance Language and, therefore, evolved from Latin, and English is full of words of Latin origin, there are many cognates in English and Portuguese that have the same origin. Some examples are displayed in Table 6.3.

Most of the words that have the same origin in English and Portuguese do not only look similar but they also share meaning. Unfortunately for language learners, this is not always true. Some words have similar forms but different meanings. These are called false cognates, and we have to pay attention to them to avoid misunderstandings when we are speaking a nonnative language. Some examples of false cognates between English and Portuguese are shown in Table 6.4. In Table 6.5 there are some common words and phrases in Portuguese.

TABLE 6.3   Cognates between Portuguese and English

Portuguese	English
necessário	Necessary
mensagem	Message
candidato	Candidate
consenso	Consensus
disciplina	Discipline
ridículo	Ridiculous
escola	School
nome	Name

TABLE 6.4   False Cognates between Portuguese and English

Portuguese	English
livraria (means bookstore)	Library
atualmente (means nowadays)	Actually
colégio (means school)	College
parentes (means relatives)	Parents
taxa (means fee, rate)	Tax
lanche (means snack)	Lunch
esquisito (means strange, odd)	Exquisite
eventualmente (means occasionally)	Eventually

TABLE 6.5   Useful Brazilian Portuguese Phrases

Portuguese	English
Oi	Hi
Como vai? Tudo bem?	How are you?
Tudo bem	I'm fine/good
Obrigado/a	Thank you
Por favor	Please
(Muito) prazer	Nice to meet you
Com licença	Excuse me
Samba	Brazilian musical genre and dance style
Bossa Nova	Brazilian musical genre
Candomblé	Afro-Brazilian religion
Capoeira	Brazilian art form that combines dance, rhythm, and fight
Feijoada	Typical Brazilian dish. It is a stew of pork, beef, and beans
Carnaval	Annual festival held for more than four days, from Friday evening until Ash Wednesday at noon, when Lent starts (40 days before Easter). It includes costumes, music, dancing, street parties, and parades

REFERENCES

Azevedo, Milton. "Vernacular Features in Educated Speech in Brazilian Portuguese." Accessed August 30, 2012. http://bib.cervantesvirtual.com/servlet/SirveObras/79117399329793384100080/p0000008.htm.

Brasil tem 305 etnias e 274 línguas indígenas, aponta Censo 2010. Folha de São Paulo. Accessed August 30, 2012. http://www1.folha.uol.com.br/poder/2012/08/1135045-brasil-tem-305-etnias-e-274-linguas-indigenas-aponta-censo-2010.shtml.

Carvalho, Ana Maria. 2010. "Portuguese in the USA." In K. Potowsky (ed.), *Language Diversity in the USA*. Cambridge: Cambridge University Press, 223–237.

Comunidade dos Paises de Língua Portuguesa. Accessed August 30, 2012. http://www.cplp.org.

Cultural Survival. "Endangered Languages in Town: The Urbanization of Indigenous Languages in the Brazilian Amazon." Accessed May 7, 2015. http://www.culturalsurvival.org/publications/cultural-survival-quarterly/brazil/endangered-languages-town-urbanization-indigenous-la.

Imigração Japonesa no Brasil: 1908–2008. Accessed August 30, 2012. In http://www.imigracaojaponesa.com.br.

Instituto de Investigação e Desenvolvimento em Política Lingüística. Accessed March 6, 2014. http://e-ipol.org.

Instituto Internacional da Língua Portuguesa. Accessed August 30, 2012. http://www.iilp.org.cv.

Matos e Silva, Rosa Virgínia. "O Português Brasileiro." Accessed August 30, 2012. http://cvc.instituto-camoes.pt/hlp/hlpbrasil/index.html.

Mattoso Câmara, Joaquim. 1976. *História e estrutura da língua portuguesa*. Rio de Janeiro: Padrão.

Modern Language Association. "Enrollments in Languages Other Than English in United States Institutions of Higher Education, Fall 2009." Accessed August 30, 2012. http://www.mla.org/2009_enrollmentsurvey.

O alemão lusitano do Sul do Brasil. Accessed August 30, 2012. http://www.dw.de/dw/article/0,1174391,00.html.

O valor do idioma. Revista Língua Portuguesa. Accessed March 7, 2014. http://revistalingua.uol.com.br/textos/72/o-valor-do-idioma-249210-1.asp.

Perini, Mário. 2002. *Modern Portuguese: A Reference Grammar*. New Haven, CT: Yale University Press.

Povos indígenas no Brasil. Accessed August 30, 2012. http://pib.socioambiental.org/pt.

Projeto de documentação das línguas indígenas. Accessed August 30, 2012. http://prodoclin.museudoindio.gov.br/.

Rodrigues, Aryon. 1986. *Línguas brasileiras. Para um conhecimento das línguas indígenas*. São Paulo: Loyola.

Rodrigues, Aryon. "A originalidade das línguas indígenas brasileiras." Accessed August 30, 2012. http://www.comciencia.br/reportagens/linguagem/ling13.htm.

Sociedade Internacional de Português Língua Estrangeira. Accessed August 30, 2012. http://www.siple.org.br.

Teyssier, Paul. 1982 [1980]. *História da língua portuguesa*. Lisboa: Sá da Costa.

Unidade e diversidade da língua portuguesa. Accessed August 30, 2012. http://cvc.instituto-camoes.pt/hlp/index1.html.

Zoppi-Fontana, Monica Graciela and Diniz, Leandro Rodrigues Alves. 2006. "Política lingüística no mercosul: o caso do *Certificado de Proficiência em Língua Portuguesa para Estrangeiros (Celpe-Bras)*." In Congresso Internacional de Política Lingüística na América do Sul, João Pessoa. Língua(s) e Povos: Unidade e Diversidade, vol. 1. João Pessoa: Idéia—UFPB, v. 1. pp. 150–156.

Family and Etiquette

Domingos Sávio Pimentel Siqueira

INTRODUCTION

Brazil, as widely known, is a continental country, super-diverse, and one of the most mixed societies of this planet. The Brazilian family reflects such diversity. In order to describe and discuss its story, there are different aspects to consider. The "typical" Brazilian family follows the common Jewish-Christian tradition, bringing in the central figures of the mother and the father. However, today, several structures do show innumerous examples of families where either one of the parents is absent, for instance, or the parents are same-sex couples. In this sense, despite societal resistance, the design of the Brazilian family has been, in many ways, transforming itself in order to respond to the winds of change produced by the so-called postmodern world.

Back in the 1950s, the average number of children of the regular Brazilian family was around four. Small families had always been rare, especially in the countryside, where the women led their lives as full-time mothers and homemakers. Authorities in the country have always equated large families with poor formal education on the part of couples who, theoretically, had no access to detailed information on contraceptive methods. This was not exactly true, of course. The problem is that information of that type was still a big taboo for different reasons, and, naturally, it never circulated so freely and fast as it does today.

In fact, having just a few children in places like Brazil turned out to be a sole exception when it came to this millennial institution, mainly in a country where the Portuguese colonial enterprise "revealed an imperialist imposition of the developed over the primitive race" (Freyre 2003, p. 45), sustaining a social process heavily marked by inequalities (DaMatta 2000). Very large families, for instance, with more than a dozen brothers and sisters had never really surprised anyone, as such a scenario had always been typical in many regions, especially in the Northeast, the country's poorest area, which comprises nine states.

In the 1960s, paying heed to the winds of the feminist liberation movement, also in Brazil, women began to seek improvement in their educational background, and, as a consequence, in the decades to follow, they set free to work outside the home,

A family of Brazilian Amerindians in the municipality of Santo Antonio de Gurupá in northern Brazil. (Hel080808/Dreamstime.com)

splitting the couple's finances and, in many cases, supporting the family on their own. As strange as it may seem, within the great majority of families in Brazil, the man usually wins a higher salary, but though slowly, things have been changing, and more and more, it has been common to see the woman playing the role of the family's major breadwinner. It is, however, next to impossible to come up with a reasonable description of the Brazilian family, even a positively stereotypical one. Families are certainly sources of emotional and financial support, and, as with all families, give Brazilians a sense of friendship and union. Nevertheless, they are also the location of serious relationship problems and all kinds of other disputes.

THE FIRST BRAZILIANS

Forming the basis of the population, we have the Portuguese, the colonizers taken as the "discoverers" of the country, African slaves trafficked from different parts of the continent, and the many indigenous peoples who, for centuries, led freely their lives on this Atlantic shore of the world. As pointed out by Ribeiro (1995), these natives were mainly from the *tupi* branch and they interpreted the arrival of the Portuguese as a fantastic mystical event, as if those strangers coming out of the rough seas had been sent by *Maíra*, their sun-god, the Creator. "It did not take long, however, for them to realize the hecatomb that had fallen over their heads" (Ribeiro 1995, p. 43).

Besides the three main ethnic origins, Brazil has had several other important influences in terms of cultural background and ancestry. This includes a significant number of European immigrants, such as 1.7 million Portuguese, who came to join the early settlers, 1.6 million Italians, 700,000 Spanish, and 250,000 Germans. Ribeiro (1995,

p. 242) identifies different periods of early immigration to Brazil, citing 1851–1885 as the first of these periods, when during that time span only 237,000 Portuguese, 128,000 Italians, 17,000 Spanish, and 59,000 Germans arrived to live in the country. There were also other smaller groups of Slavic populations and Asians, especially Japanese (230,000), who began setting foot in this country at the very beginning of the 20th century (1901–1915) due to an agreement between the two governments that aimed at stimulating massive immigration from Japan to South America.

In this context of migratory movements to Brazil, it is also worth mentioning the Jews, who first arrived on these lands in the 17th century, fleeing the Inquisition in both Portugal and Spain. The first groups settled in Recife, Pernambuco, at the time under the Dutch rule, which granted them with religious freedom. The first synagogue in the country dates from as early as 1636 in that city. Today, Brazil comprises the ninth-largest Jewish community in the world (Maio and Caiaça 2000). Their contribution to the diversification of this rich and unique mosaic that is the Brazilian population today, as much as all the others previously referred to, is widely recognized. In other words, these peoples who one day, for different reasons, came to live in Brazil bringing along with them their customs and traditions, have all contributed to give shape to what we could call, in a non-essentialist way, the Brazilian family.

Living within such a wealth of life habits, postures, behaviors, and attitudes is intrinsically challenging. For the same reason, it is extremely rewarding being in contact with so many people, making us fully aware of the important diversity that permeates different local groups. This diversity, as one might predict, unites these groups in several aspects while, in others, sets them apart.

Brazil, as highlighted by Gomes, Barbosa, and Drummond (2001), is a highly complex country, and therefore, it does not fit any formula or any single explicatory scheme. To delineate the typical Brazilian family is for sure a challenging task. When someone decides to engage in such an endeavor, resorting to a certain extent in personal experience, and in what he or she sees and lives in his or her own microcosm and, finally, is able to do so, his or her perception should be taken as partial, and not representative, in this case, of the entire complexity of the Brazilian society. It is true that, like in any other country, differences tie Brazilians together, as they also singularize their behaviors and habits, but still make everyone living on that land, above all, Brazilians.

Because of the general poor socioeconomic conditions, in the past, the Northeast of Brazil was usually taken as a common reference for the most emblematic examples of large families in the country, different from other more developed regions, especially the Southeast, where the two Brazilian megacities, Rio de Janeiro and São Paulo, are located. Brazil has begun in the Northeast, and although there was also immigration to this specific area, especially of the Portuguese, many of them former Jews already known as "new Christians," and different groups of Arab descent, more specifically, Syrians and Lebanese, these movements have not been as significant there as in other parts of the country like the South and Southeast, for instance. Despite the greater ethnic variety in these southern areas, it is possible, therefore, to say that the matrix of the great majority of Brazilian families is really the aforementioned "original three" (Portuguese, African, and indigenous), certainly, depending markedly where the Brazilian is originally from. For Ribeiro

(1995, p. 223), however, "despite all the vicissitudes the Afro-Brazilian faces, he [the African] has come to be the most creative component of the Brazilian culture, and the figure who, along with the indigenous peoples, best singularizes our people."

Despite the significant presence of the three original ethnic groups, even in faraway places where more diversified groups of immigrants were not so common, one could easily come across families that had more than the usual Portuguese and indigenous blood running in their veins. Curiously enough, these mixed families, showing opposing characteristics, since half of their children had darker skin and half of them of a much lighter complexion, caught people's attention especially because of the white ones. In many parts of the country, including the Northeast, people bearing bluish or greenish eyes, sort of red or blond hair, came to be called "galegos," even if not of Galician descent, exactly because of the large number of Galicians who settled in those areas. For many people, actually, they resembled foreigners from a distant land.

Originally, galegos (Galicians) are a national, cultural, and ethnolinguistic group whose homeland is Galicia, the green and lush northwestern corner of Spain. According to Klein (1994) and Guimarães and Vainfas (2000), they comprised the innumerous groups of Spaniards who came to the "New World" even before 1500, and their presence in Brazil, especially, was considered strategic and important toward the end of the 16th century and middle of 17th century. South America, today, has the largest number of people of Galician descent outside of Spain. Several million South Americans are descendants of Galician immigrants, mostly living in Argentina and Brazil.

THE FORMATION OF THE BRAZILIAN FAMILY

Although race has always been a sensitive topic in Brazil, it is common to be raised in families that barely discuss or show deep concern, for example, toward the skin color of siblings or even of people in general. In fact, we can have children growing unaware of these issues as, depending on where they live, this never becomes the central point of discussion when big portions of the family get together. They are just families, and this is essentially what matters, although many of their beliefs and values related to this and many other themes remain, to a certain extent, disguised or subtly understood.

This type of Brazilian family mentioned is in several ways very traditional and conservative, especially if they are born and raised in small cities in the interior of the country. Unbelievably, despite the almost global stereotype of being a very liberal country, this is one aspect that turns out to be very difficult to generalize about Brazil. It definitely depends on where you live and on your parents' and grandparents' family background. Although this country has already reached over 500 years of history, it is advisable not to disregard certain rigid features that oriented the formation of the Brazilian family, predominantly patriarchal in colonial times, where the very tight and not rarely oppressive family bonds would impose limitations to the individuals' future lives (Holanda 2005, p. 144).

To begin with, Brazil is a very religious country. Although for some sociologists in religion Catholicism in Brazil is more of a tradition than a religious practice itself, the Brazilian Institute of Geography and Statistics (IBGE) 2000 Census shows that Catholicism remains the largest religion, registering around 125 million members, that is, three quarters of the Brazilian population (73.8%) of self-declared Catholics.

Despite these solid figures, census data also reveal a significant reduction in the percentage of Catholics in the country, depicting, consequently, an important increase in the number of those who declare themselves evangelic, mainly Pentecostal and Neo-Pentecostals (Menezes 2005). In other words, if the absolute numbers reinforce this massive presence of Catholicism, a more accurate look on these results tells us that, as of the second half of the last century, there has been a progressive reduction of its members (Teixeira 2005). Certainly, Brazil is still the single country with the largest Roman Catholic community in the world. However, it is relevant to point out that there is a significant and important diversity of other religions and religious practices in the country, including those of African origin.

Under such perspective, families who profess any religion tend to adopt rigid patterns of morality. As Freyre (2003, p. 91) argues, Brazil's formation did not at all worry its colonizers in terms of race unity or purity. "During the entire 16th century," contends the author, "the colony remained wide open to foreigners, being the colonial authorities' only preoccupation that the newcomers professed the catholic faith or religion." Certainly, a lot of this religious tradition remained.

One of the aspects that excel within the families has to do with the different treatment given to boys and girls. Most of the time, boys are allowed to do practically everything they want. In fact, they are encouraged to date very early, to drive as early as possible, and to assume responsibilities a girl would probably have a little later. Girls, on the other hand, are to concentrate more on their studies, delay dating, and, depending on the family, keep their virginity intact until they marry. Although the status of women in the Brazilian society has changed significantly, some of these "traditions" within families in which the boy is taken as the "stallion," and the girl as the "little homemaker," waiting patiently to be proposed by a "prince" who, in her dreams, one day, would come to marry her, remain true. In this sense, Harrison's observations in her comparative work between Brazilian and North American behavior, *Behaving Brazilian* (1983), vividly illustrate a little bit of the open "machismo" typical of most Brazilian families:

> Men are expected to make passes at women, and a man may very well make a pass at a woman he has no interest in and whom he expects will refuse him (p. 6).
>
> Brazilian men adopt the attitude that a man must protect the woman he wants to marry, that he can go to a prostitute if he so desires (p. 8).
>
> Men are expected to "play the field" both before and after marriage, but for a married woman to have an affair is unacceptable (p. 9).

Still in this concern, being liberal or more conservative, Brazilian families do tend to establish a clear line on how their children should play their roles inside the family and in society. Despite the fact that girls today are less controlled by their parents, this condition has never paired off with the treatment being given to boys along the years. Being very explicit, boys, in general, are allowed to go out at night, for example, and come back home whenever they want. If this happens to girls, things get a little complicated, as the family will not approve of such a behavior with the frequency they do toward boys. Parents, in many families, are also lenient toward drinking when it comes to boys, but not exactly with girls. In a lesser extent, the same happens when it comes to smoking.

Certainly, changes have taken place, and they are naturally more apparent in bigger cities. As it is easily noticeable, women are much more independent today, they have more control of their sexuality, and, for sure, they are assuming responsibilities they would have only at a later age some years ago. However, this would not happen without their having to overcome several barriers, be it here or on other lands, as exemplified by Debiaggi (2003, p. 177) in her text about Brazilian families living in another cultural context, in this case, the United States:

> In Brazil, the woman was responsible for the home chores and for taking care of the children. Even when there was a maid, it had always been implicit that these duties belonged fundamentally in the female universe.

TABOO MATTERS

Another sensitive issue related to tolerance matters in the heart of Brazilian families has to do with sexual freedom and dating at home. In general, fathers talk about sex with their adolescent sons and mothers talk to their adolescent daughters, though, as we know, this can never be taken as a rule of thumb. There are indeed families that talk openly about sex, putting everybody together, boys and girls, in a round table to discuss, reinforce information gathered at school, clarify things about sex life, and, naturally, provide all sorts of advice.

However, it seems still reasonable to say that it is well accepted to allow, and even encourage, the girlfriend to come and spend the night with the son, sleeping with him in his own bedroom. It is common to hear from parents that, due to rampant violence in many of Brazil's big cities, they prefer to have their children date at home than to see the couple go out to bars and motels running therefore avoidable risks. It is important to point out that motels in Brazil are basically used for sexual encounters. Of course, there are motels of different standards, and, consequently, rates that vary according to quality and services rendered.

The same situation might also happen with girls, but it does not seem to be the norm, or something promptly accepted by parents. As found out by Borges et al. (2007, p. 1591) in a research study on factors associated with sexual debut among adolescents enrolled in a family health unit in the eastern region of São Paulo, "It is possible to consider, in a certain way, that mothers and fathers do transmit a positive idea toward the sexual initiation on the part of their sons."

When it comes to homosexual couples, once the relationship is taken naturally by the parents and the entire family, despite the lack of studies in Brazil, it can be assumed that at least the attitude of caution and protection will not be different. In such a context, it seems relevant to mention the term "chosen family," as França (2009, p. 27) clarifies:

> Well-resolved homosexual couples have an ample social network of friends who, in general, socialize amongst themselves; when this interconnection of friendship

bonds lasts for a long time, it starts to assume characteristics of a family; for this reason, the term "chosen family" is referred to.

As one can easily imagine, in such a private and conflicting environment, the expectations and demands toward who is going to be the partner chosen by the girl is much more subjected to deep scrutiny than the one selected by the boy. However, the day to introduce the new partner, especially the very first one, is always, a moment of hesitation, apprehension, and great expectation. Parents' approval will depend on several elements, from appearance to educational background to color of skin, even the preferred soccer team, and the postulant will be carefully analyzed by everyone, including the long-time maid (*empregada doméstica*) who, although slowly becoming a luxury of the past along the years, has become a Brazilian "institution." The *domésticas* are low-paid women who begin working at homes of the Brazilian middle and middle-high classes at an early age, and, depending on the relationship with the families, they may stay with them their entire lives. However, nowadays, "teen girls that used to be fodder for domestic help are required by law to stay in school, or, in really poor families, their families will lose out on receiving social welfare programs like *Bolsa Família*" (Rapoza 2013, p. 1). Besides that, as older women in their 20s on up are finding better employment in retail and manufacturing, maid service is turning into something harder and harder to find in several regions of the country.

Along the ubiquitous presence of the maid, in Brazil, "family means parents, children, grandparents, aunts, uncles, second, third, and fourth cousins" (Harrison 1983, p. 9). In this particular issue, everybody's approval is well-wished and a great sign of relief. Naturally, the final verdict will come with time, and once approved, the person (boy or girl) has to be aware that with the "seal of consent" he or she a relationship with the whole family, and not with the girlfriend/boyfriend alone.

EXTENDING THE FAMILY, CHANGING ROLES

In Brazil, "Romantic attachments," adds Harrison (1983, p. 8), "often form between members of the extended family," and if someday they turn into marriage, the situation continues on for better or worse: the person will be married to the entire family and should be ready for all kinds of interference, especially from the figure of the ever-powerful and domineering mother, turned now into the severe and overprotective mother-in-law. However, there is the bonus too. Once the new member is admitted, he or she will know that most of the time, there will be a lot of support, especially when it comes to raising the children who will be anxiously expected and sometimes even required through daily comments and supposedly "innocent" hints: "When will my first grandchild come?" Everyone entering a Brazilian family should try to be ready for those "demands" inside and outside the private circles.

Despite the different designs of families in Brazil currently (single parents, same-sex couples, etc.), for some reason, it is easy to notice that the old patriarchal heritage has not remained so vividly as we might come to believe, at least in the so-called traditional family. Depending on the region, we may be thinking of, the figure of the father is still very much respected, but it is the mother who has (almost) total control of the entire dynamics of the family. In other words, she is the one who exercises very significant (and sometimes excessive) influence on the children, especially the boys. As Harrison (1983, p. 6) reminds us from her studies, "Brazilian women often took a dominant role, particularly in family matters; [they] often have things their way." This can be easily explained when we look at facts socially and historically, mainly from the perspective of those people who live in the Northeast of Brazil. In poorer and tragedy-prone areas like this, for example, circumstances such as hunger, effects of long and severe droughts, and lack of work opportunities which would force men to move to the big cities, leaving the family behind, have contributed to radical changes on the role of the mother, seen previously as the despondent housewife, the one who would do the cooking, the cleaning, taking care of the children, staying home her entire life.

Going back in time, because of massive migration to the "El Dorados" of the South (São Paulo and Rio de Janeiro), beginning in the late 1950s, as mentioned, entire hordes of men would leave their homes in the arid and semiarid backlands of the Northeast in search of job opportunities and better life conditions. This adventure, as in a lot of the immigration experiences throughout the world, necessarily, would not include wife and children at first. The men had to go on their own and work hard day and night to survive in the so-called concrete jungle. The wife, on the other hand, could do nothing but stay with the children usually for a long time waiting for the news. Sometimes they would wait forever, as innumerous cases show that, once established in these cities, with a steady job and a house to dwell in, many of these men would never return, not rarely, settling down with another family, as it happened with former Brazilian president Luiz Inácio Lula da Silva's family. Nevertheless, instead of just conforming to the situation, in the ex-president's case, his mother decided to take her eight children with her and go after the husband who already had another family in the prosperous state of São Paulo.

As it became common, and many other men did, Lula's father abandoned the first family, letting them survive (or die) on their own. Therefore, this supposedly fragile Brazilian homemaker had to stand up and fight as mouths depended totally on her to be fed. She grew stronger and stronger, overcame many obstacles in her new life, and built up a story that, in many ways, illustrates the stories of millions of poor women in contemporary Brazil. For the simple fact of having to struggle to survive the adversities of a hard life, millions of women in Brazil were forced to change roles inside the family structure and take their lives by their hands.

THE SINGLE-PARENT FAMILIES

Single-parent families are getting extremely common in many places in Brazil, and this single parent is usually the woman. It has been shown that in low-income communities in large cities like Rio de Janeiro, São Paulo, Belo Horizonte, Salvador, and Recife, most families are raised by a single hardworking mother who daily struggles on her own to raise an average of three to four children. According to IBGE, in 2005, 49 percent of Brazilian households still depicted the couple (man and woman) as the standard nuclear family along with the children. However, there has been a significant increase in families whose main responsible figure is a female member, without the presence of the man. In the early 1990s, 22 percent of women declared to be the only reference in the family, while these numbers reached 29 percent in 2002 (Marin and Piccinini 2009).

This figure means that, more and more, these women are leaving home and usually working two to three shifts, most of the time, more than one job, to fulfill the family's needs. In the *favelas* of Rio de Janeiro, for example, this is the reality to the majority of dwellers and once the older kid is able to take care of the younger ones, he or she plays the role of the caring mother while the mother is away at work. This also includes an increasing number of grandmothers who raise their grandchildren due to the fact that young pregnancy is still very common in these communities. In other words, single mothers become single grandmothers, and the cycle goes on.

THE SAFE PORT

Members of the stereotypical Brazilian family are said to be very tied to one another. And this is absolutely true in most contexts. Along with the figure of the housemaid mentioned earlier, tardy (or forever) adolescents have become another Brazilian institution. They reach adulthood, go to the university, get a job, and still live with the parents, as to continue enjoying the comfort of an easier life, especially in middle-class, middle-high-class, and high-class families. As these families usually have at least one long-life maid, the children, since childhood, get used to being served and pampered in all senses, and practically all the time. Many of them get so used to the services of the maid that they rarely make their beds or do any kind of domestic work. Most of them do not cook, do not wash the dishes, do not do their laundry, do not iron their clothes, do not buy their food, do not help carry groceries, and so forth.

The parents' home, in many ways, may serve as this "safe port" for the rest of the children's lives, not only emotionally but also practically. Even when they get married and have children, the parents (now grandparents) will have a very strong influence on them, and, not rarely, as grandmas and grandpas help take care of the grandchildren, it is very common to have them interfere directly on the way their grandchildren are raised. Sometimes this turns into a zone of conflict, but they end up overcoming the differences, since grandparents come to be extremely important in the practicality of their children's lives. In extreme situations, it is not uncommon to have grandparents support their children financially even after they are married. Harrison (1983, p. 80) precisely summarizes this Brazilian peculiarity:

> Children remain at home until they marry, even if this means until age twenty-seven or twenty-eight. If a young couple cannot afford their own home when they marry, they live with one set of parents. When children establish their own households, they try to settle close to the parents and continue to visit them regularly, at least once a week.

Because of this attachment to the comfort of their homes and, in many situations, emotional dependence, young Brazilians do not feel very much tempted to go on adventures that will take them out of the country, the state, or even the city where they live. It is not uncommon to read and hear testimonies of Brazilian exchange students, for example, who had a hard time adapting to their new homes and realities in most countries where they go because, like everybody else, they were supposed to do home chores and become totally independent in this sense. The ones who resist, certainly, tell that in the long run it turns into a learning process for life, but many really get into trouble and, not rarely, come back to Brazil, speaking of very bad experiences, until they return to the comfortable life back home. Certainly, as previously mentioned, this is not the reality of lower-middle or poor classes, but once they ascend in life, one of the first "luxury items" they might acquire will probably include hiring a maid or at least a babysitter or a nanny. And then the stories will probably repeat themselves.

Due to this closeness, it is not an exaggeration to say that many families do their maximum to have their children around even after they become fully adults. Although study and professional mobility were not enforced by the common Brazilian family, this scenario has been changing more and more. In the recent past, because of the poor educational infrastructure in many smaller cities in the interior of the country, students at a later stage in their schooling were prepared to move to a bigger center or the capital city to enter the university. This fact, for many families, has always been seen as one more rite of passage, and there is a lot of suffering and emotional pain involved, generating what psychologists have come to call "the empty nest syndrome." Despite the fact that these adolescents are usually in the same state and their parents have total financial control of their lives, since the great majority of them do not work at all and are not forced to do so before graduation, it seems that Brazilian mothers get more affected by the effects of the "syndrome" than others from different cultural backgrounds, including those in Latin America. In her 1983 comparative study, Harrison comes to this very same conclusion:

> North Americans stress individual independence and see the son's or daughter's move out of the house, often described as "leaving the nest," as an important step towards adult status. Not so in Brazil, where living alone is not a sign of independence, but a sign that one lacks family. (p. 80)

Once boys and girls reach professional life, although it is much more usual and logical now to "go where the good jobs are," it does not come as a surprise to hear of situations in which this or that Brazilian professional turned down a job offer, be it in the country or abroad, because of "family reasons." In other words, such decisions to a certain extent are still made in a sort of family collegiate, and their opinion will certainly have a decisive weight on what the son or daughter should do. Maybe this has a little to do with what Ribeiro (1995, pp. 243–244) affirms about Brazilians leaving Brazil to live in another country:

> I could feel in the exile how difficult it is for a Brazilian to live outside Brazil. Our country has so much of a sap of singularity that makes extremely hard [for us] to accept and enjoy the experience of living with other peoples. (. . .) We just need to observe a gathering of Brazilians, from the half a million we are exporting as workers, to feel the fanaticism through which they cling to their Brazilian identity and the rejection to any idea of remaining abroad.

OPEN DOORS, PLENTIFUL TABLES

Brazilians in the collective imagination are supposed to be very welcoming, warm, friendly, happy, and gregarious. This stereotype "sold" and spread all over the world depicts some truth, of course, and the traces of these features, without a shade of a doubt, materialize within the family environment. Yes, Brazilians are happy, but also noisy, rowdy, and sometimes (very) inconvenient when in droves. In many places of the country they are always ready for partying. They gather for the most common celebrations like weddings, birthdays, anniversaries, baptisms (or similar ceremonies in different religions), Mother's and Father's Day, and Christmas and New Year's Eve, but they also meet festively to celebrate very particular and private events and situations, for example, when someone in the family enters the university, graduates, gets engaged, passes a hard exam for a highly desired job in the public sector, or, most commonly, to watch soccer matches, no matter the division the team plays in over lots and lots of beer, Brazilian barbecue, and loud music. For this latter issue, what matters is to have fun and to root for the victory.

Equally common are the "sacred" Sunday lunches, when sons, daughters, husbands, wives, boyfriends/girlfriends, partners, grandchildren, and sometimes (too) many friends come to grandparents' or in-laws' homes to spend the day, eat the indefectible "feijoada," and have fun together. In some families, it is not an exaggeration to say that this is more than a habit; it is almost a ritual, with every little detail under the command of the "super" Brazilian mama. As DaMatta (2000, p. 26) would state, this filial and familial attitude is expanded to godfathers/mothers and friends, "for whom the doors of our homes are always open and our table is always set and plentiful."

This particular Brazilian characteristic of "opening doors" certainly involves friends who are usually taken as almost brothers and sisters, and as Harrison (1983) highlights, people from other cultures get surprised by Brazilians' general warmth and enthusiasm toward their friends. This approximation, in many ways, poses some inconveniences as these friends, once invited to a social gathering, for example, may take the liberty of bringing other friends without even consulting the host. Margolis (1994, pp. 184–185), in her ethnographic work with Brazilian immigrants living in New York City, had the opportunity to register such a feature:

> Carlos, a Brazilian cabdriver, told me that his entire social life revolves around gatherings held nearly every weekend with the same group of ten or twelve Brazilian friends at one of their homes. When members of the core group invite their own friends, as many as thirty people may show up. (. . .) Clarissa, a member of the group, cooked a large *bacalhau* (codfish) casserole and told a few close friends to come over for dinner, but to her amused chagrin, there was barely enough to go around, when nineteen Brazilians appeared on her doorstep that evening.

Mealtime, actually, is an interesting source of discussion and appreciation from the Brazilian family's perspective. For example, mother and father would practically oblige the children to stop everything they are doing to have the meals together, especially lunch, the main meal in most regions of the country, in more conservative families. So children do. Of course, in these super-busy and hurried times, healthy habits within the family environment like this, it is reasonable to say, tend to be ignored, as children today are so occupied and seem not to have time at all to sit and talk over a few things with the parents, or brothers and sisters, at least at these social moments at home. In this concern, sociologists would say that this could be a sign of individualism taking its toll in the postmodern family structure. Therefore, in such situations, children end up gulping down their food, sometimes in front of the television or the computer, and then move on with all the duties they are supposed to fulfill.

Another interesting habit many Brazilian families used to incorporate was an imposition on boys since early childhood to never sit at the table bare chested. In families that held such a value, fathers would never do it either, naturally making a point of setting the example. It was, and still is for a lot people, a sign of respect, though most of these now labeled "conservative" behaviors are, in many ways, no longer attended to. Children, even in smaller cities, have become more tied to computers and smartphones than to the idea of having to come to the table and, over the three meals, enjoy and savor moments of pleasure, collective discussion, and deep learning with the members of their direct family. We can suppose this type of "tradition" falls into a code of the Brazilian family that it is explained by DaMatta (2000, p. 28) when he argues that at home in Brazil, "there is a tendency to always produce a conservative discourse, where the traditional moral values are more often defined by the elderly and by men."

SOLIDARITY WITHIN THE FAMILY

Moving to another direction, it is noticeable that there is a lot of mutual support within the Brazilian family. When one member gets into financial difficulties, the first people the person in trouble would look for in order to get some help and counseling are the closer ones to him or her in the family. The ones who are in better conditions, for sure, would do anything to contribute to the solution of the problem, providing the possible means to alleviate the stress and the suffering usually caused by such situations. However, family members also fight, interfere with one another's lives, and, in extreme situations, may break relations forever, especially if it involves very delicate and sensitive issues like money, gossip, property, internal disputes, and inheritance. Like any family in the world, disagreement is present, but Brazilians often try to solve them the best way, avoiding problems that could cause any breaks in relationships.

When it comes to choosing the career to be pursued by the children Brazilian families in general brace together and work hard toward that objective, engaging in a real collective endeavor, regardless of the profession envisioned. Certainly, as it happens in many places, traditional and highly prestigious areas like law, medicine, engineering, and technology are the most desired, and conflicts arise frequently when the child goes for something else, since parents normally project the future for their children based on their own views and beliefs. It is common for those families to expect the children to become some of those potentially successful professionals who are able to support themselves and, in a natural sequence in life, complete the cycle, forming their own families. However, in other situations, for example, when parents come from nontraditional families, even though they do value formal education, they never force their children to try to study to be a doctor, a lawyer, or an engineer. Nevertheless, it is reasonable to say that such an attitude is something still highly expected

Many Brazilian parents put a large emphasis on education, and make a point to help their children with their homework. (Jose Wilson Araujo/Dreamstime.com)

in Brazilian society. It is a motive of pride, since they are the careers that guarantee better-paying jobs. They think that they will never risk leading their lives living from hand to mouth. This may sound like a blunt generalization, but it is still very true for the average family in Brazil. It may be taken as a myth, but it is still present in our social context, especially when the father is a doctor, a lawyer, or an engineer. "Like father like son," as the saying goes.

FAMILIES GETTING OLD

It is quite visible in Brazil that its population is living much longer. Unfortunately, it seems that families in general are not prepared to take care of the older beloved ones, as the Chinese, for example, who since an early age, they are taught that one of their noblest missions in life is to care for their parents when they grow old or get sick. For sure, for different reasons, Brazilians are different from Americans, who created a national institution called nursing homes, especially on warmer lands, where the ones who can afford flock into these places in order to spend the rest of their lives comfortably and independently. Nursing homes in Brazil are few and usually very expensive, so the elderly normally come to live in one of the children's home, or, for those from middle and high classes, the elderly stay in their homes with professionals currently called "old people's caretakers." In this regard, as noticed by Harrison (1983, p. 80), "Brazilians show dismayed surprise at the commonness of retirement and nursing homes in North America and see in them a lack of concern for the aged and for the family unit as a whole." However, as the population gets more and more involved in such a situation, having their elderly live longer and longer, and, in many ways, facing several obstacles to take care of them personally, it seems that this view may go through some type of change, approximating themselves to what happens in other cultures.

The elder in Brazil may also experience a very common and ever-growing phenomenon that involves families who, for different restrictions, especially economic, opt for staying physically together after the children grow up and get married, and the time comes to raising the next generation's offspring. With the blessing and consent of the grandparents, the children expand their parents' house (one more bedroom and bathroom) or build a neck-to-neck new house or an apartment in the back or on the next floor. This completely social phenomenon has produced all over the country rough constructions that came to be jokingly called "puxadinhos" (an extra small house). Depending on the number of children, the puxadinhos can be several, and in homes where the children's finances are not so stable or not good enough, it is the older members, with the earnings of their very often-meager retirement pensions, who very frequently will help support the entire family in a cycle that will end only when they die.

Besides that, as DaMatta (2000) explains again, the Brazilian home is by nature an inclusive space. However, at the same time, it is exclusive, as it, along the years, will always have people who are not part of the family around, coming and going. These are called the "agregados" (the *aggregate*), and they can be many, bearing different profiles. Among them are, for example, a relative who often comes from the interior to see the doctor, a friend experiencing financial problems, a person who is undergoing successive marital crises, a politician who needs to talk to someone important in the city, a solitaire who does not have much of a company to chat with, an older ex-worker who does not have a place to go, a woman who escapes from her father or brother and needs some time to decide what to do with her life, a drunkard who wanders around for years on the street where the family lives, a godfather who needs a job, or even a stray dog that would know the right place to get stationed as there will always be some leftover food thrown at it. They all become members of what DaMatta (2000, p. 26) calls the "positive space of the residence."

FAMILY AND FAITH

We cannot talk or write about Brazilian families without touching on a sort of sensitive issue, that is, religion. Yes, a great number of Brazilians, as mentioned earlier, fear God and believe feverishly in the mysteries of "the spiritual world," and, naturally, this is reflected in the space of the family. The Brazilian religious experience, needless to say, is very rich, and as it happens in countries where many cultures gather, extremely diversified. Roman Catholicism, African religions, Spiritism, Evangelicalism, Islamism, Buddhism, Judaism, just to name a few, are present in the country, and families, most of the time, profess religious faith together. It does not matter which religion is more popular or congregates a greater number of worshippers, Brazilians are profoundly religious, and, in case of affective relationships, for instance, this may be an "advantage" or an "obstacle," when couples decide to make their union official. Certainly, it is possible to say that in some families, such a feature is not a reason for stress, but in others, depending on how the family sees the whole matter, a member of the couple, if he or she really wants to stay together, will need to convert to this or that religious orientation, as it certainly happens in other societies. However, as surprising as it might seem, we shall not find totally strange if we come across within family members who worship Catholic saints and, at the same time, entities from another religion, making visible the traces of a syncretism, the fusion of diverse religious beliefs and practices, which, in many ways, despite some situations of tension and intolerance, has made Brazil an example of society where religions can occupy the same space democratically.

THE FANTASY OF THE RACIAL DEMOCRACY

Among several important themes, it is of great relevance to bring up some discussion on the race issue within the Brazilian family. For some reason, maybe related to the myth of the "cordial man" (Holanda 2005), "in the sense of a man strongly determined by emotions, subjectivity and the heart" (Ianni 2005, p. 3), Brazil came to be known as a racial democracy, but we all know this is a fantasy and that several of the things people say about there not being racism in Brazil is a fallacy. There is racism in this country, and this is naturally reflected in the boundaries of the family. Brazil's racism continues being disguised and dissimulated. Although it is possible to see interracial marriages in some places, as far as we know, they are still rare. Even in regions like the Northeast, where there is a beautifully mixed population, of a dark-skinned color majority, a light-skinned girl marrying an Afro-Brazilian young man (and vice versa) is still something that calls people's attention. Middle and high classes in this area, for example, where a good portion of them tends to be light-skinned, would make a strong effort to accept the idea that someone in the family is dating a dark-skinned person, especially girls.

Certainly, generalizations should be avoided, but it seems imperative to call attention to this fact, so we do not perpetuate this dangerous state of hypocrisy we sometimes try to sell all over, though certain barriers, especially in more intellectual spaces, have been surpassed. As Schwarcz (2012, p. 52) reminds us, "Brazilian racism constitutes a type of common discourse practiced as such but very little institutionalized; [. . .] in fact, one of the particularities of the prejudice current in the country is its non-official character." So, in this concern, it is reasonable to admit that there is still a long way to go. How many times have we heard of cases of male children hiding the fact that they are in love with an Afro-Brazilian girl from their parents, so it would not cause any discomfort in the family? After a long time of planning to reveal the situation, in case the relationship continued, the family would have to be "prepared" in advance in order not to react negatively. In many ways, this is also Brazil, and it makes no sense in trying to hide it behind the inexistent and dangerous curtains of a supposedly stable racial democracy below the Equator.

CONCLUSION

As it could be seen from what was discussed throughout the chapter, like any other place in this world, the Brazilian family is varied, mixed, and, above all, highly complex. Brazilian families gather everyday, they fight, they love, they are everyone's safe heaven, and they (sort of) prepare us for the world. As mentioned earlier, any task set to define a typical family in a given society will always be a partial and fragile work. In this sense, it is important to see that what has been described in this text fills out lines impregnated with fragility, but it is a totally honest reflection. Any family is singular, and whether we grasp that concept clearly or not, each and every family will always reflect an entire country, as we all are part of countries within a country. Therefore, no generalizations are permitted. In the case of this huge "continent," which from "brasil," the ordinary wood native to a great part of these lands, has become Brazil, things are not to be different.

As we all know, Brazilians love to receive, welcome, and mingle with foreigners. Once newcomers get to the country, start moving around, and get to know people, they become family. This is in our genes, and for the better or the worse, it makes us a unique people living on a unique land. Full of problems, struggling to seriously attack its deep and cruel social inequalities, as we reflect on this particular trace of this enormously diverse nation, one of the keys to understand this society through the lenses of the family is grasping the idea that "the home congregates a complex and fascinating network of symbols which are part of the Brazilian cosmology" (DaMatta 2004, p. 15), and what is irradiated from the people who inhabit this special space clearly depicts traces of the colorful mosaic that has throughout the years enriched Brazil and the vastness of its territory.

REFERENCES

Borges, Ana Luiza Vilela, et al. 2007. "Fatores associados ao início da vida sexual de adolescentes matriculados em uma unidade de saúde da família da zona leste do Município de São Paulo, Brasil." *Cadernos de Saúde Pública*, Rio de Janeiro, 23 (7): 1583–1594.

DaMatta, Roberto. 2000. *O que faz o brasil, Brasil?* 11th ed. Rio de Janeiro: Editora Rocco.

DaMatta, Roberto. 2004. *O que é o Brasil?* Rio de Janeiro: Editora Rocco.

Debiaggi, Sylvia Dantas. 2003. "Famílias brasileiras em um novo contexto cultural." In Ana Cristina Braga Martes and Soraya Fleischer (Org.), *Fronteiras cruzadas: etnicidade, gênero e redes sociais.* pp. 174–197. São Paulo: Paz e Terra.

França, Maria Regina C. 2009. "Famílias homoafetivas." *Revista Brasileira de Psicodrama*, 17 (1), São Paulo: 21–33.

Freyre, Gilberto. 2003. *Casa-Grande & Senzala: formação da família brasileira sob o regime da economia patriarcal*, 43rd ed. São Paulo: Global.

Gomes, Graziela Luara, Lívia Barbosa, and José Augusto Drummond, eds. 2001. *O Brasil não é para principiantes: Carnavais, Malandros e Heróis, 20 anos depois.* Rio de Janeiro: Editora FGV.

Guimarães, Lúcia Maria Paschoal, and Ronaldo Vainfas. 2000. "Sonhos galegos: 500 anos de espanhóis no Brasil." In *Brasil: 500 anos de povoamento*, pp. 101–121. IBGE: Rio de Janeiro.

Harrison, Phyllis A. 1983. *Behaving Brazilian: A Comparison of Brazilian and North American Social Behavior*. New York: Newbury House Publishers.

Holanda, Sérgio Buarque de. 2005. *Raízes do Brasil*, 26th ed. São Paulo: Cia. das Letras.

Ianni, Octavio. 2005. "Types and Myths in Brazilian Thought." *Revista Brasileira de Ciências Sociais*. Vol. 1, special issue, pp. 1–11, São Paulo. Accessed September 20, 2014. http://socialsciences.scielo.org/pdf/s_rbcsoc/v1nse/scs_a03.pdf.

Klein, Hebert. 1994. *A imigração espanhola no Brasil*. São Paulo: Sumaré, FAPESP.

Maio, Marcos Chor, and Carlos Eduardo Caiaça. 2000. "Cristãos-novos e Judeus: um balanço da bibliografia sobre o antissemitismo no Brasil." *BIB—Revista Brasileira de Informação Bibliográfica em Ciências Sociais*. Rio de Janeiro, No. 49, 1st semester, pp. 15–50.

Margolis, Maxine L. 1994. *Little Brazil: An Ethnography of Brazilian Immigrants in New York City*. Princeton, NJ: Princeton University Press.

Marin, Angela, and Cesar Augusto Piccinini. 2009. "Famílias uniparentais: a mãe solteira na literatura." *Psico*, Porto Alegre, PUCRS, 40 (4): 422–429, out./dez.

Menezes, Renata de Castro. 2005. "Uma visita ao catolicismo brasileiro contemporâneo." *Revista USP*, São Paulo, 67: 24–35, setembro/novembro.

Rapoza, Kenneth. 2013. *Forbes Online*. pp. 1–2. Accessed August 20, 2014. www.forbes.com/sites/kenrapoza/2013/03/10/in-brazil-maid-service-becoming-thing-of-the-past.

Ribeiro, Darcy. 1995. *O povo brasileiro: a formação e o sentido do Brasil*. São Paulo: Cia. das Letras.

Schwarcz, Lilia Mortiz. 2012. *Racismo no Brasil*. São Paulo: Publifolha.

Teixeira, Faustino. 2005. "Faces do catolicismo contemporâneo." *Revista USP*, São Paulo, 67: 14–23, setembro/novembro.

Literature and Drama

Eduardo F. Coutinho

FIRST MANIFESTATIONS

As a consequence of a long process of colonization that lasted over three centuries and that can still be felt, though not from the same matrixes, in cultural and economic terms, Brazilian literature has always been marked by a tension between the mere incorporation of a European tradition and the attempt to create a new one of a local or native coinage. Although these two trends have oscillated throughout the history of Brazilian literary production and have never been completely abandoned, the search for a proper profile and the effort to create a local tradition to replace the one brought from Europe has become so significant in the country that it has been often referred to by critics as a growth of Brazilian consciousness. With a look at the cadre of Brazilian literature, from its first manifestations to the present times, it is not difficult to observe that it has gradually broken away from the influence of Portuguese writings and has begun to deal with the native themes and types, and to express itself in the new Brazilian idiom. The development of this consciousness had its origins in colonial times, but it reached a considerable relevance at two special moments: the Romantic period, right after the independence of the country, and the Modernist movement of the 20th century, where Brazilian literature seems to have reached a solid and proper tradition.

At the beginning of colonization, the literary manifestations that appeared in Brazil were seen as extensions of Portuguese literature. Such is the case of Caminha's letter of discovery–the first document written about the new land—and of the reports that followed: accounts of the first travelers, descriptions of the land and its inhabitants, and the chronicles of missionaries and soldiers. It was the literature of Portugal that generated the nascent Brazilian literary spirit. It served as a vehicle for the inheritance of European, Western, and Christian ideas that laid the foundations of Brazilian consciousness. It brought about classical values, literary techniques, and aesthetic models that were adapted to the new environment. Yet it also gave rise, from the very beginning, to a yearning to create something endowed with a Brazilian sense. Portuguese literature brought to Brazil the medieval and Renaissance heritage. From the Middle Ages came Brazilian old poetic yardstick, in the form of popular lyrics and courtly versions of troubadour ballads, traditional dramatic forms that had grown up in the plays of Gil Vicente and the Jesuit theater, whose legacy can be seen in the voice of Anchieta and his colleagues, particularly Nóbrega. And to the Renaissance atmosphere Brazilian literature owes the boastful lyricism of exaltation of local things and countryside, the cycle of the literature of expansion and the prestige of the classical languages, particularly Latin, through the three first centuries of colonization, and also of Greco-Roman culture.

THE BAROQUE AND ARCADIAN PERIODS

But if the first great influence exercised on the nascent literature of Brazil was that of the Portuguese, whose great authors–Camões, Gil Vicente, Sá de Miranda, the chroniclers, the poets of the *cancioneiros*, and the prose-writers of the 16th and 17th centuries—constituted a constant presence in the Brazilian psyche, the Baroque period opened up a new and important source of influence, the Spanish. The Baroque movement was introduced into Brazil by the first Jesuit writers, but it penetrated the 17th and 18th centuries, appearing in the prose and poetry of *ufanismo* (a logical exaltation of the land and countryside), in the native poetry of Gregório de Matos, in the exhortations of Father Vieira and his successors, and in the poetry and prose of the literary academies. In spite of the fact that the major manifestations of Baroque art in Brazil are to be found in the plastic arts—in painting, sculpture, and architecture—particularly in the works of Aleijadinho and in the extraordinary art forms and collections of Bahia and Minas Gerais churches, it was under the aegis of the Baroque, defined not only as an artistic style but also as a cultural complex, that Brazilian literature was actually born. The literary genres most cultivated at the time were the dialogue, lyric and epic poetry, and the theater, along with historiography and pedagogical meditation. In most cases, literature was used to serve the religious and pedagogical ideal of conversion and catechism; the theater, for example, was an extraordinary vehicle in the process of expanding Catholicism.

What mainly characterizes Baroque art in general is the fusion of opposites, such as natural and supernatural, light and shadow, and good and evil, and the contradictions emerging from these opposing terms found a fertile ground in Brazil, due to the tension already existing between the European and the nascent native tradition. As a result of this, the Baroque style became a kind of a modus vivendi in the country and was at the basis of the long process of interbreeding, which came to be one of the most significant traits of Brazilian culture. At the time Baroque art reached its zenith in Brazilian literature, the 17th century, two figures deserve particular mention: Father Antônio Vieira (1608–1697), whose sermons had great impact on the colonizing process, contributing, among other things, to avoid the slavery of the Amerindians, and Gregório de Matos (1623–1696), whose poetry constitutes a formidable satire of every single aspect of the colony's way of life. Matos's poetry was dominated by Baroque dualism: a mixture of religiosity and sensory perceptions, mysticism and eroticism, earthly values and spiritual aspirations. He is a good example of the Baroque soul, in his polar situation, his state of conflict and spiritual contradictions.

In the 18th century, the Baroque taste for hyperbole, for ostentation and the outstanding, was followed by a search for the classical qualities of measure, convenience, discipline, simplicity, and delicacy. This constituted the Rococo or Arcadian, or, to use a more general term, Neoclassicism. In Brazil, the 18th century was a moment of great importance for it constituted a phase of transition and preparation for independence. From the discovery and possession of the land, from the deeds of the *bandeirantes* as they pushed back the Western frontiers, and from the defense against the invader, there came naturally the formation of a common consciousness, a national feeling, which replaced the description of nature and the Indian. The lyrical sentiment of the previous century was replaced by a kind of national pride. There emerged the figure of the "Brazilian" half-breed in blood and soul, the local type, a product of miscegenation, who spoke a language that differed considerably from peninsular Portuguese in accent, intonation, lexicon, and syntax. The literary product of this cultural complex was the Arcadian movement, which flourished among the poets of the so-called Minas School—Cláudio Manuel da Costa, Basílio da Gama, Santa Rita Durão, Alvarenga Peixoto, Tomás Antônio Gonzaga, and Silva Alvarenga—and its beginning was marked by the publication of Cláudio Manuel da Costa's *Obras poéticas* (1768). It was a movement of European origin but marked by a strong nationalist feeling composed of a group of poets who used to call themselves shepherds and lived in straight contact with nature. Among these, Gonzaga was the highest expression.

ROMANTICISM

Romanticism, the movement that followed the Arcadian, in Brazil assumed a particular makeup, with special traits and characteristics of its own, along with the broad elements that linked it to the European movement. Besides, it is a movement of extraordinary relevance, for the country owes to Romanticism an acceleration of the evolution of the literary process as never before. The period between 1800 and 1850 shows a great leap forward in Brazilian literature as it passed from a mixture of decadent neoclassicism and nativist exaltation into an artistic manifestation by which a whole group of high-quality poets and prose-writers was brought together. Some critics even state that this period consolidates in Brazilian literature the autonomy of its national tonality and of its forms and themes, as well as the technical and critical self-awareness of that autonomy. Here, José de Alencar can be pointed out, in prose-writing, as the symbol of the literary changes that have taken place. In poetry,

Gonçalves Dias and Castro Alves followed the same revolutionary process. The task of introducing Romanticism in Brazil fell, though, to Domingos José Gonçalves de Magalhães, in whose manifesto, presented in the magazine *Niterói* (1836), his revolutionary attitude for the renovation of the country's literature was expressed.

Having started in the decade that followed the independence of the country, Brazilian Romanticism had a strong political and social coloration: the writer was not only a man of letters but also an intellectual who had a role in social and political reforms, an educative function, a note that can be clearly felt both in the nationalism of José de Alencar and Gonçalves Dias, present, for example, in their idealization of the Amerindian, and in Castro Alves's struggle for the abolition of slavery in his poetry. Along with this came a break with obligations to the tradition of the language and a yearning to create a new sensibility, represented by the search for different aesthetic forms. Literary genres acquired a distinct makeup: lyrical poetry predominated over the epic or the bucolic and the novel was established, having been widely successful from its first manifestations. The genre favored an atmosphere of sentimentalism, idealism, and sense of the picturesque, as well as historical and social preoccupation, and as such, it conquered the sensibility of the bourgeois society that was being developed particularly in the cities. Summing up, Romanticism adjusted itself perfectly to the spirit of the people, hence the importance it had in Brazilian literature and culture.

THE REALIST ERA

After Romanticism, three great literary movements in prose and poetry flourished during the second half of the 19th century and reached into the 20th century: Realism, Naturalism, and Parnassianism. These currents occupy a cultural period of great relevance in Brazil, in which national and international historical circumstances coincided with the advent of bourgeois, democratic, industrial, and mechanical civilization. In Brazil, three important questions agitated people—the slavery, the religious, and the military question—and in all of them one could feel the influence of those ideas that formed the spirit of the time: laicization, materialism, rationalism, anticlericalism, naturalism. Realism, Naturalism, and Parnassianism, as revolts against Romantic subjectivism, shared the same spirit of precision and scientific objectivity, of exactness in description, of an attention to detail, and of the cult of the fact. Fiction evolved in Brazil toward Realism and Naturalism, and around 1880 the first fruits began to appear. Afterward, it is sometimes the line of Realism, sometimes that of Naturalism, by which short story or novel is written. Poetry, at this time, is under the aegis of Parnassianism; it is a descriptive type of poetry, with exactness and economy of images and metaphors.

Two directions marked the evolution of Realism in Brazil: the social current, attracting social problems, urban and contemporary themes, the common material of everyday life, and the regionalist movement that brings to the fore local color and the importance of the land, which is the real protagonist of this literature. Ever since Romanticism, the growing importance of regionalism was a fact of great significance in Brazil, but whereas in the former movement the region was portrayed from a picturesque and idealized perspective, in the Realist and Naturalist periods Romantic nostalgia and escapism were replaced by a picture of contemporary existence and its environment. Realism taught the Brazilian writer to deal aesthetically with native material rather than sentimentally, and with it literature put roots down into the native soil and attained what Modernism was able to ratify later. As far as the genres are concerned, prose-fiction constituted the best means of literary realization now, and the great figure representative of the whole period was undoubtedly Machado de Assis, followed in the case of Naturalism by Aluísio Azevedo, and in that of Parnassianism by Olavo Bilac, Raimundo Magalhães, and Alberto de Oliveira.

Machado de Assis

Joaquim Maria Machado de Assis (1839–1908) is usually considered by critics as the greatest writer of Brazilian literature. A master of the novel and short story, a refined poet and playwright, and a sensible literary critic, Machado began his production in Romanticism and ended it in Impressionism, but his major works are mostly associated with Realism. Yet he never attached himself to any type of doctrine. Instead, he absorbed aspects of various styles and incorporated them into his own aesthetic ideal. His brilliant and subtle voice set him apart from his 19th-century contemporaries and pointed the way to Brazilian literature of the 20th century. Machado felt that the secret of art lies in classical equilibrium, and he learned that originality and invention are not opposed to tradition. His works can be said to constitute the most genuine and thorough expression of the Brazilian spirit in literature.

Machado was a writer who constantly corrected and perfected himself. The short story was the most fruitful laboratory for his experiments, and he published more

Brazilian novelist Joaquim Maria Machado de Assis in 1896, when he was 57 years old. Machado de Assis is often considered Brazil's greatest writer. (Fundação Biblioteca Nacional)

than 200, some of which are masterpieces of the genre in any language. Yet his major success came from the publication of his novels, especially *Memórias póstumas de Brás Cubas* (1881), *Quincas Borba* (1891), and *Dom Casmurro* (1899). *Memórias póstumas* (*Epitaph of a Small Winner*) was a revolution in terms of narrative technique. By writing his memories from his grave, and with a sharp humor, the narrator offers a critical view of Brazilian life in the mid-19th century; *Quincas Borba* (*Philosopher or Dog?*) marks a development of the philosophical principles that dominate the former novel; and *Dom Casmurro* is the story of a supposed adultery that puts into question the novels based on this theme, on account of the ambiguity with which the issue is treated.

Machado's works reflected his time and environment, and one of his greatest abilities is that of creating characters. He deals with the problems of these people, their customs, preoccupations, and ideals. For him, the national element is not opposed to the universal. He was certainly influenced by his readings of foreign authors, but the elements he borrowed from them were transformed, and by so doing, he attained his originality. "One can use an alien spicery, he used to say, but [he] must season it with a sauce of his own making" (Coutinho 1989).

THE TRANSITION PHASE:
SYMBOLISM AND IMPRESSIONISM

Around 1890, the Romantic elements that were latent during the Realist-Naturalist period reappeared in Symbolism, as a revenge of subjectivity over objectivity, of the interior over the exterior life, of the individual on society. Yet, a movement of idealistic origin, Symbolism in Brazil had to face opposition and hostility by the Realist and Positivist worldview that had been dominant since 1870. The prestige of Parnassianism, which even conditioned the foundation of the Brazilian Academy of Letters in 1897, left no margin for the recognition of the Symbolist movement. Brazilian Symbolism promoted a significant transformation, especially in poetry, and inspired artists of later periods: several Modernist writers, for example, of the 20th century began as Symbolist poets. As well as Symbolism, Impressionism also played an important role in the development of Brazilian literature. Here, though, unlike in the former case, the major transformations were in the field of prose-writing. Around 1890, prose-fiction in Brazil was leaving the Naturalistic aesthetic behind and was heading toward Impressionism, which is in a way a form of Realism as well, the difference being that now the reproduction of reality in an impersonal and objective manner gives way to the recording of the impression that reality provoked in the artist at the very moment when the impression was made. In Brazil, the first great example of Impressionism can be found in Raul Pompeia and later in Graça Aranha and Adelino Magalhães.

THE MODERNIST MOVEMENT

At the beginning of the 20th century, Brazilian literature plunged into a phase of transition and syncretism in which elements of Parnassianism, Symbolism, and

Impressionism were blended together. But the importance of this period is undeniable, for it brought about the transformation that resulted in Modernism. The period witnessed, among other things, the integration of intellectual life, culture, arts, and letters into Brazilian reality. By this time, the word of order in Brazilian literature was to think about the country, to interpret it, and to emphasize the values of Brazilian civilization and the regional qualities of the country's ethnic, social, and cultural traits. A feeling of Brazilianess became a central theme at that time and came out to be one of the highlights of the Modernist movement. This movement, though prepared long before, had its outbreak in 1922 with the realization of the Modern Art Week in São Paulo, and was characterized by its revolutionary tone.

THE MODERN ART WEEK

The Modern Art Week, which took place February 13–17, 1922, at the Municipal Theater in São Paulo, was a landmark of the Modernist movement in Brazil. It was an event originated in the suggestion made by the painter Di Cavalcanti to Paulo Prado that they organize a week of scandals in São Paulo, just like the series of scandals of Elegance Week in Deauville. With the literary spirit long prepared for renovation, some vanguard artists and intellectuals gathered together and laid plans for the battle in which they would assault the bastions of tradition. The innovations that culminated in the Modern Art Week were being prepared long before. So, the Week was rather a coronation than a starting point; it was a result, a point of convergence of forces that had been struggling to manifest themselves. And these manifestations comprised every form of artistic expression—literature, theater, music, painting, sculpture, architecture, and others—and they affected every single aspect of Brazilian intellectual life and culture. But if there was a certain grouping of participants around the Week and a common aim, based on the idea of putting tradition into check, the movement never had the homogeneity and unity of a doctrine. The central idea of the Week was that of demolition, and its main direction was critical. So, everything that made up the patrimony of Brazilian literature was rejected: oratorical emphasis, eloquence, Parnassian practice, the cult of rich rhymes, perfect and conventional meters, classicizing language. After the Week, Modernism began to be redefined by different groups and currents. But the seeds of transformation had been planted, and their consequences would be gradually felt throughout the century.

The Week, a suggestion made by the painter Di Cavalcanti to Paulo Prado, was a destructive blow against the old order, or rather against what its participants called "the bastions of past-ism." Mário de Andrade, one of its most active participants, outlined, 20 years later, what he thought the initial directions of the movement had been: the break with academic subordination, the destruction of the conservative and conformist spirit, the demolition of taboos and prejudices, and the permanent adherence to three basic principles, that is, the right to aesthetic investigation, the updating of Brazilian artistic intelligence, and stabilization of a national creative consciousness (Coutinho 1969, p. 222).

A few years after its outbreak, the Modernist movement began to divide up into groups and divergent currents, and several manifestoes, program-articles, prefaces, and even books of doctrine-poetry were published.

CARLOS DRUMMOND DE ANDRADE

The author of an extraordinary production in poetry, short stories, essays, and chronicles, Carlos Drummond de Andrade (1902–1987) is one of the highest expressions of Brazilian literature in the 20th century and one of the greatest Brazilian poets of all times. He began writing poetry in the 1920s and soon became one of the best-known Modernists in the country, maintaining his position of preeminence until his death, when he was widely respected both in Brazil and abroad. His poetry expresses a preoccupation with the mystery of human existence and with the perplexity of contemporary man before the problem of his own identity.

Drummond's poetry is characterized by the coexistence of an intimate type of lyricism with a strong social concern. The problems that afflicted the world, and particularly his native land, were always present in his poems, and were treated with a high dose of humor. He was very critical of the conventions of society and intolerant of any sort of authoritative measure. Drummond devoted a special affinity toward the poor and simple people, and obtained great pleasure from the observance of daily life. The language employed in his poems was colloquial, retaining the spontaneity of oral communication.

Drummond's prose-writing, his short stories, and chronicles contained the same themes that conquered full expression in his poetry. They were poetical fragments of everyday life, to which he added a reflection about man and his existence, a criticism about capitalist society, or a meditation about his own life, his family, or his childhood in Minas Gerais. The language used in these texts is even more colloquial, and the style is more informal, but at the same time revealing a full domain of narrative techniques. For all this, Drummond is undoubtedly a clear representative of the modern spirit in Brazil.

But the movement lasted until the mid-20th century, and it is usually subdivided into three main generations, those of 1922, 1930, and 1945. The first was a revolutionary generation both in art and in politics. Its objective was the demolition of a fictitious social and political order and of a type of art and literature produced by imitation of foreign models. It was predominantly a "poetical" phase in which the main formal and aesthetic conquests of the movement in the field of poetry were established. The second generation repeated the results of the preceding one, replacing the destructive nature with a constructive intent for the restructuring of values and the configuration of the new aesthetic order. Poetry follows the task of purification of means and forms that had been earlier initiated, and prose broadens its area of interest to include new preoccupations of a political, social, economic, human, and

Brazilian author Jorge Amado in Salvador, Bahia, in 1995. Amado was Brazil's best-selling author. His work has been translated into 49 languages and published in 55 countries. He was 88 when he died on August 6, 2001, in Salvador, Bahia, Brazil. (AP Photo/Luiz Prado-AE)

spiritual nature. It was mainly in prose that its greatest effect resides, since 1928, with the publication of *A bagaceira* (*A Bagaceira*), by José Américo de Almeida, and *Macunaíma*, by Mário de Andrade, as well as by figures like Graciliano Ramos and Jorge Amado.

JORGE AMADO

One of the most popular Brazilian writers of all times and one of the most well-known abroad, Jorge Amado (1912–2001) is perhaps the only man of letters in his country who made a living from his writing. His works have been translated into 49 languages, and many have been adapted for television, cinema, and the stage. Although he was primarily a novelist, he also wrote poetry, drama, political pamphlets, memoirs, short stories, travel accounts, and books for children.

Having started his career as a committed writer, Amado later evolved into a more complex author, aware of his literary *métier* (craftsmanship) and revealing a full domain of narrative techniques. Yet he never set apart his sympathy for the working classes and the poor, who form the axis of his narrative universe.

Most of these people are the protagonists of his stories, which often place an emphasis on Brazilian popular culture and folklore, including Carnival, Bahia's cuisine, the *capoeira*, and the Afro-Brazilian religious cult of Candomblé.

Amado is also known for being an extraordinary storyteller, and his stories are usually of two sorts: a saga of a realistic tone, based on his own knowledge and life experience, and the imaginary narratives that he places on the mouth of his many characters. In the first case, Amado has created a regionalist narrative cycle, represented by novels like *Gabriela, cravo e canela* (*Gabriela, Clove and Cinnamon*, 1958). In the second case, his stories are mostly set in the city of Salvador and are marked by a high lyrical tone, the use of a colloquial language, and a significant dose of humor, sometimes complemented by some touch of the fantastic. Amado has conquered a relevant space in Brazilian literature and has significantly contributed to the projection of this literature in international terms.

The third generation is known especially by a greater precision in form, an effort at disciplinary recuperation, emotional containment, and severity of language. In prose, there is an attempt to revitalize the short story by means of new experiments on the level of language. The great names in this phase are those of Guimarães Rosa and Clarice Lispector.

The critical conscience that Brazil's Modernism developed was undoubtedly one of the movement's major contributions to Brazilian cultural life. Having arisen under the sign of a national consciousness, Modernism continued in this direction and produced a true rediscovery of Brazil, creating a powerful awareness of the country's particularities. Formerly, Brazilian intellectuals had lived with their eyes turned to Europe. With Modernism, this mentality changed, turning artists to experience their native land and give it artistic representation. European contribution continued to be appreciated, but it came to be approached with a critical gaze and to be selected rather than blindly imported as before. Furthermore, these European elements were not the only ones appropriated by the movement. Brazil's Modernist writers also looked back at their nation's literary tradition with a similar critical filter, as it can be exemplified by the image of "anthropophagy," or rather, cannibalism, which stands out as an emblem of the movement. This attention to Brazilian land and environment brought a new preoccupation with regionalism, traditionalism, and folklore. Amerindian and African traditions, regional legends, and popular language with Amerindian and African contributions quickly became common in literature, both in poetry and in fiction. There was also a deep investigation into the country's life, not only in literature but also in its historical, social, ethnographic, and linguistic aspects. Brazilian music and plastic arts were given special value, for example, and there arose a strong preoccupation with what has been called a "Brazilian language" as opposed to peninsular Portuguese.

João Guimarães Rosa

One of the most distinguished authors of 20th-century Brazilian literature, *João Guimarães Rosa* (1908–1967) is best known for the innovations he introduced into Brazilian narrative language. Yet Rosa's importance as a writer is not restricted to this aspect. Having drawn a deep criticism of any commonplace worldview, he offered his readers an oeuvre of such great philosophical dimensions that it is only comparable in his country's literature to Machado de Assis's novels. Despite their apparent difficulty, his works are some of the most widely read and frequently reprinted in Brazil, and they have been translated into several languages.

A meticulous writer, Guimarães Rosa used to rewrite his works often, and he exploited to the utmost the potentialities of linguistic signs and the devices of narrative technique, thus creating a style that was his own and that could be described as an aesthetic whole formed by the various dialects spoken in Brazil, blended with contributions either invented by himself or borrowed from other languages. Rosa is mainly a short-story writer, but he also wrote novellas and one single novel—*Grande sertão: veredas* (*The Devil to Pay in the Backlands,* 1963)—considered his masterpiece.

The setting of Rosa's fiction is the Brazilian backlands and his characters are usually the inhabitants of that region. Yet he is not a regionalist writer alone. His characters transcend the regionalist types due to their existential dimension, and his backlands are a microcosm of the world. Rosa conceives reality in an all-encompassing manner. Myth and fantasy are for him a part of reality as well as rationalistic logic, and he treats these elements in a similar way in his fiction. Rosa's works are at the same time regional and universal, mimetic and self-conscious, "realist" and "antirealist," and it is precisely this paradox that best characterizes them.

Clarice Lispector

Born in Ukraine but raised in Brazil from her early childhood, Clarice Lispector (1925–1977) is one of the most distinguished women writers of her time and one of the foremost Brazilian writers of the 20th century. Her novels, short stories, and chronicles have been translated into several languages and are studied both in Brazil and abroad. Considered as an introspective kind of fictionist, she was mostly preoccupied in examining the self, and the account of her existential or metaphysical speculations was transmitted in a style peculiarly her own. For this reason, she made ample use of modern technical devices. The search for a new type of language to translate her inner life was also at the center of her aesthetic preoccupations.

Lispector extracts most of her fictional material from daily life, and her themes form a wide spectrum, but they usually bring about a deep questioning on issues like death, God, the writing craft, people's attitudes and behavior, and women's role in society. She is not a feminist writer in the strict sense of the term, but the majority of her novels and short stories focus on the world from a feminine point of view.

Lispector's short stories are usually masterpieces; it is in their construction that she reveals her full domain of narrative technique. Two other genres widely explored by Lispector were the chronicle and the stories for children. The former is in her case of a predominantly biographical sort, and the latter are a kind of narrative with a good dose of humor and some philosophical touch that pleased both children and adults. Some of Lispector's works were adapted for the stage and made into films with much success, as it is the case of *A hora da estrela* (1977), *The Hour of the Star*, her last novel.

THE MODERNIST LEGACY AND POSTMODERNISM

As a consequence of the conquests achieved by Modernism, in terms both of the density and depth of the Brazilian theme and of the pursuit of technical perfection of the literary art, literature reached a position that showed the degree of maturity and conscious integration of the Brazilian mind and soul. From the second half of the 20th century to the present, Brazilian literary scene has been mostly characterized by a plurality of trends, which on the one hand follow the lines established by the Modernist movement and on the other hand express the multiple set, based on micro-narratives, that has been often designated by critics as Postmodern. In both cases, however, foreign imports are now combined with a consciousness of Brazilian cultural aspects, and the result has been the emergence of different currents, all marked by a kind of dialogue between a foreign influence and a local touch. Such is the case in poetry of most well-known movements like Concretism, Neoconcretism, Praxis Poetry, Postal Art, and Tropicalism, and in prose-writing of the journalistic fiction, the memoirs or testimony, the excursions into the fantastic, and the self-conscious, particularly feminine, line of narrative. All these currents—to which we can add more recently the production of the so-called minority groups, like the Amerindian, the African, and those resulting from 19th- or 20th-century European immigration, and a type of popular literature called *literature de cordel* (cordel literature)—crammed with aspects like the constant presence of the media, the fragmentation of the text, the abundant use of a polyphony of voices, an emphasis on stylistic eclecticism, a strong intertextuality, parody, and the frequent use of metalanguage, are efforts to represent the new lifestyle of a country where sophisticated computers are found together with a high measure of misery and illiteracy.

THE BRAZILIAN DRAMA

As well as literature in general, theater in Brazil was born under the sign of the Baroque. Brought to Brazil by the Jesuit priests, it constituted a vehicle of extraordinary effect for pedagogical and moralizing reasons, and as such, it played an important role in the process of conversion and catechism. By using grandiloquence and sumptuousness, luxury and pomp, or other devices that intimidated and impressed the senses, plays were staged to infuse into people's spirit a gloomy and negative concept of earthly life, of the contrast between good and evil, between spirit and flesh, salvation and damnation. The idea was to spread Christian faith at all cost, and they even made use of aspects of the Indian cultures in order to intimidate and dominate them. These plays broke with the rules of Renaissance poetics and introduced novelties such as the dislocation of the center of interest or gravity, the multiplications of points of view and protagonists, disproportion, and ornamental pomp. Also introduced were such operational elements as the division of the stage from the audience, the darkening of the theater, and the use of the devil as a character and of devices as thunder, lightning, fire, and smoke to give an impression of death or hell. Besides, it was not only the actors who took part in the play, but the audience too, which was pulled into the dramatic event and involved in it.

After this period of indoctrination in which the first theatrical manifestations were found in Brazil, theater went through a kind of hibernation only to reappear in Romanticism under the form of drama. During that time, however, there were some profane performances, usually put forth to celebrate an important event or an illustrious personality, and the staging of some European plays, mostly French, that had been translated into Portuguese. Two figures had some relevance at that period, namely, Manuel Botelho de Oliveira and Antônio José da Silva, o Judeu, but it was only in the 19th century that Brazilian theater would give a decisive step, thus establishing itself as an important genre. At that time, there appeared not only several playwrights of certain scope, among whom Martins Pena, the creator of popular comedy in Brazil, but also many established poets and prose-writers ventured into the drama, producing a number of plays that met with success from the public. The most common theatrical performances were the Romantic drama, historical in basis, bringing together social, political, moral, psychological, and religious problems; broad questions; and a great number of characters. This type of theater renounced the classical unities of time and place, and turned toward the national past and modern history, in search of a new form. Besides, local color and customs formed the basis of the reality represented. Finally, it mingled verse and prose, and brought about a union of the noble and the grotesque, the grave and the burlesque, the beautiful and the ugly.

In the second half of the 19th century, the period of predominance of the Realist and Naturalist movements, theater experienced an important development. Now, the role played by João Caetano, who had been the great actor of the Romantic period, was no longer unique, and the number of companies had multiplied. Yet two tendencies held the scene: on the one side, the setting of European plays, and, on the other side, a taste for the local color, expressed mainly by comedy, based primarily on episodes from daily life. Theater did not receive a doctrinaire aesthetic outline, but it often brought about themes in vogue at the moment, some of them of a certain political tonus. Among the playwrights of the period, Artur Azevedo was the most popular, especially due to the portraits he offered of the Brazilian man of his time. After him, a few plays by different authors met with success from their audiences and were appreciated by criticism, but it was only around 1920, when the Modernist movement began to spread its rays upon the intellectual milieu that theater went through a new flourishing period, regaining its prestige among the public.

The renovations introduced by Modernism were seen in the aspects not only of dramatic creation itself but also of set designing, staging, and the formation of actors. New techniques inspired by Symbolism, Expressionism, and Surrealism were explored, new light and sound devices were used, and several schools for the formation of actors appeared, granting the theater with a professional stamp it had never met before. Many groups and companies were formed, encouraged by the growth that theater was undergoing, a number of new theater houses were built up all over the country, and a constellation of good actors came to hold the public's admiration. Among the playwrights of the time the main tendency was toward a confluence of foreign experiences with Brazilian themes, be them of a social, historical, or cultural order. The greatest expression of this new moment of Brazilian theater was Nelson Rodrigues, who broke abruptly with traditional conventions and gave way to themes so far forbidden, as that of sex. Also significant, though representing different trends, are names like those of Juracy Camargo, Henrique Pongetti, Dias Gomes, Ariano

Suassuna, Jorge Andrade, Gianfrancesco Guarnieri, Oduvaldo Viana Filho, Augusto Boal, and, in the case of the theater for children, Maria Clara Machado.

Modernism was highly beneficial to dramatic literature, as it had been to literature in general in Brazil, for it was the moment in which the theater found its way and established itself as an important genre in the cadre of the country's literary production. Yet, in spite of the number of good playwrights, directors, actors, and theatrical apparatus, Brazilian theater cannot be said to have attained the same level as that of the poetry and prose-fiction produced in the country. From that moment on, what can be observed is a pluralization of trends, as in case of the other genres, and a gradual improvement in the quality of productions and in the variety of themes. Theater continues, in the late decades, to hold a significant place in Brazil, and despite the vertiginous growth of other genres as the cinema and the television, it is going through a phase of great vitality, perhaps one of the most fruitful of its entire history.

REFERENCES

Coutinho, Afrânio. 1969. *An Introduction to Literature in Brazil.* Trans. Gregory Rabassa. New York: Columbia University Press.

Coutinho, Afranio. 1989. In Carlos A. Sole, and Maria Isabel Abreu, eds. *Latin American Writers.* 3 vols. New York: Charles Scribner's Sons, Vol 1, pp. 257–258.

Coutinho, Afrânio, ed. 2004. *A Literatura no Brasil*, 7th ed. 6 vols. São Paulo: Global Editora.

Coutinho, Afrânio, and J. Galante de Sousa, eds. 2001. *Enciclopédia de Literatura Brasileira*, 2nd ed. 2 vols. São Paulo: Global Editora.

Coutinho, Eduardo F., ed. 1991. *João Guimarães Rosa*, 2nd ed. Rio de Janeiro: Civilização Brasileira.

Coutinho, Eduardo F. 1997."Postmodernism in Brazil." In Hans Bertens and Douwe Fokkema(eds.), *International Postmodernism. Theory and Practice.* pp. 327–336. Amsterdam/Philadelphia: John Benjamins Publ. Co.

Coutinho, Eduardo F. 2007. "Brazilian Modernism." In Astradur Eysteinsson and Vivian Liska(eds.), *Modernism.* Vol. 2., 2 vols, pp. 759–768. Amsterdam/Philadelphia: John Benjamins Publ. Co.

Rector, Mônica, ed. 2005. *Brazilian Writers* (*Dictionary of Literary Biography*, v. 370). Detroit and New York: Thomson Gale.

Solé, Carlos A., and Maria Isabel Abreu, eds. 1989. *Latin American Writers.* 3 vols. New York: Charles Scribner's Sons.

Art

Alice Heeren

The story of the "discovery" of Brazil by Portugal in the 16th century has been a greatly disputed topic. Firstly because the notion of discovery implies a civilizing Europe attempting to recuperate the lost souls it found in the land we now call Brazil.

It disregards the previous history of the geographical area and its inhabitants and can be considered an act of violence in itself. Second, the narrative that in 1500 Pedro Álvares Cabral officially "discovered" Brasil, arriving on the shores of present-day Porto Seguro in an attempt to reach the Indies through the Atlantic Ocean, has been questioned by scholars. Research on the period has suggested that before Cabral, and even the Spaniard Vicente Yáñez Pinzón, set eyes on the Brazilian coast, Portugal had already been to the region. This early discovery had led the Portuguese Crown to ask the pope for a change in the demarcation that previously divided the world into Spanish and Portuguese territories. This new treaty, which put Brazil on the Portuguese side of the border, was the Treaty of Tordesillas of 1494.

Nevertheless, and despite the political maneuvers to guarantee the right to Brazil, between 1500 and 1534, Portugal paid very little attention to its South American colony, more concerned with the already profitable Eastern colonies. The Portuguese Crown turned its attention to Brazil only in 1534 when it divided the land into territories giving to rich Portuguese merchant families. Later, in the early 1550s, with the constant threats of invasion by French, Dutch, and other European maritime powers, Portugal finally began to actively populate the "empty" Brazilian territory, particularly the coast. The Portuguese Crown named Tomé de Souza as the general governor and Salvador the capital of the colony.

Although this Eurocentric version of history has been widespread in history books inside and outside the country, the study of the art of Brazil showcases how limited this narrative really is. A look at Brazil's vast and prominent archeological patrimony reveals that there were other peoples who expressed themselves artistically in what is known as Brazil today long before the Portuguese arrived in the 16th century. The idea of an empty Brazil waiting for the Portuguese to colonize and explore it, as well as to civilize it, assumes that those that lived there previously did not constitute people with cultural and material practices, social structures, and ownership over the territory they inhabited. The justification for colonization by European nations at the time was in most cases bringing civilization to the "savage" indigenous peoples from around the globe and converting them to Christianity. In Brazil, it was not different. This excuse was used to justify the extirpation of the natives' territories, their imprisonment, and their murder. The genocide of the indigenous people of Latin America is yet to be fully acknowledged, but most important for the purpose of this discussion, the material culture of the indigenous peoples of the region allows us a glimpse into the societies and artistic lineages that were interrupted with the arrival of the Portuguese to Brazil.

INDIGENOUS ARTS OF THE BRAZILIAN REGION

Since the Brazilian territory was inhabited by several native indigenous societies upon the Portuguese's arrival, as well as had been the home of many other groups of peoples before, which moved away before the 16th century to other regions of Latin America or became extinct, it is not surprising that Brazil has a large and important archeological patrimony. Nearly 10,000 sites have already been unearthed, such as Sambaqui do Pindaí (state of Maranhão), Parque Nacional da Serra da Capivara in São Raimundo Nonato (National Park Serra da capivara) in São Raimundo Nonato (state of Piauí), and Lapa da Cerca Grande in Matozinhos (state of Minas Gerais).

The paintings found in the walls of caves and shelters in the Lapa da Cerca Grande and Parque Nacional da Serra da Capivara, as well as many sites in the Amazonian region, for instance, are remnants of groups that inhabited this area as early as 6000 BCE.

Analyses of patrimony from the preconquest era (pre-1500) contribute to a reevaluation of the official history of Brazil giving a voice to the indigenous peoples of the region. Many of the objects that survived were those looted from Brazil during the colonization era, as well as those that were gifted to the Europeans by the Indians themselves. Other sources important to the study of this material culture are images and accounts produced between the 16th and 18th centuries by European travelers such as Andrè Thevet and Hans Staden, who spent considerable time in the Brazilian region and had direct contact with the indigenous groups of the area.

Ceramics, spears, clothes, and featherwork are among the most important remants of native peoples from the Brazilian region. These objects had both ceremonial and utilitarian purposes, but whatever their practical end, they depict singular aesthetic ideals, which reflect the group that produced them, illuminating not only the habitual, everyday practices these objects were part of but also the cultural and aesthetic fabric of the native Brazilian. Important examples are the ceremonial spears from the Munduruku tribe, the ceremonial vests from the Tupinambás, and ceramic objects for funerary or domestic use, spoons, statues, and body adornments of the Marajoara and Santarém peoples.

The ceramic works of the Amazon region developed from crosshatched patterned ceramics of the 1000 BC Annatubas peoples to the complex polycromic designs of the later Marajoara phase. The Marajoara peoples, named such because of their arrival to the Marajó island in the state of Pará around 400 of the current era, had already started to disappear after 1350, their culture absorbed by the migrant Aruã peoples who came to settle in the area and later encountered the Europeans. Nevertheless, archeologists have recovered several objects from the Marajoara group in the Marajó region, and aspects of their style have survived and are still influential today. Their stylized representation of the human body and simple, geometric decoration using red, black, and white and their use of both incision and excision, as well as modeling in the production of the pieces, show the technical prowess of these indigenous peoples. Not only pots, pans, plates, and spoons but also thongs, funerary urns, and figurines are part of the archeological material recovered from the marajoara phase.

Another cultural group, known as the Tapajós, had a wide production in ceramics, which have been recovered throughout archeological sites in Brazil. These are called Santarém ceramics and are more complex in terms of ornamentation than the simple, geometric decoration of the Marajoara pieces. Human figures support the Santarém vases. The Santarém ceramics is populated by figurative motifs, with zoomorphic designs being the prevalent ones. The use of floral motifs are extremely rare, and even the animals are highly antropomorphisized in the Santarém designs.

The Tapajós inhabited the area where the rivers Tapajós and Amazonian meet in the state of Pará. They were first mentioned in the documents of the colonial period when, in an expedition inland to the Amazon, the Spanish Orellana encountered them in 1542. Later, the Portuguese captain Pedro Teixeira, in 1626, would make contact, estimating their numbers at 240,000. The Tapajós resisted the *conquistadores*

Fifth-14th–century funerary urns uncovered on Marajó Island in the Northern region of Brazil during an archeological investigation. (Carlos Moral/Dreamstime.com)

(conquerors) and the Christian missionaries well into the 18th century. Again we find pots, pans, thongs, and figurines, but not funerary urns since the Tapajós did not keep the ashes of their dead.

The featherwork of the Tupi peoples is another part of the material culture of preconquest era Brazil, to which we have access today. It is important to realize first that the term "Tupinambá" is a European construction and was applied generically to different peoples living along the 400-kilometer Brazilian coastline during the 16th century. The Tupi featherwork, one of the main artistic manifestations of those peoples, includes capes, cloaks, headdresses, bonnets, and ankle bracelets, made from a variety of colorful feathers taken from different animals found in the Brazilian territory or obtained by the Tupi through trading with other indigenous peoples. The feathered cloaks and capes were markers of status in the Tupi society and used by both leaders and shamans: the two most important figures in the social hierarchical organization. Several techniques were used to make these pieces, including different knots and processes to modify the feathers. A bonnet from the era now found in the Copenhangen Nationalmuseet Etnografisk Samling recalls a newly hatched bird. This is accomplished by the use of brightly colored yellow and blue-hued feathers. The soft plumage gives the impression of the head of a bird just coming out of its egg. Tupi featherworks were important collectibles in the 16th and 17th centuries, making their way to Europe and filling curiosity cabinets and houses of merchants or aristocrats. These pieces were wildly exchanged, procured, and gifted by the Jesuit priests who were in direct contact with the Tupi peoples in Brazil. The peoples, now known under the umbrella Tupi, were nearly extinct by the 18th century because of a combination of diseases, slavery, war, and exodus to the interior of the country to evade the Europeans.

The remnants of the indigenous Brazilian cultures, their aesthetic and Symbolism, have been reappropriated in subsequent centuries in attempts to construct a national cultural and political identity. The instrumentalization of indigenous (and later African) cultural expressions, to construct the ideal of an autochthonous artistic practice, will be a topic revisited in several sections of this chapter.

DUTCH ARTISTIC INFLUENCE IN COLONIAL BRAZIL

In the 1580s, Brazil lived through a period of turbulences suffering repeated attacks of the English, French, and Dutch. As these European nations settled for short periods in the Brazilian territory, they helped artistic movements to flourish in the colony. One important example is the period between 1630 and 1654, when part of the Brazilian Northeast (now state of Pernambuco) was occupied by the Dutch who attempted to circumvent the Spanish effort to block its participation in the sugar market. Mauricio de Nassau was sent to Brazil to govern the region. Nassau recognized the importance of the arts as a colonial tool, how it could give the Dutch a foothold by promoting good relations with the inhabitants of the area—mullatos, Amerindians, and second-generation Europeans—as well as its importance in the development of a Dutch colony in the Americas. Nassau was known for his religious tolerance in a moment when the Catholic Church was extremely intolerant, and he built the first Brazilian Synagouge in the city of Recife, even though he was a Calvinist himself. Nassau looked to transform Recife into a modern city and created an urban design that still has left traces today.

The artists, writers, draftsmen, and scientists of the Dutch India Company, who were part of Nassau's entourage, were especially interested in the inhabitants of coastal Brazil and their environment, Albert Eckhout and Frans Post most prominently. They depicted these individuals in paintings and novels, and used them in scientific studies.

Eckhout produced a unique oeuvre of 26 oil paintings, among them figural paintings of Amerindians and still lifes of local flora later gifted to Frederick III, which remain today in the Nationalmuseet in Copenhagen, as well as over 400 drawings published by Christian Mentzel in the volumes *Theatri Rerum Naturalium Brasiliae* e *Miscellanea Cleyeri*. Eckhout remained in Brazil for seven years, between 1637 and 1644, and his most renowned works of this time are the four pairs of ethnographic paintings of Brazilian peoples of African and Amerindian descents now at the Nationalmuseet in Copenhagen. The fragile balance between reality and fiction, in situ representations, and wish fulfillment of European assumptions about Brazil is evident in these oil paintings. The compositions are typical of ethnographic representations—a genre being consolidated as a style during the period Eckhout was working: frontal, full body views in the center of the composition as close as possible to actual human size. Eckhout also included in the background of his pieces symbolic objects he associated with the ethnic group he was looking to represent. Both in Eckhout's figural paintings and in his still lifes, the influence of the then-flourishing Dutch style is evident. His focus on textures and the material quality of what he represented, the stark light and dark contrast of his compositions, and the many perspectives presented at once in his paintings show the influence of the prominent northern artists of the time.

Frans Post was also prolific during his stay in Brazil. The most common themes in Post's work of the time were civil architecture, landscapes, and scenes of naval and terrestrial battles. He paid special attention to the inhabitants of the region while also depicting the natural and man-made monuments of the Brazilian Northeastern coast. Post remained eight years in Brazil and although his production may have been large, only few examples survive, such as *Vista da Ilha de Itamaracá* (*View from the Isle of Itamaracá*, 1637), *Vista dos Arredores de Porto Calvo* (*View of the Surroundings of Porto Calvo*, 1639), and *Forte Hendrik* (*Hendrick Fort*, 1640). Post's style is observable in these examples, as is the influence of the Dutch landscape painting tradition. The low horizon lines creating large expanses of sky, the juxtaposition of the tropical vegetation and the buildings designed in European styles, and the peoples who navigated between these spaces are recurring elements of his work. He nevertheless maintained the sober palette of the Dutch traditional landscape painting instead of turning to the expansive colors of the Brazilian flora.

Post's *Indians in the Forest* (1669) is particularly interesting in the environments it condenses into the pictorial space: the natural landscape of Pernambuco, the Jesuit mission villages, and the market place around the town, as well as the people around the market: the Europeans followed by their loincloth-clad slaves and the Indian Tupis who perform a dance to the right middle ground. This image presents the complexities of Brazilian society at this time with its hierarchical nature, its histories of domination, and its bountiful natural and ritual diversity.

The complexity of the production of Eckhout and Post cannot be underestimated, and the way they constructed the notion of the exotic New World and their peoples was largely influential in Europe at the time as it is still today. Together with the artifacts the Dutchmen took back to their home country, these works are still some of the most important archeological and historical evidence of life in Brazil during this period.

BAROQUE AND ROCOCO: THE 17TH AND 18TH CENTURIES IN BRAZIL

Continuing through the 17th century and further into the 18th century in the art history of Brazil, it is important to note how much architecture and art were intertwined to the point that a history of art in this period without a history of architecture is almost impossible to write. The changes in the artistic manifestations of this period are marked by the instauration of the religious *confrarias* and *irmandades*—groups of powerful nonreligious patrons, which worshiped specific saints and promoted their cults through building, art practices, and processions. These groups had money and influence and were especially aware of the growing Baroque and Rococo movements in Europe, which ideals they imported to Brazil.

The Baroque style in Brazil was brought by the Portuguese artists and artisans and hit its height in the 18th century with its variation into the Rococó. It was a mixture of the Baroque style of the Iberian Peninsula and that of France and Italy. Developed in the lay guild groups, the early traces of the Baroque in Brazil are particularly visible in the *portões*, large carved portals set in the facades of religious buildings. Built of stucco on masonry with orange cornice and pilaster with front pieces of gray soapstone elaborately carved around great wooden doors with heavy bosses painted apple green or violet, the churches of Ouro Preto, Mariana, Recife, Salvador, Rio de Janeiro, among others, had elaborate interiors, and, particularly in Rio and the Northeast, made use of the traditional Portuguese blue tiles.

The first three decades of the Baroque movement in Brazil reflect the incorporation of the European models, but from 1730 onward the movement took a new form in the country. Valentim da Fonseca e Silva (Mestre Valentim) and Antônio Francsico Lisboa ("O Aleijadinho"—The Crippled) were two of the most important artists/builders of the period. Mestre Valentim was active in the region of Rio de Janeiro, designing the tower at the entrance of the Praça XV de Novembro in Rio, and Aleijadinho was the most renowned artist of the Minas Gerais region.

The syncretic mixture of the Portuguese styles like the Joanine and the Pombaline, the Baroque already developed in Brazil, and the newly arrived rococó references is nowhere more fully realized in the country than in the region of Minas Gerais. During the mid-18th century, Vila Rica—known later as Ouro Preto—the main city of the Minas Gerais region supersedes the old coastal centers as the locus of wealth in the Brazilian colony. The Rococó of Minas Gerais appears most prominently in churches like the São Francisco de Assis of 1766 attributed to Aleijadinho. Its sculptural *portões* in soapstone (typical material of the region) and its sinuous architectural plans show the influence of European rococó. The chaotic and dissonant elements join to create a total harmony, which has been recognized as the trademark of Aleijadinho's work.

The theatricality of the sculptural figures and the interplay between the gold-gilded woodcarving and the ceiling design by Manoel da Costa Ataíde are central aspects of the design of the São Francisco de Assis chapel. The web of hollow supports in the rococó-style ceiling substituted the heavy perspectival models of the Baroque, and the Assumption of the Virgem painting in Ataíde's ceiling in the nave of the chapel is the most successful instance of perspectival painting in colonial Brazil. Ataíde's ceiling, painted between 1801 and 1812, is a delicate composition favoring tons of whites, blues, reds, yellows, and browns with figures and ornaments framed by painted architectural elements like columns and arches. Ataíde is also responsible for the series of six panels imitating blue Portuguese tiles that are also found in the chapel. The designs are copies of engravings representing the life of Abraham by Dermane found in an illustrated French Bible of the period. The difficulty in exporting tiles to Minas Gerais because of its inland location caused the use of tiles, most prominent in Northeast Brazil, to be reduced in Minas Gerais, this false tile ensemble by Ataíde being an interesting counterexample.

Another collaboration between Aleijadinho and Ataíde is the complex at the sanctuary of Bom Jesus dos Matosinhos in the city of Congonhas do Campo in the region of Minas Gerais. Built as payment for a promise made by the Portuguese Feliciano Mendes, the sanctuary in Congonhas is modeled after the sanctuary of Bom Jesus in Braga, Mendes. While Tomás de Maia Brito was responsible for the architectural project of the sanctuary complex, Aleijadinho was responsible for the 12 sculpted prophets in the central churchyard and the 64 sculptures for the *via sacra* for which Ataíde executed the polycromy and gold leafing.

Regarding the production of religious sculpture during the late 18th century in Brazil, it is important to highlight the importance of the state of Bahia because of both the abundance of pieces and the specificity of their style. The internal market for religious images and small personal oratories was very large in Brazil, since a personal relationship with images was a character of luso-Brazilian Catholic practices. The religious images produced in Brazil as a whole were characterized for having only the most basic iconographical markers and a strong expressional force. The Bahian style in particular was marked by delicate gestures and attitudes, vibrant colors, and gold leafing, while the Minas Gerais sculptures were more discrete and simple, but with a closer attention to expression and movement.

If Aleijadinho was the main figure in Minas Gerais, in Rio de Janeiro it was Mestre Valentim's works that were most prominent. Particularly, the Passeio Público complex in Rio and the tower in the front of the Praça XV de Novembro are important examples. The Passeio Público, part of the project of viceroy D. Luis de Vasconcelos for the embellishment and sanitation of Rio de Janeiro, began in 1779 and was inaugurated in 1784 inspired by the Pombaline movement in architecture in Portugal, as well as the example of leisurely gardens across Europe. The decoration with sculptural elements like obelisks, ancient gods, and animal sculptures, as well as fountains called upon neoclassical French landscape design and was framed by the careful choice of local Brazilian flora, particularly palm trees.

Nevertheless, Mestre Valentim's hybrid style, encompassing elements of Baroque, rococo, Pombaline, Joanine, and French neoclassicism, is central to the understanding of this complex. The Baroque contortion of the iron portal opening to the Rua do Passeio, the joanine pilasters topped by classical urns, the iron rococo design of the main entrance with its floral and curvaceous ornaments, all point to Mestre Valentim's abilities to condense styles: the trademark harmonious syncretism of his designs. European styles aside, like Aleijadinho, Mestre Valentim's African origins are yet to be fully explored by scholars of the period.

Another prominent mulato artist in Brazil during the second half of the 18th century was the painter Leandro Joaquim, who painted as part of his important series of images of Rio de Janeiro, the *Vista da Lagoa do Boqueirão e do Aqueduto de Santa Tereza* and the *Revista Militar no Largo do Paço* (*Sight from the Boqueirão Lake and the Aqueduct of Santa Teresa and Military Swipe of the Largo do Paço*). These paintings, at the surface, appear to depict a harmonious marriage of the Brazilian religious, political, and social spheres. Nevertheless, once examined more carefully, the images denounce the very instability of these ideals during this period. Joaquim's series of six oval panels dates from the end of the 18th century and were meant to decorate one of the pavilions of the Passeio Público in Rio de Janeiro. These paintings were elaborated in a period when, following the Haitian Revolution, European powers like Portugal were concerned with the growing restlessness in their colonies, particularly due to their harsh exploitation of the African and Indian descendants. The propagandistic nature of the painting production of the period show the attention of governors, like viceroy Vasconcelos, to these instabilities, and their attempts to maintain the status quo through utopian discourses. It is thus even more interesting to see how this instability transpired in Joaquim's work, even if the paintings remain faithful to the neoclassical genre.

Joaquim's attention to the people who inhabited the city is of particular interest as this six-panel series commissioned by viceroy Vasconcelos was meant to exalt the natural resources of Rio de Janeiro, while simultaneously calling attention to the urbanist restructuring promoted by the viceroy's government. The aqueduct of Santa Teresa was a major sanitation project and the Largo do Paço—a major civic monument based on the Praça do Comércio of Lisbon—was depicted in Joaquim's painting on the day of its inauguration. Nevertheless, the compositional strategy of *Vista da Lagoa do Boqueirão e do Aqueduto de Santa Tereza*, as observed by historian Rafael Cardoso, points to how the aqueduct was not Joaquim's main concern in this painting. In fact, the aqueduct itself does not completely appear in the composition. Although the aqueduct was a key endeavor of the colonial government, in Joaquim's painting it is overshadowed by the people depicted in the first plane of the composition engaging in everyday activities. Also, while a shift in angle would have allowed the painter to represent the aqueduct as a whole as well as the fountains for which it provided water, Joaquim chose instead a part of the city where such resources were not available, where there were no fountains or carefully designed landscapes. In the painting we stare at the activities taking place on a lake in the outskirts of Rio, where black people push cow-led vehicles and carry packets in their heads, and a handful of children play next to animals refreshing themselves. There are six figures in the forefront of the composition: six slaves—as perceptible by their engagement in labor like carrying clothes and merchandise in their heads or playing instruments in the street, as well as the fact that they are barefoot—who appear prominently close to the viewer. This juxtaposition, of the slave labor and the poor peoples of the community, below the monumental aqueduct surrounded by the lush immensity of the Brazilian landscape pointed to the unstable relationship between the institutions of colonial Brazil and the people they exploited. So more than a propagandistic support to the governmental policies of viceroy Vasconcelos, what Joaquim shows here is the underlining social hierarchy that was implicit in these measures for the beautification of Rio de Janeiro and its sanitation strategies, which like Baron Haussmann's Paris, attempted to cleanse the city of unwanted communities pushing them to the periphery and setting in their place monuments that favored only the high-class white families of the capital.

Similarly, in *Revista Militar no Largo do Paço* (*Military Swipe of the Largo do Paço*), Joaquim represented an important monument in the colonial imaginary and destabilized it. In this second case, the Praça XV de Novembro square is depicted on the day of its inauguration. Joaquim paints the square as viewed from the sea with its new neocolonial buildings and the military parade in honor of its opening. This military presence, the spectacularization of colonial power depicted in the very spot—in the waterfront of the Praça XV de Novembro—where slave ships traditionally arrived and their cargo was sold, creates a tension within the painting that is extremely poignant. The Praça XV de Novembro, initially known as Terreiro do Paço or Largo do Paço, was until 1770 the main entrance of slaves into the Southeast Brazilian territory, and the juxtaposition of this spot with the military parade—exposing both the violent power of the colonial government and its aspiration for order and cleanliness—depicted in the piece results in a destabilizing image of the central gateway into Rio de Janeiro. Again Joaquim subverts the intents of a public commissioned work meant to commemorate the beautification of Rio de Janeiro and the force of the colonial government by inserting the memory of those exploited by this social and political structure.

NINETEENTH-CENTURY BRAZIL: THE FRENCH MISSION AND THE ROOTS OF MODERNITY

From the end of the 18th and throughout the 19th century, the French and their neo-classic or Beaux-Art styles began to dominate Brazilian art and architecture. After Emperor Dom João VI, fleeing Napoleon's army, ran to Brazil with his court in 1808, Rio de Janeiro, the capital of the colony, became the center of the Luso-American Empire. With the opening of the Brazilian ports and the end of the monopoly of Portugal over Brazilian exports, the country entered transatlantic capitalist routes and was rapidly influenced by other European powers.

Looking to solidify the status of Rio de Janeiro as a modern city and its image as capital of the empire, as well as guarantee the influence of the Portuguese Crown over the Brazilian territory, in 1816, the emperor brought a group of artists, architects, and intellectuals to shape the minds of the Brazilians and the space of Rio de Janeiro; this has been known as the French Artistic Mission. This group, inclined to leave France after the return of the Bourbons to power and led by Jacques Lebreton, included artists Jean-Baptist Debret, the brothers Marc and Zépherin Ferrez, Nicolas Taunay, Auguste-Marie Taunay, and architect Grandjean de Montigny. Even though the arrival of the French Mission meant the professionalization of the art and architectural profession in the country, and its separation from the realm of craft, these individuals did not represent one artistic and stylistic current. While on one hand there were those dedicated to the neoclassical style such as Grandjean de Montigny, on the other hand, there were those, like Taunay, distanced from the rigors of neoclassicism and inclined toward Romanticism, which would also have a substantial impact on the country.

Most of the individuals who arrived with the French Mission later taught at the Academia Imperial de Belas Artes (Imperial Academy of Beaux Arts), inaugurated in 1826 and one of the main projects of the artistic mission, influencing the next generation of academic artists. The Museu Real and the Biblioteca Real were also institutions founded in the country at the time. The question of the nation, its structure, goals, and ideologies, became an important question in Brazil of the early 19th century. With the country's newfound independence and the severe economic, cultural, and political changes later in the century, the artistic field began to shift. Rio de Janeiro had already been impacted by a series of progressive viceroys—particularly Vasconcelos—in the 18th century with the construction of civic buildings in a monumental style and investment in urban services such as water supply. Also, painters, feeling the impact of contemporary currents in Portugal, focused on secular topics with classic influences, as in the work of Mestre Valentim and Leandro Joaquim. These previous changes, in many ways, set the tone for the impact of the Academia in Brazil and the change in the role of art and artists.

The academy advocated for art as the expression of ideal beauty, valued classic themes like historical painting and portraits, highlighted the importance of technique over experimentation, and promoted the use of noble materials like oil, marble, and bronze. Ultimately, the academy was supported by the royal family and produced works under their aegis. It generated the institutionalization and standardization of artistic practice and education across the Brazilian territory, a process further emphasized by the employment of imigrés from Europe to lead regional schools like Recife and Salvador.

While sculpture and architecture, led by the Ferrez brothers and Grandjean de Montigny in Brazil, more consistently followed the French academic precepts with their attention to symmetry and reference to Greek-Roman models, painting, from Debret and Nicolas Taunay, as well as his son Félix-Émile Taunay, manifested a larger array of artistic options. Although Nicolas Taunay quickly returned to Europe, the five years he worked in Brazil resulted in pieces that, although faithful to the neoclassical attention to drawing, also manifested the artist's sensibility for local colors. Furthermore, Taunay focused on landscape painting, which was less important within the academic doctrine, thus leading the development of this genre in the country without much resistance.

Jean-Baptiste Debret's production on the other hand is characterized by two lines: First, he accompanied many travelers throughout the country during the 19th century, thus registering the plant and animal life of Brazil, as well as the everyday practices of those settled in the country at that point; second, he created large oil paintings commissioned by the Portuguese royal family, which included portraits and paintings of important historical events. In the first branch of Debret's work, one finds a mix of exoticism and concern for scientific cataloguing, as well as a more pronounced formal freedom and spontaneity. Also, this part of his practice shows his political interests revealing themes approached by the French Revolution: absolutism, religion, and slavery. Debret's portraits of slaves and the slave trade in Brazil are specifically poignant with his print *Slave Market*, Rio de Janeiro, Brazil, 1816–1831, which appeared in the collection of his works published as *Voyages Pittoresque et Historique au Bresil* (*Historical and Picturesque Travels to Brazil*). Underlined by the caption "Boutique de la Rue du Val-Longo" ("Boutique at the Val-Longo Street"), this print shows adult slaves sitting against the walls of a small room while children play on the floor. To the left of the composition sits the seller negotiating with someone looking to purchase. Other works of Debret found in the *Voyages* highlight the realities of the Atlantic slave trade in Brazil, such as *Pelourinho* (or Negros no Tronco, ca. 1820s), where a slave is found chained to a pillar (a *pelourinho*, a type of pillar found in the squares in Brazil during this time) and flogged. The contrast of the watercolors, prints, and drawings created by Debret depicting the realities of the slave trade in Brazil and the oil paintings of the royal family commissioned from the artist in this period are important social and political, as well as artistic, documents of the Brazilian 19th century.

The second branch of Debret's work, his historical paintings and commissioned portraits, presents a style more in line with French neoclassicism in terms of technique, themes, and composition. His *Retrato of Dom João VI*, 1817, is a case in point, as the artist echoes Hyacinthe Rigaud's famous *Portrait of Louis XIV* of 1701. Debret paints the Portuguese monarch in the same pose and within the same composition as Rigaud's painting of the French king. These similarities between the works are meant to express a clear message: showing Dom João VI's hierarchical position while highlighting, through the differences between him and Louis XIV, the more populist and least ostentatious quality of the former. It is also important to note that this

A Brazilian family at dinner on a coffee plantation, circa 1820, from Voyage Pittoresque et Historique au Bresil *by Jean-Baptiste Debret. The woman of the house feeds scraps to a small child while a female slave cools the air with a large fan. (Hulton Archive/Getty Images)*

painting by Debret was later reproduced as a print by Carles-Simon Pradier. Taking under consideration the objective of a portrait like this, the production of a print is not surprising as it allows a wider circulation of the image and its message.

From the Regency (1831–1840) onward, groups of intellectuals connected to the state began to gain more power, and the concern over the creation of a Brazilian identity disconnected from a Portuguese one became a central theme. Significantly, in 1838 the Instituto Histórico e Geográfico Brasileiro (Brazilian Historic and Geographical Institute) was founded; it was responsible for constructing a Brazilian history, a national memory that valued the autochthonous aspects of the country. The middle of the century saw the rise of some of the students of the French Mission artists and a new concern with the image of the monarch. Pedro II slowly began to present himself as a liberal politician with a love for the arts and sciences. With this and a growing concern with the mythification of the nation, historical paintings became a key genre in the country and Romanticism an adequate style. Two painters received the major commissions from the state during this time: Vítor Meireles and Pedro Américo. The approximation of the scene to the spectator, the more dynamic compositions, and the concern with showing the descendants of the Portuguese, the native peoples of Brazil, and the African slaves side by side, as well as the tropical flora of the country, marked the work of these artists, among the most prominent being Vitor Meireles's *A Primeira Missa no Brasil* (*First Mass Performed in Brazil*) of 1860, exhibited at the Salon the following year being widely praised by the critics.

Although landscape painting continued to be considered inferior by the Academia, Nicolas Taunay's legacy led by his son Félix-Emile Taunay remained a strong reference in Brazil. The younger Taunay turned to the forest and the unexplored landscape of Brazil, producing paintings with Romantic tones.

On another hand, Realism slowly became a movement with little, but significant influence in Brazil during this period. After its rise in France following the political agitations of the mid-19th century, Realism would make an impact on Brazilian art. Important is the case of painter José Ferraz de Almeida Júnior with his focus on regional themes and the concern with the colors of everyday life. Almeida Júnior studied at the Academia and later traveled to Europe studying at the École National Supérieure des Beaux-Arts in Paris. He returned to Brazil in 1883 and settled in São Paulo. His paintings focused on the man of the countryside of Brazil, and the realism of his works took after French artists like Gustave Courbet and Jean-Baptiste-Camille Corot. This is apparent in pieces such as *Caipiras Negaceando* (*Caipiras Negotiating,* 1888) and *Caipira Picando Fumo* (*Caipiras Cutting Tobacco*, 1893) of remarkable quality of color and calling attention to the atmosphere of the interior of São Paulo, moving away from a traditional neoclassical style to focus on the specificity of Brazil and its people. Almeida Júnior presents his characters in their poor and isolated environments with sad or pensive expressions without mythologizing them or elevating them to the status of heroes. He is valued by intelletuals like Monteiro Lobato as a painter of the national reality and authenticity of Brazil, leading to his popularity with the rising bourgeoisie.

The rise of photography in Brazil in the 19th century is another key aspect of the period. The new medium was received in the country with a lot of enthusiasm, particularly by Emperor Pedro II, who in 1840 acquired a daguerreotype, setting the stage for the rise of photography in the country. With the techniques becoming more reliable in the late 19th century, several studios opened in Brazil, particularly in Rio de Janeiro, Recife, and Salvador. José Cristiano Júnior, Marc Ferrez, Auguste Stahl, and Georges Leutzinger are just some of the key names emerging during this time. They registered landscapes and scientific expeditions and the process of modernization of Brazilian cities, as well as the rising bourgeoisie and the large slave and Native Indian population. The photographic archive of the 19th century reveals important social, aesthetic, and political characteristics of the period.

Finally, in the 1880s the two important movements began to develop: the abolicionist and the republican. Even if the economic structure of the country, based on agricultural exportation, did not change, the influence of positivist thinking led to significant changes in the later decades of the century, including the abolition of slavery in 1888 and the proclamation of the republic in 1889. With these political and ideological shifts, the Academia also faced a reformulation becoming the Escola Nacional de Belas Artes (National School of Fine Arts) in 1890 and changing its faculty from old masters like Pedro Américo and Vitor Meireles, who were very much in line with the French Mission of the beginning of the century, to a lineup of new artists such as Rodolfo Amoedo, Modesto Brocos, and its new director, Rodolfo Bernardelli—who would stay until 1915.

This new leadership of the now Escola Nacional de Belas Artes marked an important change for art in the country. Rodolfo Amoedo, Eliseu Visconti—one of the few artists to have been influenced, even if moderately, by Symbolism in Brazil as his works *Gioventí* (1898) and *A Dança das Oréades* (*The Dance of the Oréades*, 1899) show—and the Bernardelli brothers formed together the "grupo dos modernos" (modern's group) and were responsible for the project for the Escola Nacional de Belas Artes in 1890. Amoedo was furthermore a pioneer of conservation and restauration in the country and taught one of the great Brazilian artists of the 20th century: Cândido Portinari. Visconti, having also studied with Amoedo, experimented with several techniques, genres, and media in his career, and between 1905 and 1939 he created a group of works for the recently inaugurated Teatro Municipal, including the ceiling of the dome and the foyer. He also had a brief excursion into Art Nouveau and elaborated posters, including the one for the Companhia Antártica. Finally, early Fauvist and Expressionist influences can be seen briefly in manifestations like *Baile a Fantasia*(*Costume Ball*) of Rodolfo Chambelland of 1913.

In the same period, a group of artists gathered around the German émigré George Grimm, after the latter had an impactful exhibit at the Liceo de Artes e Ofício in Rio de Janeiro in 1882. Grimm advocated for painting in situ, which he ingrained into his students like Hipólito Caron, as Caron's *Praia da Boa Viagem* (*Boa Viagem Beach*) shows. Although traces of Romanticism remained, there was a more pronounced concern with luminosity and atmospherical alterations. The academic styles endured well into the 20th century, when Brazilian artists who had spent time in Europe, like Anita Malfatti and Tarsila do Amaral, brought home the avant-garde influences from the Old World. Nevertheless, the neoclassicists also absorbed their share of the vernacular colonial art and architecture, as well as influences from black and mulato artists, leading to a syncretic manifestation of neoclassicism in the country deep into the 20th century.

THE RISE OF THE MODERN: THE 20TH AND 21ST CENTURIES IN BRAZIL

In a country influenced by the growing nationalism after World War I and at the beginning of its industrialization, the Week of Modern Art in São Paulo of 1922 was the birth of a new movement that looked to break paradigms and engage in the creation of a national art. It was the experiences of young Brazilian artists such as Anita Malfatti, Tarsila do Amaral, and Di Cavalcanti that impulsed changes in Brazilian art at the beginning of the 20th century. Having traveled to Europe and experienced the works of the Impressionists, Expressionists, and other groups, these artists returned with the hopes of changing the character of Brazilian art. They found resistance and harsh criticism in an older generation of artists and critics. An iconic example was the critics of Anita Malfatti's *Exposição de Pintura Moderna* in 1917 in São Paulo, where the artist showed works with a clear Expressionist vein such as *O Homem Amarelo* (1915–1916), next to works from international artists such as Floyd O'Neale, Sara Friedman, and Abraham S. Baylinson. In *O Homem Amarelo* (*Yellow Man*, 1915–1916), Malfatti engages with Modernism in the freedom of the brushstrokes, the new relationship between the foreground and background and the new Expressisionist character of the color. The intellectual Monteiro Lobato in the newspaper *O Estado de S. Paulo* in December 1917 critiqued this new path taken by Malfatti, pointing to its hermetic qualities and exposing his concern with the input of international styles. This ignited a fierce debate between Lobato and other intellectuals such as Oswald de Andrade, Mário de Andrade, and Menotti del Picchia, who spoke up in favor of Malfatti and against what they called the provincialism of Brazilian academic art. This debate fueled the group of aspiring Modernists, who would conceptualize the Week of Modern Art of 1922.

The Week of Modern art was the birthplace of important concepts and the beginning of a new movement for the development of a national art. It was a step toward the "decolonization" of Brazilian art and culture without needing to isolate from European artistic currents. The leaders of the Week and the modern movement that followed, Oswald de Andrade, Mário de Andrade, Menotti del Picchia, Tarsila do Amaral, Emiliano di Cavalcanti, and Anita Malfatti, advocated for free experimentation and for the absorption of international styles of art in order to impulse new forms of national production. Neither the Week nor the group that spearheaded it ascribed, or proposed, one clear aesthetic program. They did very much the opposite and looked to erupt the existing conservative art structure, showing a clear futurist ideal of raising the field for the construction of a new modern movement. Ultimately, the Week marked a change in the discourse of art in the country and the beginning of the search for a new style.

In 1928 Oswald de Andrade's "Manifeto Antropofágo" ("Anthropophagi Manifesto") would come to give another push to modern art in Brazil. With its aphorism and humor influenced by the thinking of Karl Marx, Sigmund Freud, Francis Picabia's *Manifeste Cannibale* (*Cannibal Manifesto*) of 1920, Jean-Jacques Rousseau, Michel de Montaigne, and Hermann Keyserling, de Andrade's Manifesto became a central text. Highlighting the work *Abaporu* by Tarsila do Amaral of the same year, de Andrade advocated for the incorporation and distillation of European ideas to Brazilian themes. *Abaporu* shows the combination of an abstraction of the body to psychological expression with an increasingly simple background in a palette of vibrant pure colors, which approximates this work to Henri Matisse's and Paul Klee's. Do Amaral's earlier work, *A Negra* of 1923, already shows a highly expressive body with a background of abstract geometric lines à la Piet Mondrian.

Emiliano Di Calvancanti was another key artist of the period. Having started working as an illustrator in 1914, he had experience in graphic design and was the creator of the poster and catalog of the Week of Modern Art. Traveling to Europe after the exhibition, Di Cavalcanti was influenced by artists such as Georges Braque and Pablo Picasso. His work after returning to Brazil showed the result of these encounters in his simplified forms and the strict pallete, even if this pallete ranged from warmer and sharper tones as opposed to the cool grays and browns of the European cubists. Di Cavalcanti's later contact with German Expressionism and his concern for political issues of the country led his work to become more connected to the everyday marginalized groups in Brazilian society: prostitutes, *malandros*, the black community, and fishermen. The results are pieces such as *Três Raças* (*Three Races*) of 1941 and *Carnaval no Morro* (*Carnival at the Hill*) of 1963. In 1932, he founded with Antonio Gomide and Flávio de Carvalho the group Clube dos Artistas Modernos (Modern Artist's Club, CAM), which would plant the seed for the rise of modern institutions like Museu de Arte Moderna de São Paulo and Rio de Janeiro (MAM-SP/RJ) in the late 1940s.

The 1930s/1940s, however, saw internationally what has been known as the return to order in the art sphere. It was a backlash to the experimentalism of the early avant-garde movements felt in the interwar period in Europe. Ideals of order and tradition marked this attempt at reconstructing Europe after the atrocities of World War I. In Brazil this backlash was also felt, and prominent artists of the first Modernist wave, such as Tarsila do Amaral, in the 1930s returned to formal traditions, classical composition, and lines in their works. Nevertheless, they continued to focus on nationalist themes already important in the art of the early modernists. Groups such as Santa Helena, led by Rossi Osir and Vittorio Gobbi, and Família Artística Paulista (The Paulista Artistic Family) led by Alfredo Volpi and Mário Zanini worked between modernism (manifested more strongly by other groups of the time like CAM and Sociedade Pró-Arte Moderna (Society Pro-Modern Art)—SPAM also founded in 1932 and led by Lasar Segall) and a lingering academicism. In hindsight, one finds that although the styles grew more eclectic and without a clear aesthetic line, they were overall characterized by an anti-experimentalism and anti-vanguardism.

The mid-20th century in Brazil was marked by political, economic, and cultural changes. The country had entered a period of intensive industrialization based on the national developmentalist policies of Getúlio Vargas. The modern movement led by CAM and SPAM continued to blossom, and the government looked to art and architecture to become the symbol of this "new era." Overall, two strong currents emerged in the late 1940s. On one side, the reintroduction of figurative art with a social vein influenced by the Mexican muralists and international artists such as Edward Hopper was spearheaded by Cândido Portinari and the Grupo Santa Helena, and on the other, an abstract group influenced by Constructivism emerged around Samson Flexor. These two lines of work were important because they marked the schools that would consolidate and become extremely influential later in the 1950s. Flexor's *Ateliê Abstração* (*Abstraction Atelier*) opened in 1951 was the first hub of abstraction in Brazil. It was where the ideals of Cubism, Neoplasticism, and Concrete art were gradually introduced and experimented with by Flexor's students such as Alberto Teixeira and Nelson Leirner. This space prepared the terrain for the rise of abstraction in the country during this decade.

The mid-1950s saw the appearance in São Paulo and Rio de Janeiro of two abstract groups: Grupo Ruptura (São Paulo, 1952) and Grupo Frente (Rio de Janeiro, 1954). Grupo Ruptura, led by Waldemar Cordeiro with Lothar Charoux, Anatol Wladyslaw, Luís Sacilotto, and Geraldo de Barros, was heavily influenced by the Italian Art Club. Lygia Clark, Lygia Pape, Aluísio Carvão, Hélio Oiticica, among others, were part of Grupo Frente led by Ivan Serpa, which saw geometric abstraction as an open field for experimentation and for questioning art paradigms. Brought forth by the I Bienal Internacional de São Paulo (I International Biennial of São Paulo) and the work of Max Bill—*Tripartide Unit*, which won the first prize at the Bienal—this period marked the rise of the geometric veins of abstract art in the country. The Ruptura and Frente groups rose in a moment, internationally, where concrete art flourished because of a growing optimism and will to overcome the difficulties of the two great wars.

In 1956, the Museum of Modern Art (MAM) São Paulo showed the First National Exhibition of Concrete Art, which put the two groups side by side. While it revealed the great amplitude that Constructive Art had achieved in Brazil, it also highlighted the discrepancies between the artists from the two centers. Waldemar Cordeiro defended Concrete Art in the models of Theo Doesburg and was contrary to all that was external to painting itself. He looked for the creation of bidimensional paintings without traces of sculpture, well inside the ideas of American critic Clement Greenberg. He found resistance in the work of artists such as Lygia Clark that saw the experimentation in the core of her creative process and had already at that time engaged in questioning the frame and other concepts of the bidimensional world, verging toward sculpture and architecture and engaging with participatory ideals.

These conflicts between the groups caused the dissolution of the Rio-based group Grupo Frente, with some of its members gathering around art critic and poet Ferreira Gullar and founding the Neoconcrete group. Lygia Clark, Hélio Oiticica, Lygia Pape, and Willys de Castro were some of the prominent artists involved with the Neoconcrete movement. They rejected the rationalism of concrete art and valued sensibility and expression over mathematical structures. Ultimately, they imposed the primacy of experimentation (both formal and thematic) over the limiting precepts of Concretism. Although the Neoconcrete group had already dissolved in 1961, it served to reinforce the will of these artists to continue experimenting.

Clark began this shift with her apprehension of the organic line, changing her titles—from lifeless inorganic to organic living things—in the end of the 1950s. Folding out the plane into itself, Clark created the *Casulos* (*Cocoons*, 1959), her first breach into the "world." From this breach she worked toward her most iconic pieces: the *Bichos* series (*Critters*, 1960–1966). The question of the experience was the central aspect of Clark's later works; it was the ideal of a relational event in space and time that took full form later in the 1960s as she breached the gap between therapeutic practices and art objects in pieces such as the *Cloth-Body-Cloth* series of 1966. The experimentations of the late 1950s also came to fruition in the participatory production of Hélio Oiticica. It was Oiticica's move to the slums and his engagement with the underprivileged people of Rio de Janeiro that really marked a shift in his work. It was then that he created the *Parangolés*: multisensory objects/garments made up of common materials: colored pieces of fabric, plastic, and, sometimes, pigment. The sensorial and participatory language of Clark and Oiticica's work was also manifested in the works of other artists of this period like Lygia Pape.

Art Informel (Informal Abstraction) had a rather small participation in the history of art in Brazil. It was widely criticized in the 1950s, when Tachism began to make an appearance at the São Paulo Bienals. Nevertheless, as a style, which did not define itself as a group of neat tendencies, but rather a focus of individual free expression, informal abstraction in Brazil clearly stands apart from European tachism or U.S. Abstract Expressionism. Thus, the works of artists such as Manabu Mabe, Antônio Bandeira, and Mathieu have been important and the principle that can most clearly define their works is a nonrepresentational expression of internal subjectivity without definitive relations between plastic elements.

Together with a small impact of Informel, the legacy of the Concrete and Neoconcrete movements was central for a new generation of artists such as Antonio Dias, Rubens Gerchman, and Carlos Vergara, and the 1960s saw the emergence of the new figurative movements: among them pop art, new realism, CoBrA, and Art Brut. In Brazil, critic Mário Pedrosa in the end of the 1960s already highlighted that artists began to work against the extremely formalist concern of other moments and engaged with figuration and iconography more thoroughly. In this field of new figuration, two lines emerged: on one hand was a return of Surrealism, influenced by the European Neodada and with a renewed attention to everyday objects—the *object trouvés* (found objects); on the other hand were the new realism and pop art manifestations, where art was seen as a form of intervention into the real social and political scenario. The Rex group, concerned with the systems of production and consumption of art and influenced by both Futurism and Fluxus, took a stand against cultural imperialism in Brazil, which it saw in institutions such as the Bienals of São Paulo, MAM-SP, prominent galleries, and the art criticism found in large newspapers. They engaged with antiart manifestations and had as their most prominent representatives Wesley Duke Lee, Nelson Leirner, and Geraldo de Barros.

Opinião 65 (Opinion 65) was a key event of this period. An exhibition organized in MAM-RJ by Ceres Franco and Jean Boghici, it brought together the new figuration artists of the School of Paris and a group of Brazilian artists, among them the veterans Hélio Oiticica, Ivan Serpa, and Waldemar Cordeiro, as well as the newcomers Duke Lee, Gerchman, Dias, and Vergara. It is important to note that the umbrella new figuration at this moment in the country encompassed a range of styles. Vergara's Expressionist tones, performance art tropes in the work of Oiticica, the pop art with political criticism of Duke Lee and Waldemar Cordeiro, and the neorealist pieces with urban concerns are just a few among many. Overall, the shortcomings of aestheticist tendencies and the insularity of abstraction were brought to the fore and criticized as other exhibitions followed in the lines of Opinião 65, among them Opinião 66, Proposta 66, and the emblematic Nova Objetividade Brasileira.

Brazil lived in a complicated moment. After the 1960s opened with the developmentalist catered by the Juscelino Kubitschek government, in 1964, after the rise of leftist João Goulart, the country suffered a military coup, which resulted in 21 years of military government. Neorealist manifestations, fueled by a leftist resistance to the military government, were central up to the 1970s. The role of artists and art at this moment was one of militance, a resistance significantly hampered by the AI-5 (Institutional Act number 5), which increased censorship and violence in the country. With the military government gaining power, even artists who previously defined themselves as apolitical assumed a critical position, among them Duke Lee.

In the midst of this appeared in Brazil the Nova Objetividade (New Objectivity) movement—also the name of the exhibition organized in March in 1967 by this group—engaging with art committed to political militance. More than a cohesive group, those who presented in the exhibition were concerned with similar problems and distanced themselves from traditional easel painting, turning to ready-made objects, tactile and sensorial manifestations, and performative propositions. Ultimately, the exhibition and the manifesto written by Oiticica were attempts to sum up the artistic currents of the moment. These artists, including Dias, Vergara, Gerchman, Pape, Glauco Rodrigues, and Carlos Zilio, distanced themselves from the neosurrealist tendencies of the early neofigurative years in Brazil manifested in Opnião 65. There was also an underlining constructivist vein to this group.

This shift from abstraction to neofigurative tendencies with a social political vein influenced by the military control becomes apparent in the work of artists such as Cordeiro. After years of critiquing the Frente artists, Cordeiro and the Ruptura group reevaluated their reliance on rationalism and mathematical rigor in the production of artworks and turned to other significant international movements. A politically engaged artist, Cordeiro was influenced by pop art and Neorealism. A pioneer in the appropriation of the banal into his artwork, the artist created works that would leave radical open veins of experimentation for the artists of the 1960s and 1970s such as *Popcreto para um Popcrítico* (*Popcrete for a Popcritic*, 1964). More than the introduction of pop to concrete art, the Popcretos are the use of concrete language for a highly complex and semiotically sophisticated work. The sharp red and abstract nature of the backdrop of *Popcreto para um Popcrítico* coupled with the hoe posed in the foreground demands a semiotic work from the spectator that is shaken out of the passive space of contemplation ignited by the rationalism of concrete art.

The 1970s saw the continuation of these neofigurative tendencies, but also the insertion of conceptual art into the scene. Cildo Meireles, Artur Barrio, among others, began working around this time and produced installations using different materials dealing directly with the legacy of the ready-made and of sensorial currents of Neoconcretism. The exhibition *Do Corpo à Terra* (*From Body to Earth*) of 1970 in Belo Horizonte presented Barrio's *Trouxas Ensanguentadas* (*Bloody Bundles*) and Meireles's *happening* "queima de animais vivos" ("burning of live animals"). The first work was made up of 14 bundles of bones and flesh thrown into the Ribeirão Arrudas River in the center of Belo Horizonte, while the second was an event where the artist burned live chickens in the Parque Municipal, the central park of the city. Both works called attention to the growing wave of repression and censorship following the AI-5 and externalized the anxiety of this period causing uproar in the streets and the engagement of the police and the fire department.

On another vein, the street art movement born in the 1960s was a reaction by the younger generation to the social and political reality of a country under military dictatorship. Already in the 1970s, *Pixação* or "wall painting" became a way of reclaiming public spaces in increasingly bolder and irreverent ways. The urban poetry of the graffiti's word play impacted the visual landscape, and these manifestations haunted the authoritarian regime and expressed the silently disturbing repression lived by the majority of the population during the violent years that followed the institution of the AI-5. The political function of street art in the country is still today one of its trademarks. During the 1980s and 1990s the movement gained space and began to develop both in style and in boldness. Groups like 3NÓS3 and Os Gêmeos created signature styles and inundated the walls of cities like São Paulo and Rio de Janeiro with their artworks. As political and social commentary, graffiti in Brazil brought questions to the fore through interventions in the urban environment that disrupted everyday life with thought-provoking images displayed on the city walls. These interventions called attention to the architectural structure of the city as the carrier of its history and the peeling and cracking walls and dirty streets of the Brazilian metropolis as backdrop where art, life, and memory bump into one another. Graffiti artists engaged the people who inhabited these spaces and echoed questions surrounding them: violence, pollution, industrial progress, and the poverty that still pervade a large amount of the Brazilian population.

The 1980s saw the return of questions of subjectivity and painting as a genre, of gesture and pictorial concerns, and of romantic theatricality. The artists of this generation made use of art history and the canon in ironic ways. Maneirism, Baroque,

Symbolism, and Romanticism returned as a backlash to the conceptual dematerialization of the 1970s opening to a new expressive painting with clear individualistic and pictorial concerns. Image took a stand against concept, and in Brazil the works of artists such as Nuno Ramos, Adriana Varejão, Daniel Senise, Leda Catunga, and Leonilson are among the most important. The Escola de Artes Visuais do Parque Lage (Parque Lage School of Visual Arts) in Rio de Janeiro and Faculdade de Artes Plásticas da Fundação Armando Álvares Penteado (University of Plastic Arts of the Armando Álvarez Penteado Foundation) in São Paulo were particularly important hubs for this development. Brazil lived through a transitional moment with the end of the military control in 1985, and the opening of the country to the international market and exhibitions such as Como Vai Você Geração 80? (How are you doing, 80s Generation?) marked the rise of a series of new questions, particularly regarding the relevance of this new production to the international artistic scene. Artist Jorge Guinle pointed at the time to the rupture of this new generation with the cerebral art of Constructivism, pop, and conceptual art, the main exponents of the earlier decades.

The rediscovery of art history and the avant-garde were central themes of the 1980s movement, as was the emptying of the ideal of newness proposed by the artists of this generation. It is a mistake however to deny this generation's engagement with the neofigurative, neoconcrete, pop, and conceptual tendencies of the 1960s and 1970s. The attention to the question of color, tactility, and the sensorial in the works of Varejão and Milhazes, the political and social questions in the ironic works of Leonilson, and the ephemerality of the materials found on the canvases of Catunda are just a few among many other vestiges of the earlier period. It was particularly the appropriation of art history and a return of concerns with the making of art that were the keys to the proposition of the artists of the 1980s, many of whom still work in the contemporary moment. The 1980s movement in the visual arts also needs to be understood as a nationwide phenomenon and not just one happening in the São Paulo–Rio de Janeiro axis. Belo Horizonte and the south of the country, led by Amílcar de Castro and Iberê Camargo, also saw an intense production in this period.

The recent decades also present the insertion of the Northeast of the country into the artistic scene. Previously distanced from the artistic axis and as such little explored by critics and historians, the Northeast as a region had a small participation in the country's museum collections and art market. Mestre Didi's work is a case in point. He began to be recognized in the national scene in the 1960s. His sculptural pieces draw on the legacy of the Yoruba peoples brought to Brazil during the Atlantic slave trade. Aside from these traditional themes and materials, there is also contemporariness to the work of Mestre Didi in his engagement with the sacred within the secular space of the museum.

Finally, the 1990s and 2000s saw the continuation of several of the tendencies of the 1960s, 1970s, and 1980s, with Varejão, Milhazes, Meireles, Barrio, and other artists such as Tunga, Eduardo Kac, Rivane Neuenschwander, and Rosângela Rennó having great impact on the production. The continued reevaluation of history includes, in 1998, the first Bienal de São Paulo completely devoted to a Brazilian theme, and later in 2000, the *Mostra do Redescobrimento: Brasil 500 anos* (Rediscovery Exhibition: Brazil 500 years). Aside from the rather amnesic naming of the latter, which invalidates the period prior to the colonization of Brazil by Portugal, its attention to art history in the country from 1500 until the contemporary moment was extremely important. Around the turn of the millennium, other tendencies also gained prominence, including video art, bio art, and environmental art. A new generation of artists such as Cynthia Marcelle, Marcius Galan, Jonathas de Andrade, Marilá Dardot, and André Komatsu have gained prominence since the turn of the century and have evoked questions of urbanity, high capitalism, the ubiquity of the media, and the legacies of exploitation in the country.

The art of Brazil remains an important exponent in the international museum scene and art market, with many artists gaining great international prominence and showing their work in large exhibitions across the globe, among them Sebastião Salgado, Vik Muniz, and Romero Britto. Salgado's photography is marked by its social engagement and the use of the media as a tool to expose the extreme conditions of groups such as the peasants in Latin America and the descendants of native tribes of Brazil. Salgado's strong contrasting photographs, as well as his highly developed sense of composition, have landed him among the great names of contemporary photography.

Vik Muniz is also known for his photography. A São Paulo native, Muniz moved to New York in the 1980s and after beginning his career as a sculptor quickly realized his interest in the photographic reproductions of his works rather than the installations themselves. His use of unusual materials like chocolate syrup, dirt, and coffee powder, as well as his engagement with iconic images of the history of art, has led him to be recognized internationally in several large exhibitions and art fairs. His recent film project WASTE LAND has gained much international attention. In this film director Lucy Walker and codirectors João Jardim and Karen Harley registered the everyday life of individuals who survive by picking and selling recyclable materials from the largest garbage dump in the world: Jardim Gramacho in the outskirts of Rio de Janeiro. As Muniz engaged with this community, his initial project of "painting" them using the garbage they work through became a collaborative project, where the individuals themselves began to rethink their lives and trajectories as they made the large-scale artworks using their own images.

Finally, Romero Britto can be considered the most successful commercial artist in Brazil today. Using the formal resources of pop art, vibrant colors, serial patterns, and everyday themes, Britto has ranked among big names of pop such as Andy Warhol and Keith Haring, being chosen in 1988, together with the latter, to design for Absolut vodka's "Absolut Art" advertisement campaign. He was also appointed recently as the official artist of the 2010 World Cup, as well as ambassador to the 2014 FIFA World Cup, which took place in Brazil. His exuberant and colorful works have populated visual culture worldwide in large-scale installations such as his sculpture for Hyde Park in London, the largest sculpture in that capital's history, as well as design products such as bottles of Evian water or Coca-Cola.

The recent boom of Brazilian art worldwide is noticeable in the recent large-scale exhibitions of Brazilian artists in important international museums such as the *Hélio Oiticica: Body of Color* at Tate Modern in 2007 and *Lygia Clark: The Abandonment of Art 1948–1988* at the Museum of Modern Art in New York; however, studies of lesser-known artists and movements, as well as a constant reevaluation of the Brazilian artistic canon and art history, inevitably remain a concern. It is difficult to write a chapter in the history of art of any region as the time frame, peoples, and distances are large and uncontainable by any narrative. Therefore, this narrative is a small glimpse into the arts of Brazil and its history, but in no way exhausts or summarizes it, but only opens the door for more research and new discoveries and a history of art built in collaboration with its reader.

REFERENCES

Ades, D., G. Brett, S. L. Catlin, R. O'Neill, and South Bank Centre. 1989. *Art in Latin America: The Modern Era, 1820–1980.* New Haven: Yale University Press.

Aguilar, N. 2000. *Mostra do Redescobrimento.* São Paulo: Associação Brasil 500 Anos Artes Visuais.

Alpers, S. 1983. *The Art of Describing: Dutch Art in the Seventeenth Century.* Chicago, IL: University of Chicago Press.

Amaral, A. 1998. *Constructive Art in Brazil: Adolpho Leirner Collection.* Lloyds Bank.

Amaral, A. A., P. Herkenhoff, and National Museum of Women in the Arts (USA). 1993. *Ultramodern: The Art of Contemporary Brazil.* Washington, DC: National Museum of Women in the Arts.

Bardi, P. M. 1970. *New Brazilian Art.* New York: Praeger.

Barnitz, J. 2001. *Twentieth-Century Art of Latin America.* Austin: University of Texas Press.

Bayón, D., and M. Marx. 1992. *History of South American Colonial Art and Architecture: Spanish South America and Brazil.* New York: Rizzoli.

Bazin, G. 1964. *Baroque and Rococo.* New York: Praeger.

Beezley, W. H., and L. A. Curcio. 2000. *Latin American Popular Culture: An Introduction.* Wilmington, Del: SR Books.

Belluzzo, A. M. M., S. Gledhill, and M. J. Cerqueira. 1995. *The Voyager's Brazil.* São Paulo: Metalivros.

Bowron, A., N. Aguilar, F. P. Venancio, Museum of Modern Art (Oxford, England), and BrasilConnects (Firm). 2001. *Experiment = Experiência: Art in Brazil, 1958–2000.* Oxford: Museum of Modern Art.

Brazilian Ministry of Foreign Relations and Ministry of Education and Culture. 1980. *Brazilian Indian Feather Art.* Brasília: Brazil Pró-Memórial National Foundation.

Brito, R., G. Bueno, S. Salcedo, C. Alves, Palais des beaux-arts (Brussels, Belgium), and Europalia. 2011. *Art in Brazil 1950–2011.* Brussels: Europalia International.

Butler, C. H., O. L. Pérez, A. S. Bessa, L. Clark, and Museum of Modern Art. 2014. *Lygia Clark: The Abandonment of Art, 1948–1988.* New York: Museum of Modern Art.

Calirman, C. 2012. *Brazilian Art under Dictatorship: Antonio Manuel, Artur Barrio, and Cildo Meireles.* Durham, NC: Duke University Press.

Castedo, L. 1964. *The Baroque Prevalence in Brazilian Art.* New York: C. Frank Publications.

Chamberlin, L. 2012. *Street Art: Rio.* S.l.: Blurb Creative Publishing.

Colección Patricia Phelps de Cisneros, Pérez-Barreiro, G., A. Locke, S. Lea, and Royal Academy of Arts (Great Britain). 2014. *Radical Geometry: Modern Art of South America from the Patricia Phelps de Cisneros Collection.*

Denis, R. C., and C. Trodd. 2000. *Art and the Academy in the Nineteenth Century.* New Brunswick, NJ: Rutgers University Press.

FotoFest 1992, W. Watriss, and L. P. Zamora. 1998. *Image and Memory: Photography from Latin America, 1866–1994: FotoFest.* Austin: University of Texas Press.

Fundação Bienal de São Paulo, and Associação Brasil 500 Anos Artes Visuais. 2000. *Mostra do redescobrimento*. São Paulo: Fundação Bienal de São Paulo.

Gabara, E. 2008. *Errant Modernism: The Ethos of Photography in Mexico and Brazil*. Durham, NC: Duke University Press.

Lemos, C.A.C., J.R.T. Leite, and P. M. Gismonti. 1983. *The Art of Brazil*. New York: Harper & Row.

Martins, L.L. 2013. *Photography and Documentary Film in the Making of Modern Brazil*. Manchester; New York : Manchester University Press.

Martins, S.B. 2013. *Constructing an Avant-Garde: Art in Brazil, 1949–1979*. Cambridge, MA: MIT Press.

Mostra do Redescobrimento, N. Aguilar, Fundação Bienal de São Paulo, and Associação Brazil 500 Anos Artes Visuais. 2000a. *Mostra do Redescobrimento: Arqueologia = archaeology*. São Paulo: Fundação Bienal de São Paulo.

Mostra do Redescobrimento, N. Aguilar, Fundação Bienal de São Paulo, and Associação Brasil 500 Anos Artes Visuais. 2000b. *Mostra do Redescobrimento: Arte do século XIX = 19th century art*. São Paulo: Fundação Bienal de São Paulo.

Mostra do Redescobrimento, N. Aguilar, Fundação Bienal de São Paulo, and Associação Brasil 500 Anos Artes Visuais. 2000c. *Mostra do Redescobrimento: Arte barroca = Baroque art*. São Paulo: Fundação Bienal de São Paulo.

Mostra do Redescobrimento, N. Aguilar, Fundação Bienal de São Paulo, and Associação Brasil 500 Anos Artes Visuais. 2000d. *Mostra do Redescobrimento: Artes indígenas = native arts*. São Paulo: Fundação Bienal de São Paulo.

Mostra do Redescobrimento, N. Aguilar, Fundação Bienal de São Paulo, and Associação Brasil 500 Anos Artes Visuais. 2000e. *Mostra do Redescobrimento: Arte popular = popular arts*. São Paulo: Fundação Bienal de São Paulo.

Museu de Arte Moderna do Rio de Janeiro, P. Vasquez, Maxwell Museum of Anthropology, and Houston Foto Fest. 1988. *Brazilian Photography in the Nineteenth Century*. Rio de Janeiro, Brasil : Museu de Arte Moderna do Rio de Janeiro.

Museum of Fine Arts, Houston, H. Olea, M.C. Ramírez, and Haus Konstruktiv (Zurich, Switzerland). 2009. *Building on a Construct: The Adolpho Leirner Collection of Brazilian Constructive Art at the Museum of Fine Arts, Houston*. Houston, TX: Museum of Fine Arts.

O'Brien, E., et al. 2013. *Modern Art in Africa, Asia, and Latin America: An Introduction to Global Modernisms*. Chichester, West Sussex: Wiley-Blackwell.

Ramírez, M.C., A. Leirner, and Museum of Fine Arts, Houston. 2007. *Dimensions of Constructive Art in Brazil: The Adolpho Leirner Collection*. Houston, TX: The Museum of Fine Arts.

Renshaw, A. 2013. *Art and Place: Site-Specific Art of the Americas*.

Sansi-Roca, R. 2007. *Fetishes and Monuments: Afro-Brazilian Art and Culture in the Twentieth Century*. New York: Berghahn Books.

Sterling, S.F., B.M. Sichel, F. Pedroso, and National Museum of Women in the Arts (USA). 2001. *Virgin Territory: Women, Gender, and History in Contemporary Brazilian Art*. Washington, DC: National Museum of Women in the Arts.

Sullivan, E.J. 1996. *Latin American Art in the Twentieth Century*. London: Phaidon Press.

Sullivan, E.J., Solomon R. Guggenheim Museum, and Museo Guggenheim Bilbao. 2001. *Brazil: Body and Soul*. New York: Guggenheim Museum.

Whistler, C., C. Avila, A.J.R. Russell-Wood, and Ashmolean Museum. 2001. *Opulence and Devotion: Brazilian Baroque Art*. Oxford: Ashmolean Museum.

Wood, M. 2013. *Black Milk: Imagining Slavery in the Visual Cultures of Brazil and America*.

Architecture

Doriane Andrade Meyer

Brazilian architecture was initially influenced by a variety of cultural, geographical, and social factors. Among them, the initial support of the natives, the physical and topographical conditions, the Catholic Church, the influx of African slaves, the discovery of gold, and, later, the immigration of different ethnic groups stand out. In addition, political, social, and economic changes in Brazil shaped the way Brazilian architecture adopted the architectural styles in vogue in each of its periods.

Architecture was in the spotlight in European culture between the end of the 15th and the beginning of the 16th centuries. The Gothic style was beginning to decrease its influence, and it started being replaced by the Renaissance style, which was disseminated by Italy, inspired by Leonardo da Vinci's *Vitruvian Man*. In Spain, the architect Lorenzo Vazquez introduced the Italian renaissance, making its principles evident in the details of the Santa Cruz Palace, in Valladolid. In Florence, Italy, Michelangelo worked in his masterpiece *David*, which is one of the best-known examples of Renaissance sculpture. In France, the Chateaux on Loire Valley—a series of castles along the Loire River—were being built. In Africa, the natives worked with terracotta (a reddish clay that is used for pottery and tiles) and glass. In Portugal, the prevailing architectural style was the *Estilo Manuelino* (also known as late Portuguese Gothic style), which had significant Islamic influence and was about to be replaced by the individualistic vision of Mannerism. In this context, King Manuel decided to explore the new lands demarcated by the Treaty of Tordesillas (which divided the newly discovered lands on the western side of Europe between the Portuguese Empire and the Spanish Empire) and sent the explorer Pedro Alvarez Cabral to the unknown western lands.

COLONIAL BRAZIL

When the Portuguese arrived in the Northeast of Brazil on April 22, 1500, they found that the indigenous people were divided into tribes, and each tribe had different architectural styles and practices. Nevertheless, it was common for the natives to live in *ocas* (tepees), which were big collective houses, usually organized in the shape of a circle, built with a wooden frame, and covered with straw. The tepees were usually surrounded with palisades to protect the tribe. The indigenous people only had a pictorial written language. They used natural pigments and made baskets and pottery. Everybody had to work together for the maintenance of the tribe. Women worked in agriculture and men hunted and fished. Although the Portuguese did not find the same resources and structure as they had in Europe in the new land, they were able to learn how to adapt to live in a tropical land from the natives.

The architecture of the colonial era can be classified as civil, religious, or military (defense). The civil architecture reflected the housing needs and the political and social organization of the colony. The religious architecture helped to provide spiritual support, as well as social and educational assistance. The military architecture arose from the need of the colony for defense from foreign invasion.

Civil Architecture

Because colonial Brazil did not count with the technology and materials available in Europe, its architecture and building methods adapted to the available materials and resources. Initially, the Portuguese built *tejupares* (from the Tupi-Guaraní *tejy*: people and *upad*: place) that were made from straw and were similar to the ones built by the natives. Soon they began to build houses of wattle and daub (*pau-a-pique* or *taipa-de-mão*, in Portuguese), which had interlaced bamboo walls covered with mud. This is a technique still used by the indigenous people and found in some rural regions in Brazil. The walls were built with pieces of wood that were initially covered with straw and later with clay tiles. Each house had only one door and one window. More-over, contrary to what was customary in Portugal (where the stove was inside the house and provided heat on cold days), because of the natives' influence, the kitchen was built outside, as an independent structure, to prevent the heat and smoke from entering the house. They placed large stones directly on the floor to support the pots made of clay by the Indians, and then they lit a fire underneath them.

Since the dense forest made access to the interior of the country difficult, the buildings were constructed on the coastline. In 1534, Portugal divided the land into large lots and donated them to Portuguese nobles, who, in exchange for exploring the land, would protect it from invaders. These lots were called "Capitanias Hereditárias" (hereditary captaincies), since they would pass from father to son. The Portuguese

A wattle and daub house in Brazil. Wattle and daub (or wattle-and-daub) is a building material used for making walls, in which a woven lattice of wooden strips called wattle is daubed with a sticky material made of some combination of wet soil, clay, sand, animal dung, and straw. (Paura/Dreamstime.com)

started to build villas, which were commonly used to ascertain their domination over the colonies. The houses were built with adobe or gravel-clay wattle made with clay, vegetable fiber, and water. To protect the walls from the rain, the houses were semidetached (glued to each other), with two roofs inclined to the front and back of the house, and eaves to protect the walls from the weather.

Around 1540, the first churches, city halls, and jails were being built at the center of each village, following the pattern of cities and towns in Portugal. The first city hall and jail were built in the city of Salvador in 1549, when its first governor, Tomé de Souza, founded the city. Like the houses, the first jail was built initially with wattle-and-daub walls and covered with straw. In 1551, the walls were rebuilt with stones covered with clay tiles. The jail occupied the ground floor, and the population fed the prisoners, because the government did not want to afford this expenditure. Like the one in Salvador, many city halls and jails were built in other cities in the colony.

Around this period began the building of urban houses, which followed standardized blueprints and always aligned to the predetermined grid of streets. These houses occupied all the area and were predominantly white due to the use of lime. For contrast, the frames were painted with bright colors such as green, blue, and yellow. The streets did not have sidewalks and were paved with stones. These urban houses were similar. "The ground floor consisted of a store with an adjacent storage area, and rooms for the slaves or guests; . . . an entrance hall, and a staircase. The first floor had a large front room that led directly to the front balcony. There was a central hall with rows of rooms or alcoves, and a large dining area and living room in the back whose stairs led to the backyard. The kitchen was attached to the back room" (Smith 1979, p. 236; translation is mine).

A street in the district of Pelourinho, in Salvador. The streets and houses date back to the colonial period. The sidewalks and the façades, painted with colors, were added later. (Photo by Doriane Meyer)

In addition, it is in this period that slaves first arrived officially in Brazil. They mainly worked in the sugarcane plantations, a monoculture with an increasing importance in the economy of the colony. Gradually, urban houses began to change according to economic status. The houses of the wealthiest class were usually built in stone masonry or gravel-clay wattle, their floors were made from wood, and the facades displayed a variety of decorative elements, whose most distinctive example is the threshing floor and border (decorative elements from the eaves, or overhangs, of roofs). This later led to the popular saying, "sem eira nem beira" (without threshing floor and border), referring to a person who does not have material goods (which is sometimes referred to as "not having two pennies to rub together"). Popular buildings were constructed with wattle and daub, adobe, and gravel-clay wattle, which are materials with good thermal properties to provide insulation from the heat during the day and help the houses remain warm during the night. The floor was made out of clay. The exterior walls were covered by clay tiles, which were made by slaves using their thighs to mold them, with no measure of quality or standard size or shape—this is the origin of a popular Brazilian saying that refers to poor quality of a product or service, "made in the thighs" (*feito nas coxas*). These tiles are known as *colonial* tiles or *Moorish* tiles, possibly because the Moors introduced them to Portugal. The use of lime on the facades produced a significant amount of humidity. This was one of the reasons for the Portuguese to start using tiles to cover the facades. Moreover, tiles protected and embellished the houses and are easy to clean. We can see a large collection of tiled facades in the city of São Luiz, Maranhão state, also known as "city of tiles" (cidade dos azulejos).

In rural areas were the sugarcane farms, usually composed of a big house with a chapel, the slave rooms (*senzalas*), the house of the foreman, and the mill to grind the cane. The master of the mill and his family lived in the big house (casa grande). They were spacious houses with living rooms, bedrooms, service sector, verandas, and usually an attached chapel. They were built with stone masonry or gravel-clay wattle covered with wood and clay tiles. The verandas were large and usually surrounded the big house. They served to cool the house, as well as receiving visitors and conducting business meetings, which was convenient because the visitors had no access to the house. Women were kept in almost complete isolation and were not allowed to have contact with strangers. Single women slept in so-called alcove bedrooms, which were rooms at the core of the house, so they were windowless and their only ventilation was through the unlined tiles of the roof.

With the arrival of the first slaves in 1549, African beliefs and culture were introduced in Brazil. Since then, they have been spread and disseminated through generations and have had a significant influence on Brazilian architecture, especially regarding the distribution of residential spaces. The home of the slaves were the *senzalas*, which were small rooms in the back of the master′s house. S*enzalas* were usually built of wattle and daub and covered with tiles. The floor was made of clay. They were joined together, forming a single rectangular building.

In the region of São Paulo, in the southeast of the country, the architectural style that developed was called "bandeirista." This style is named after the *Bandeirantes*, who were men responsible for the capture of runaway slaves. Unlike other parts of the country, in which the buildings adapted to the slopes of the land, the *Bandeirantes* leveled the terrain and built houses that rested on a flat foundation. The houses were rectangular and had a hip roof divided into four sides, all of them inclined downward. The hip roof extended in wide overhangs to protect the walls from the weather. They had a front porch, which led to the chapel and the guest room. The porch was a kind of balcony designed to cool the house and protect it from the sun and rain. The houses usually had two porches: the front or social porch, which led to the guest room and the chapel, and the service or interior porch, which led to the back of the house, served as a breakfast room for the family, and was commonly called "veranda." The guest room led only to the front porch, so it had no access to the inside of the house. Generally, the rooms of the slaves were near the kitchen and bathroom, which were outside of the houses. In the interior, the main hall was situated in the center, and the rooms were distributed surrounding it. In Rio de Janeiro, there is a well-known example of a typical Bandeirista house, which is well preserved: the house of the farm *Mato de Pipa*. The walls are built with the so-called rammed earth technique, which was introduced in Brazil by the Portuguese and is widely used in Bandeirista architecture, which Bandeiristas spread in the colony in their travels.

In the state of Minas Gerais, only a few houses were built of gravel-clay wattle or using the rammed earth technique because its architecture closely followed Portuguese architecture, which used wooden structures.

The sugar trade helped building several *solares*, which were huge houses where wealthy families lived. Notable examples of *solares* are the *Solar do Unhão*, *Solar Berquó*, and *Solar Ferrão*, in the city of Salvador—all of them from the 17th century and still well preserved. Located in the Bay of All Saints, the *Solar do Unhão* received in its pier the sugar production to be stored and exported; the storage area was on the right side of the *solar*, the *senzalas* were on the left of the *solar*, and the chapel in the background. After being refurbished by the architect Lina BoBardi in 1960, the *Solar do Unhão* started to be a museum and cultural center (Museum of Modern Art of Bahia). The *senzalas* are now a café-restaurant.

Religious Architecture

The architectural styles of the colonial period were Mannerism, Baroque, and Rococo, and their characteristics are best expressed in religious buildings. The "Portuguese" Mannerist style was the simplified expression of the Italian Renaissance, which arrived late in Portugal. This was the style adopted in Brazilian religious buildings during the first century of the colony. The first chapel in Brazil was the "Igreja de São Francisco de Assis do Outeiro da Glória" (Church of Saint Francis of Assis on the Hill of Glory). It was built in 1503 by Franciscan friars who arrived in the first Portuguese expeditions. It was made of adobe and grass in the town of Porto Seguro. In 1526, the "Igreja de Nossa Senhora da Misericórdia" (Church of Our Lady of Mercy) was built in Porto Seguro, and it is the only church from that period that is still preserved and in use. The Jesuits arrived in Brazil in 1549 and applied the Mannerist style consistently in their works, which made the Mannerist style to be also known as Jesuit style in Brazil. It is marked by triangular pediments and scarce decoration. Although the first cathedral was the "Sé Primatial" (demolished in 1933), the most important cathedral built by the Jesuits was the Basilica Cathedral of Salvador, which was inaugurated in 1672 to be the Jesuit College. It was considered the most important religious building in colonial Brazil due to its monumentality and artistic collection. Many churches were built with stones, as it was done in Portugal. By the early 17th century, the Baroque became the dominant style, as part of the Catholic Church's response to Luther's Protestant Reformation. The Baroque, therefore, is a representation of the Catholic Counter-Reformation and the reaffirmation of the cult of images, displaying its entire splendor in religious architecture. The Baroque style, marked by strong religious appeal of visual impact, was introduced in Brazil from Europe by the Jesuits. The Baroque churches usually had round side towers, stilted pediments, and a bull's-eye on top of the portal. Brazil has the largest collection of Baroque works in the world, which can be seen especially in the cities of Salvador, Olinda, and Ouro Preto—where Baroque fully developed with the discovery of gold. Until their expulsion, the Jesuits were the most significant architects of Brazil, and their legacy can be seen all throughout the Brazilian coastline, where they built several villages. In their villages, in addition to the church, there were schools and hospitals.

Besides providing spiritual support for the Portuguese and the African slaves, the purpose of the churches was to catechize the natives, or Amerindians, and to teach the Catholic religion. With the goal to seduce its visitors and parishioners, the interiors of the churches were richly decorated with carvings, paintings, and tiles. There are three outstanding examples of this rich decoration in Salvador. The Church of Saint Francis (*São Francisco*) has prominent painted woodwork and a tiled cloister, and is mostly covered with gold. The Franciscan monastery next to it is decorated with Portuguese tiles painted with biblical motives. The Third Order of Saint Francis Church, built between 1702 and 1705, displays similar rich ornamentations, and it is one of the most important Baroque expressions of Brazil and one of the Seven Wonders of Portuguese Origin in the World.

In the second half of the 18th century, the Rococo style arose, with simpler and softer expressions than the Baroque. This style appeared at the end of the Baroque style, and for this reason, Rococo is often confused and mixed with the Baroque. One of the most renowned figures of this period is the sculptor and architect Antonio Francisco Lisboa, who was colloquially known as "Aleijadinho" ("Little Cripple") after he suffered a debilitating disease. Aleijadinho's works are from the transition period between the Baroque and the Rococo, but even though the influence of these two styles can be seen in his works, Aleijadinho created his own, unique style.

Antônio Francisco Lisboa was born in the city of Ouro Preto, in the Minas Gerais region, when this area was at the peak of its economic prosperity because of the discovery of gold and precious stones. He drew attention to himself with his simple work on a sketch of a fountain that had been assigned to his father, the Portuguese architect Manuel Francisco Lisboa. Even after being stricken with a terrible disease that left him deformed and mutilated, Aleijadinho left numerous works. Among his extensive work, an outstanding example is the Sanctuary of "Nosso Senhor do Bom Jesus de Matosinhos" in Congonhas (Our Lord of the Good Jesus), where he carved the 12 full-length prophets in soapstone. Another example is the church "São Francisco de Assis" (Church of St. Francis of Assisi) in Ouro Preto, designed and decorated by him in Rococo style with paintings by Master Ataíde—a renowned artist, famous for his religious paintings, who specialized in perspective works on church ceilings. The church has curved facade with details in soapstone. Aleijadinho

Façade of the Third Order of St. Francis Church, in Salvador. The church was built between 1702 and 1705, and it is one of the most important baroque-style pieces of architecture in Brazil. St. Francis Church was named one of the Seven Wonders of Portuguese Origin in the World. (Photo by Doriane Meyer)

is today one of the most highly regarded Brazilian artist. As modernist author Mário de Andrade said, "In Aleijadinho we see the genius of a sculptor joined with the genius of an architect" (Andrade 1993, p. 66; translation is mine).

Military Architecture

Because the captaincy system was not working well, in 1549 Portugal set a governor-general, seeking to centralize the administration and organize its colony, which it considered a valuable business. This marked the beginning of the military architecture. Once the general government was established in Salvador, the first governor-general, Tomé de Souza, brought the military architect and master brick layer Luís Dias to design the capital of the colony. In the same way as Duarte Coelho founded Olinda (state of Pernambuco) on a hillside in 1535, Salvador was built on top of a high ground to better be monitored and make the attacks of pirates and Amerindians more difficult. Salvador's urban plan was similar to the one used in Lisbon. The main axis of the highest portion of the city was the *Terreiro de Jesus*, a square around which the main churches and best homes would be built.

A particularly distinctive feature of Salvador is an escarpment (a steep slope or long cliff) that separates it into an upper town (*cidade alta*) and a lower town (*cidade baixa*). In the lower town are the port, warehouses, workshops, and the pier, and this is where the slave market used to be. In the 18th century, landfill was used to expand the lower town. In 1873, the Lacerda Elevator was installed to connect the two sections.

This elevator was the first one installed in Brazil and helped carry heavy loads to the upper town. It has been and continues to be the greatest symbol of the city of Salvador. Before the installation of the Lacerda Elevator, goods and merchandises had to be carried by slaves and animals through the hillsides that connect both sections.

More than 30 forts were built in and near the capital, and some of them still exist today, such as Forte Santa Maria; Forte de Santo Antônio da Barra; Forte Monte Serrat; and Forte São Marcelo, which is the only fort in Brazil with a circular shape. Forte São Marcelo, built at the entrance of the Bay of All Saints, was responsible for protecting the port against attacks from invaders. The number of forts built in Brazil was over 350 most of them on the coastline, with few exceptions, like the imposing Forte Príncipe da Beira, in Rondônia. It was built in the middle of the Amazon forest, in a place still inaccessible, with the goal to defend the northern border and assist in the expansion of the land. On the north coast of Bahia, in Tatuapara, is the Fort Garcia D'Ávila. Garcia D'Ávila, who arrived in Brazil with his father, Tomé de Souza, received the lands in exchange for defending the coast. He lifted one of the first large buildings in Brazil, a castle/fort built with medieval influence in a former sugar plantation. Today it is in ruins, but its hexagonal-shaped chapel remains almost intact. The forts had great importance during the colonial period because they ensured protection against the invaders. Without forts, probably Brazil would not be the size that it is today.

After a failed attempt to invade Salvador in 1624, in 1630 the Dutch, led by Maurice of Nassau, invaded the captaincy of Pernambuco with the goal of taking possession of adjacent fields and captaincies. With the invasion, they virtually destroyed the city of Olinda. The Dutch decided to build next to Olinda, in a fishing village, a city protected by reefs, which gave it its name, Recife (which means "reef"). "In Recife, the Dutch, contrary to what they did in other American colonies, no longer exclusively followed their own building traditions. . . . partly because of the climate, partly because of the Luso-Brazilian architecture influence" (Smith 1979, p. 251; translation is mine). In Recife, Nassau built the Friburgo Palace and demonstrated huge interest in urban planning. The Dutch brought engineers and architects; built bridges, markets, zoos, and botanical gardens; and paved streets, which surpassed the work of Brazilian settlers by more than a century. The Dutch remained in control of a large portion of northeastern Brazil during 1630–1654, and Recife was the center of their government, as well as the port for export of sugar. In 1654, the Portuguese, the Africans, and the Amerindians joined against the common enemy and won the second battle of Guararapes, casting out the Dutch from Brazil.

In 1763, due to the lower production of sugarcane and the discovery of gold and precious stones in Minas Gerais, the capital was transferred from Salvador to Rio de Janeiro, which became the center of political and economic power of the colony. In this period, the world was going through significant changes that would influence the future of the colony. In 1776, England lost its main colony when the United States proclaimed independence and elected George Washington as its first president. In France, the revolution of the people joined with the bourgeoisie stormed the Bastille and guillotined the king and queen. In 1804, Napoleon proclaimed himself emperor of France. In 1807, the industrial revolution boosted England's economy. Spain and France signed the Treaty of Fontainebleau, intended to divide Portugal between them. Feeling increasingly depressed and unable to face Napoleon—who wanted to prevent the influence of England over Portugal—and with the help of England, the prince regent of Portugal, Dom João, decided to transfer the Portuguese royal court from Portugal to Brazil.

IMPERIAL BRAZIL

Neoclassical

Upon arriving in Brazil, Dom João docked first in Salvador for a strategic visit, attempting to form a coalition with the people of Bahia who were dissatisfied about the change of the capital to Rio de Janeiro. During his visit to Salvador, Dom João, among other things, opened the ports for imports and approved the creation of the first medical school in Brazil. On March 8, 1808, the royal family settled in Rio de Janeiro and the city went through a huge change that included social behavior. Before, women were kept recluse, but at this point, the trend was to encourage the social lives of women, which included being hostesses in their homes. This created the need to redecorate the houses, which required a dance room. In court, the prevailing style was the French decor. The wealth, which was hidden before, was to be displayed. The population was euphoric with the arrival of the court. The houses suffered transformations; the windows were opened to the balconies. The court's presence had a significant effect on architecture. After the French Revolution, neoclassicism was embraced as a more appropriate bourgeoisie style, so it replaced the Baroque style in Europe. In Brazil, through the influence of the court, neoclassicism also became the prevailing style. Neoclassicism, as its name suggests, is inspired in the classical arts of ancient Greece and Rome, which was characterized by simplicity and reduced ornamentation. The city was transformed, adopting the neoclassical style, squares and houses were renovated, and the streets were paved. As part of the change, the eaves of roofs disappeared and were replaced by parapets and triangular pediments.

The city needed to adapt to the growing population. The entire administrative structure had to be organized for the functioning of the state. The court also needed to get settled; that is, they needed adequate housing for everyone who came to the city. The viceroy of Brazil, D. Marcos de Noronha e Brito, gave his residence, the Palace of the Governors in the Palace Square (Largo do Paço), to the royal family, after which it was called Paço Imperial. Nevertheless, soon after that, Dom João VI moved into the Palace of Boa Vista, in São Cristovão, which became his residence, leaving the Paço Imperial for official use only.

At that time, the houses started having side spaces, which improved ventilation and lighting, as there were windows and the front door on the side as well. They also started to have front gardens, ensuring the family's privacy. Dom João ordered the removal of *muxarabiês* (balconies enclosed with lattices) of the facades of the houses—which was a hallmark of Arab influence in Portuguese architecture—leaving the interior of the houses exposed to the public eye. Houses built in this period had balconies with iron balustrades and fixtures hanging on the facades.

During that time, narrow sidewalks were built along the houses. They planted trees and implemented outside lighting. The sidewalks had rectangular-shaped stones. People started to stroll with their families in the squares and gardens, so that social life was no longer restricted to residential space. The Botanical Garden in Rio de Janeiro was constructed in this period.

During the colonial and imperial periods, the slaves were responsible for the construction and operation of the houses. It was the slave who did all the services. As Brazilian architect Lúcio Costa puts it, "[the slave] was sewage, water, and fan," alluding to the multiple roles that the slaves played in the homes (Costa 1952, pp. 9–10; translation is mine). There was not even urban sanitation yet. The first water inflow systems were the fountains and aqueducts. The waste (urine and feces) was placed in barrels to be collected by the so-called "tigers," who were slaves who threw the waste in established dumps, beaches, or rivers. They had that name because the waste full of ammonia and urea leaking from the barrels rolled down their bodies and marked their skin with light stripes, giving them the appearance of a tiger.

Dom João, a great admirer of art, invited a French mission of artisans, sculptors, painters, and architects to come to Brazil in 1816. With this mission, the architect Grandjean de Montigny arrived in Rio. De Montigny's goal was the creation of the first school of fine arts in Brazil. It was later called "Imperial Academy of Fine Arts." It offered classes in drawing, painting, sculpture, and architecture. The French mission had a strong influence in the adoption of neoclassicism, also known as academicism, for being a resumption of the principles of Greco-Roman antiquity. Colonial art and its religious themes were left behind and, in this period, art began to focus on secular aspects.

The academicist architecture was characterized by constructive clarity and simplicity of form. The Baroque and Rococo arts were at that point considered as exaggerated. The curves were replaced by rectilinear shapes. However, the houses were not entirely reformed to follow the new style. Instead, houses had a kind of makeup. They received new paint, French wallpaper, and a transformation in their facades. The new houses began to have partially underground basements and external stairs with landing and balustrade. The facades began to be decorated and the top of the parapets to have sculptures.

Meanwhile in Europe, the Portuguese were pressing for the return of Dom João VI to Portugal. In 1821 Dom João decided to return, leaving his son, Dom Pedro I, prince regent, in Brazil, to run the colony. However, after political differences, and the Portuguese demanding his return to Portugal, Dom Pedro I proclaimed the independence of Brazil, on September 7, 1822.

THE REPUBLIC OF BRAZIL

Independence created the need for public buildings, to provide administrative services. Almost all public buildings were built in neoclassical style. In the beginning of the 1830s, the coffee production surpassed the sugarcane production and became the main export product, financing the urbanization of cities like Rio de Janeiro and São Paulo.

The houses of this period had ventilation and space surrounding them in at least one of the sides, with entries transferred to the sides; the houses had high basements and the wealthiest homes also had gardens. Even the simplest buildings had decorative details, intended to show the social mobility aspirations of the owners.

Eclecticism

In the late 19th century, three facts influenced new changes in the architecture: the abolition of slavery, the mass influx of immigrants, and the proclamation of the republic.

With the abolition of slavery in 1888, the society had to be reorganized. As Paulo F. Santos explains, the slave had a triple meaning: in economics, it was a production factor; in the family, it meant warmth and cohesion; and in the house, it meant maintenance and operation (Santos 1981, p. 75). Without them, both the family structure and the organization of the house had to be changed, and this had an impact on architecture and the cities.

In urban areas, there was an accumulation of residents due to migration of former slaves from rural areas in search of work and housing and because of the arrival of immigrants. As a response to all these housing needs, specialized dwellings called "cortiços" arose. They were multifamily units with bathrooms and service area for community use. In Rio de Janeiro, many of these multifamily units were in the houses abandoned after the abolition of slavery, when the owners were no longer able to keep them without the services of slaves. At that time, works of sanitation and re-urbanization were carried out. With salaried labor and the techniques brought by immigrants, the construction techniques were improved. Railways and railway stations were built in this period. Annateresa Fabris explains that "iron and masonry are mixed in its composition, the first representing the technical and functional aspect, the second providing the decorum and ornament, as shown by the example of *Estação da Luz* in São Paulo, fully planned in London and whose components all came from Great Britain" (Fabris 1993, p. 140; translation is mine). The Estação da Luz (Railway Station), built to transport coffee from the city of São Paulo to the Santos port, also helped the immigrants to arrive in the future-biggest metropolis in South America. It was built with a mix of styles, which was characteristic of eclecticism. The building shown has simple and straight lines, which were typical of the rococo style, and neoclassical arches. It is also worth noting the use of iron in composition with the masonry in the building. The watch tower was inspired by the Big Ben in London.

In 1889, after the proclamation of the republic, new nationalist sentiments and references to other cultures appear, in an attempt to sever ties with the Portuguese past. This mixture of different cultures naturally sees eclecticism as a new architectural language. Eclecticism, a style that is a fusion of different historical styles, with its grandeur influenced almost all the reforms of expanding cities in the early 20th century. For instance, in 1903 the Central Avenue (currently Rio Branco Avenue) in Rio de Janeiro was built, inspired by the then-recent French urban reform with the creation of wide avenues. The Central Avenue quickly became the most frequented place by the Rio society. In 1905 the construction of the Municipal Theater of Rio de Janeiro was started, with classical lines but Baroque decoration. The eclectic building was inspired by the building of the Palais Garnier opera house in Paris.

At the same time, the infrastructure of cities began to improve, with sanitation and lighting. With the work of embellishment and "cleansing" of Rio de Janeiro, buildings and houses were demolished, many of which were tenements that lacked the most basic sanitary installations and thus were a significant factor in the spread of many diseases. Some of these people, without available housing options, moved to the outskirts of the city to live in the suburbs. Others, wishing to continue near the city center, and taking advantage of its topography, built their "shacks" (wooden houses) in the hills, which is the origin of the *favelas*—slum areas with no sanitation, public services, or urbanization.

Meanwhile, in the northern region of the country, Belém and Manaus had a strong economic growth because of rubber exports. Construction of new buildings followed an eclectic style and took advantage of the return of ships that carried rubber to Europe to bring back imported construction materials. A great example is the Amazonas Theater, in Manaus, founded in 1896, of eclectic architecture, already with Art Nouveau features, a style that was beginning to take effect in Europe. Professionals and various materials from Europe were brought for its construction, as the Carrara marble and the iron parts.

In the Southern region, with the arrival of new immigrants, new cities were founded that had significant influence from foreign cultures, especially German and Italian. The architecture of these new cities emerged spontaneously from the traditions brought by these immigrants. They used existing materials in Brazil but construction techniques brought from their homelands. An example is the city of Gramado, which has a strong Italian influence, where we find the use of stones and wood in several buildings. The colonists built all with their own hands—nothing was industrialized. The Germans brought the "Enxaimel" architecture, whose main characteristic is its hip on gable roofs with great inclination. The houses were built with double wood-work, and their walls were filled with materials available in the region. German architecture can be found in highest concentration in the state of Santa Catarina.

Art Nouveau

After World War I, Art Nouveau (French for "new art") appeared as a reaction to academicism and romantic sentimentalism, aiming integration to everyday life, and modern-life social changes, emphasizing beauty and craftsmanship. Seeking to make nicer industrialized projects, the artists looked for inspiration from nature, for example, making extensive use of flowers, insects, and curved lines. It was a unique style, independent of any other, that had a wide influence on interior architecture. It used modern materials such as glass and iron, and it also influenced the typography, clothing, jewelry, lighting, and home furnishings. With Art Nouveau, Paris continued its strong influence on Brazilian architecture.

One remarkable example is the *Confeitaria Colombo* in Rio de Janeiro, which opened in the end of the 19th century and represents Paris's *Belle Époque*. In 1905, it was redecorated in Art Nouveau style. Many important people, such as Queen Elizabeth of England in 1968, visited the *Confeitaria Colombo* in Rio de Janeiro. It is an important historic place in the city and has the same sophisticated ambiance until today.

In São Paulo, there is a building that is considered one of the most representative examples of Art Nouveau: the Vila Penteado, which was designed by architect

Carlos Ekman in 1902. Today it houses the headquarters of the Graduate School of Architecture and Urbanism of the University of São Paulo.

The kitchens in the homes built in this period were closer to the living rooms. With access to artificial lighting and electricity, in the 1920s, the radio became popular and families used to gather around it in their living rooms. Another significant feature of homes built during this time is the gardens that surround them.

Parallel to the Art Nouveau emerged the neocolonial movement, which sought inspiration in Brazilian colonial architecture and was opposed to academicism and its international orientation. The neocolonial movement aimed to the resumption of the use of elements of the colonial era, such as large eaves with clay tiles.

Art Deco

In 1925, the Exposition Internationale des Arts Décoratifs et Industriels Modernes (International Exhibition of Decorative Arts and Modern Industries) took place in Paris and started the Art Deco style there. Unlike the Art Nouveau, Art Deco was a less labored style of a decorative art, with a predominance of abstract design, straight lines, and geometric shapes.

It was an architectural style not committed to general theoretical principles and was characterized by simpler ornamentation. It was influenced by the tendencies of that time, such as symmetry in blueprint distribution and a decrease in construction costs. It was an important step for the development of modernist architecture.

Due to the abandonment of the old *solares* and the absence of slaves to help in the organization of the house, the architecture started to verticalize, and the first buildings appeared. Even though at first they were met with public resistance, they were soon embraced, initially by the middle class and then by the upper class. The architecture of this period was largely inspired by American ideals and the American way of life—bathroom facilities, modernist chandeliers, and elevators were adopted from the United States. These made possible the skyscrapers of São Paulo and Rio de Janeiro, which produced a total change in the urban landscapes. The urban houses started to have garages for the cars, while rural homes and houses in the suburbs had backyards, gardens, and balconies.

In Salvador, the Oceania building, built in 1942 under the total influence of Art Deco, set a new standard of luxury housing in the city. Another fine example in the city of Brazilian Art Deco architecture is the Elevador Lacerda (Lacerda Elevator) idealized by the engineer Antonio Lacerda to help transport between uptown and lowertown, which used to be done by using the roads on the slopes of the surrounding hills.

The Art Deco left great examples throughout Brazil, as the Theater José de Alencar, in Fortaleza; the *Viaduto do Chá* in São Paulo; the Hotel Copacabana Palace; the Tower of Central Brazil Station's clock; and the statue of Christ the Redeemer, the highest Art Deco statue in the world—all of them in Rio de Janeiro.

This was a time of great public works, which included the construction of train stations, factories, and exhibition halls. The Brazilian city that best represents Art Deco is Goiania because it was designed at the apex of this architectural style.

However, despite having generated a great change in the architectural scene in the country, Art Deco did not last long and soon gave way to the modernist style. In São Paulo, the *Edifício Esther*, one of the first residential buildings, marked the transition to modernism. As Carlos Lemos points out, "The Art Deco style, popular since the late 20s, masked the architectural panorama of the time, as it also had modern traits. . . . With one foot in the functionalism of Le Corbusier, and another on the compositional spirit or decorative Art Deco is the *Edifício Esther*, by Alvaro Vital Brazil and Adhemar Marinho, 1936" (Lemos 2005; translation is mine).

The *Edifício Esther*, one of the landmarks of modern architecture in Brazil, was also cited in 1943 by Mário de Andrade as an example of the new architecture in the country, in opposition to the neocolonial movement. This building displays distinctive characteristics of the new architecture, ranging from the rational use of materials and construction methods to the lack of ornaments. This building has a typical standard blueprint of that period. There are two apartments on each floor. They have living rooms, bedrooms, bathrooms, kitchens, and a service area with a sink and storage. They also share a common service hall. Strongly showing the social division, the elevators were designed to operate in separate halls, social and service, evidencing the gap between employers and employees.

Modernism

Although it had not had much effect on the architecture, the 1922 Modern Art Week with demonstration of new ideas and artistic concepts marked the arrival of modernism in Brazil. Modernism was widely discussed and promoted by Brazilian intellectuals. The renowned poet Oswald de Andrade wrote *The Cannibal Manifesto*, whose main argument is that Brazilian culture was formed by "cannibalizing" other cultures and returning to its indigenous and African roots. The movement did not have much influence on architecture because the main Brazilian architects were influenced by foreign architects such as Le Corbusier, Mies Van Der Rohe, Walter Gropius, and Frank Lloyd Wright. Le Corbusier directly influenced Lúcio Costa's and Oscar Niemeyer's works with his visit to Brazil in 1933. Le Corbusier's influence propelled the transformation of Brazilian architecture, opposing the existing academicism and moving away from any ideas or elements that were reminiscent of other styles.

In 1925, the Russian architect naturalized Brazilian Gregori Warchavchik published the manifesto *Acerca da Arquitetura Moderna* (*On Modern Architecture*), where he went against the academic eclecticism when he stated that "the modern architect must study the classical architecture to develop their aesthetic sense. . . . The modern architect should not only fail to copy the old styles, but also stop thinking about the style. . . . [He or she] should love his time" (Fracalossi 2013; translation is mine). In 1928, Warchavchik built the first modernist house in Brazil in São Paulo, ignoring the whole concept of the colonial era to a residence in a tropical country.

In 1930, Getúlio Vargas assumed the presidency of Brazil in a spirit of change and renewal. As a result, he created the Ministry of Education and Health. The new ministry needed headquarters that represented the new spirit of modernity. It was an open competition for the project. The winning project, however, was still linked to academicism, contrary to the new spirit desired by the new government, which made the current minister, Gustavo Capanema, intervene, paying the prize but discarding the winning project and inviting the architect Lúcio Costa to implement his proposal, which won the second place. Lúcio Costa assembled a team composed of architects Oscar Niemeyer, Jorge Moreira, Carlos Leão, Affonso Reidy, and Ernani Vasconcellos. The new design was inspired by the five points of the Franco-Swiss architect Le Corbusier. Its basic characteristics were pilotis—pillared system to raise the building floor and allow free access; free plan—absence of supporting walls, providing independence and diversity in the internal spaces; facade free—independence of the structure giving freedom to create; terrace garden—providing use of toppings for domestic purpose; and horizontal window—longitudinal openings by cutting the entire length of the building bringing more light and providing better visibility of the external space.

At the request of Lúcio Costa, President Vargas authorized the coming of the Franco-Swiss architect to Brazil. During his 30-day stay, Le Corbusier attempted to move the construction of the building to another location that was not approved by the ministry. Based on Le Corbusier's sketch, the team of Brazilian architects developed a new project for the construction site. After Le Corbusier and the ministry approved the project, the construction of the building started, which marked the end of academicism in Brazil and made Brazilian architecture to be admired and recognized internationally. *The New York Times* stated that the Education and Health Ministry project is "the most advanced architectural structure in the world" (James Brown, *Harvard Magazine Review* of Latin America/spring/summer 2010).

The Ministry of Education and Health (MES) was housed in two separate buildings, one with 10 floors and the other, smaller, perpendicular to the larger one. Its south glazed facade was what was most modern. "In 1938, with the building of the ministry already under construction, there was not in New York any skyscraper with glass facade yet—the 'curtain-wall' or 'mur rideau'—emerged later" (Costa 1995, p. 122; translation is mine). The north facade had a high incidence of sunlight, so in front of the windows, it used *brise-soleil*, which was designed by Le Corbusier and consists of pieces with rotary movement to be positioned to deflect sunlight and decrease heat gain. It was the ideal solution for a tropical country like Brazil. The main building, with its open plan interior, is supported on stilts. The smaller building used tiles for its outer covering and stilts—a suggestion of Le Corbusier—returning to the tradition of the colonial period. The terrace has a landscape designed by the renowned landscape architect Roberto Burle Marx. Murals painted by Candido Portinari and sculptures by Celso Antonio, Jacques Lipchitz, and Bruno Giorgi provided the finishing touch.

During the construction of the MES, Lucio Costa won the competition to design the Brazilian pavilion at the 1939 New York World's Fair and invited Oscar Niemeyer, second place in the competition, to join him. The lightness and elegance of the Brazilian pavilion drew attention, helping Oscar Niemeyer and Lúcio Costa to gain international recognition.

In the 1940s more than 50 buildings were built in Rio de Janeiro with the use of reinforced concrete. The bathroom had drains and used ceramic tile on all the walls to facilitate cleaning. Buildings of this period used concrete, steel, and glass. The use of concrete allowed the architects to play with the shapes without limits. Affonso Eduardo Reidy projected the *Pedregulho* Residential Complex, aimed at public low-income employees. This project proposed the integration between housing and the exterior, provided by community living spaces such as schools, sports courts, swimming pools, and leisure facilities. It had sinuous lines on stilts following the topography of the land, and *cobogós* on the facade to provide better ventilation of the apartments. *Cobogó*, a traditional element user all over Brazil until today, is a prefabricated hollow element made of cement or pottery, created in the state of Pernambuco, in the Northeastern region of Brazil and gets its name because of its inventors' surnames—Coimbra, Boeckmann, and Góis. The *cobogós*, which refer us to the ancient *muxarabiês* in the colonial period, help ventilating the corridors and service areas and keeping the interior privacy. In the same period, the use of awnings to moderate the internal temperature of the apartments by decreasing the incidence of direct sunlight became popular. The set *Pedregulho* is one of the icons of modernist Brazilian architecture. The project was landscaped by Robert Burle Marx and had panels painted by Candido Portinari.

Almost in the same time, Lúcio Costa projected the housing set *Parque Guinle* (Guinle Park) aimed at the upper class. The landscaping of this housing project was made by Burle Marx. Its blueprint was inspired by colonial houses with social balconies in front and familial convivial balconies behind. As Reidy, Costa used *cobogós* in the facades and projected stilts as supports for the buildings to deal with the irregular topography of the terrain. In the future, when designing Brasília, Lúcio Costa designed the "superblocks," which were inspired by the Guinle Park.

In 1940 Juscelino Kubitschek, then mayor of Belo Horizonte, invited Oscar Niemeyer to design a casino in Pampulha. Kubitschek wanted to create a leisure district in Pampulha unlike anything that existed in the country. Niemeyer also designed in Pampulha the Yacht Club, with Cândido Portinari panels; the Church of St. Francis of Assisi, with Picasso painting in tiles; the Ball House; and the Golf Club. The shape of the Church of St. Francis of Assisi shows the plasticity of the reinforced concrete.

With the economic independence that Brazil achieved as a result of its industrial development, the government began to modernize cities, culminating in the construction of Brasília, which was a dream of President Juscelino Kubitschek. He said that the Senate agreed to finance his project only because it thought that it would never be actually built. Kubitschek was of great importance for Brazilian architecture because Brasília was internationally recognized as a masterpiece, gaining respect and being imitated.

After Oscar Niemeyer declined Kubitschek's invitation to design Brasília's urban plan, it was decided to make a competition for selecting the new capital's pilot plan. Niemeyer, however, agreed to design the government buildings (Niemeyer 1961, p. 12). Lúcio Costa explains that Kubitschek already had Niemeyer in mind when he became president of Brazil: "The architect had been previously chosen. This means that the competition was only for the city's urban planning, the Pilot Plan. I drew up the master plan. I did not intend to participate in the contest, but in the middle of the prompt, I had an idea, I thought it was worth it and I ran" (Oliveira 2005; translation is mine).

Lúcio Costa decided to participate in the contest for the new capital project only two months before the deadline, and he delivered his proposal on the last day: "Born of the primary gesture of marks a place or taking possession: two axes intersecting at an straight angle, that is, the cross sign itself" (Costa 1995, p. 284; translation is mine). Costa's proposal consisted of two crossing axes, with the government buildings on the main—or monumental—axis and the "superblocks," divided by the main road axis, on the parabolic—or residential—horizontal axis. At the meeting of the two axes is the cultural center. The superblocks consist of building sets of six floors supported on stilts. Because they are supported by stilts, the buildings seemingly float and leave the areas below them for community use of the residents. Each set of buildings is in an area delimited by vegetation. Between the superblocks are educational and commercial buildings. Traffic was designed so that the paths do not intersect, avoiding in this way traffic jams and the need for traffic lights.

The federal government gave the architects complete freedom of design for the projects and for the corresponding budgets. Brasília was planned as a symbol of Brazil's modernity. Niemeyer designed monumental buildings, such as the Cathedral of Brasília the Palácio da Alvorada (the residence of the president), the Pálacio do Planalto (the official workplace of the president), and the Itamaraty Palace (the headquarters of the Ministry of External Relations). As Ottaviano C. De Fiore remarks, "Few architects in history have had the opportunity to construct an entire city with complete freedom of design, unlimited funds, national sympathy and the unconditional support of the nation's President" (De Fiore 1985, p. 23).

Although they had received criticism from several architects who accused them of a return to academicism with the monumental buildings of Brasília the architects were not intimidated. "Monumentality never frightened me when a stronger topic justified it. After all, what remains in architecture are the monumental works, the ones that mark history and technical evolution—those that socially justified or not, still touch us" (Niemeyer 2000, p. 176). On a visit to Brasília in 1962, Le Corbusier said, "Brasília has been built . . . It is magnificent in its invention, courage, and optimism; and speaks to the heart. . . . In the modern world, Brasília is unique" (Costa 1995, p. 141; translation is mine).

The Brazilian government had significantly contributed to the dissemination of art for the public building museums such as the Museum of Modern Art (MAM) of Rio de Janeiro and the Museum of Modern Art of São Paulo (MASP). The Museum of Modern Art of Rio de Janeiro has landscaping by Burle Marx in harmony with the Affondo Reidy project. The building draws attention because of its bold structures. In its garden, Burle Marx used native flora. For the MASP project, the government invited the Italian architect, Lina Bo Bardi, who lived in Brazil. The project had to preserve the view of the city center and the Serra da Cantareira (a mountain in the city of São Paulo that has one of the biggest urban forests of the world, a piece of Atlantic forest). Lina achieved these objectives with a design that suspended the building on prestressed concrete pillars, leaving the required clearance below.

In the last decades of the 20th century, the use of residential environments underwent changes and adapted to technological innovations. In the 1960s, the houses were being built in series. The buildings were already a sales success. Television joined the families in the living rooms and the gardens began to be forgotten. In the 1970s, home appliances became more sophisticated, and homes had game rooms, saunas, and swimming pools. Because of the increased need for security, luxury-gated condominiums, with pool and sports facilities, became popular. In the 1980s, nearly each bedroom gained its own television. With pools and grills in the home, outdoor social life returned, as it was in colonial times. With the computer and the Internet, many self-employed professionals started working at home. The old houses that had been turned into tenements almost ceased to exist—they were demolished to make way for luxurious condominiums. The urban planning of big cities began to change, and great slums began to grow next to high-end luxury condominiums. The government started to build affordable housing to meet the needs of proletarian class. In the 1990s loft apartments—apartments with a large adaptable open space, without dividing walls, and where only the bathrooms have privacy—appeared. The target audience for loft apartments is young professionals, usually unmarried, who are attracted to the feeling of freedom in this broad and integrated space. The loft apartments were copied from an American concept popular in New York City, in which large spaces were renovated and adapted for other uses. For instance, buildings that were previously used by light industries were transformed into loft apartments, becoming popular among young artists and self-employed professionals.

Verissimo and Bittar comment that Brazilian homes began to turn away from international tastes and values and into more specifically Brazilian values. They claim that "the so-called 'modern' is looking for plastic elements that remind them of colonial architecture" (Veríssimo and Bittar 1999, pp. 43–44; translation is mine).

Modernism left a huge legacy for Brazilian architecture. Among the main figures are Alvaro Vital Brazil, the Roberto brothers (Marcelo, Milton, and Mauricio), Vilanova Artigas, Oswaldo Bratke, Carlos Leão, Rino Levi, Hélio Uchoa, Sergio Bernardes, Lina Bo Bardi, and many others, some of whom have already been mentioned in this chapter. The following are some of the main architects of modernism with additional information on their lives and work.

Lúcio Costa—A French-born Brazilian architect and urban planner, Costa graduated from the Academy of Fine Arts of Rio de Janeiro in 1924. In his first projects, he began following the eclectic academicism of neocolonialism, which was an attempt to return to Brazilian roots inspired mainly by the Barroco style. However, in 1922, after a travel to Diamantina, Minas Gerais, a city that to this day still retains architecture of the colonial era, he came to the conclusion that there was a contradiction and exaggeration in the ideals of neocolonialism. Returning to Rio, he adopted the values and principles of modernism and the ideas of Le Corbusier, from whom he applied the five key points in the MSE project (current Capanema Palace). During the work of the MES, Costa was the winner in the competition for the construction of the Brazilian flag of the NYC Fair in 1939. In 1948, he made the project of the Guinle Park, "the first set of buildings built on stilts and the harbinger of Brasília's superblocks" (Costa 1995, p. 205; translation is mine). In the Guinle Park, he referred to the colonial period, designing social balconies in the front and familiar convivial balconies behind. On the facades of the buildings, he placed *cobogós*, made from different types and sizes of blocks, interspersing them with windows and *brise-soleils*. During his career, he made several projects and wrote several books; the last one, released in 1995, is almost an autobiography. He was one of the precursors of modernism in Brazil, having several works consecrated internationally, such as the MSE and Brasília, which cast him forever in the history of modern world architecture.

Gregori Warchavchik—A Russian architect naturalized Brazilian, Warchavchik was born in Odessa, Ukraine. He graduated from the Regio Istituto Superiore di Belle Arti, Rome (College of Italian Architecture) and settled in Brazil. A follower of Le Corbusier, he was credited with building the first modernist house in Brazil, in the city of São Paulo. He published in 1925 the manifesto *Acerca da Arquitetura Moderna* (*On Modern Architecture*), where he went against the academic eclecticism. In the early 1930s he was a work partner of Lúcio Costa for a short time, during which they designed the first modernist popular building in Rio de Janeiro. Warchavchik was always worried about the functionality and aesthetics of his projects.

Oscar Niemeyer—He is without a doubt the biggest name of Brazilian architecture so far. He worked initially as a trainee in the office of Lúcio Costa and Carlos Leão. There he had the opportunity to work with Le Corbusier for four weeks, during his trip to Brazil in 1936, from whom he basically absorbed all the concepts and principles of modern architecture. He was part of the team that designed the MES and along with Lúcio Costa designed the Brazilian Pavilion at the New York World's Fair in 1939, which made him known in the international market. In 1945, along with other architects, he was invited to design the United Nations building in New York. Oscar Niemeyer gave new directions to Brazilian architecture after the various projects in the neighborhood of Pampulha in Belo Horizonte and was definitely recognized after the designs of the Brasília buildings. No one knew how to mix traces of the colonial era to its modernist buildings like him. Niemeyer was master in using the mixture of architecture with painting and sculpture in his projects. In his own words, "It was important for me to forge a link with the architecture of old colonial Brazil and the elements common to those days by expressing the same plastic intentions, the same love of curves, rich and refined form, that are so characteristic of the colonial style" (Peter 1994, p. 240). When the military took power in Brazil in 1964, Niemeyer exiled himself in France and did several projects outside the country, including the headquarters of the French Communist Party and the Cultural Center of Le Havre. After his return to Brazil, he worked on the projects of the JK Memorial in Brasília, the Sambadrome in Rio de Janeiro, the Memorial of Latin America in São Paulo, among several others. In 1988, he received the Pritzker Prize; the purpose is to honor a living architect or architects, whose work demonstrates a combination of talent, vision, and commitment, which has produced consistent and significant contributions to humanity and the environment built through the art of architecture. When he was 102 years old, he did the project of the Administrative City of Minas Gerais, one of whose buildings, the Tiradentes Palace, has the largest clearance in the history of architecture. Today he is considered one of the great modernist architects in the world, followed by many architects in Brazil and in the rest of the world. On December 27, 2007, the architecture critic Nicolai Ouroussoff wrote in his article in *The New York Times*: "The force of that vision reverberated across the United States and Europe. Lincoln Center in Manhattan, Empire State Plaza in Albany, the Los Angeles Music Center—all owe a debt to Niemeyer."

João Filgueira Lima (*Lelé*)—As soon as he finished his university degree, Lelé worked in the construction of Brasília, where he learned the techniques of precast, often with precarious support structures. He continued to improve and dominate these techniques, and they became a hallmark of his work.

Lelé has always been a professional concerned with the cost and social role of his works. In the projects of the Sarah Hospitals chain, he used natural light and ventilation. The Sarah Hospitals chain is aimed at rehabilitation of people with physical-motor problems and the integration that exists in the hospitals between humans and the environment contributes greatly to the recovery of patients. Later he built the Centro de Tecnologia da Rede Sarah (Sarah Chain's Technology Center), where he developed hospitals and equipment for several Brazilian cities.

He lived in Salvador, where he did much of his work. Invited by the state government, he designed several buildings in the Administrative Center of Bahia (CAB), whose road plan was authored by Lúcio Costa. At CAB, Lelé designed the church, the exhibition center, and several buildings where he preserved the existing Atlantic Forest, suspending the buildings by pillars. He also developed the factory community facilities (FAEC) in Salvador, with which he created the colorful walkways of the city, as well as several schools and kindergartens. There are several colorful walkways in the city of Salvador, which are a part of the landscape of the city. Although built in reinforced concrete and covered with coloring metal, they have smooth appearances.

He was responsible for creating the CIEPs—integrated center for public education in Rio de Janeiro, designed by Oscar Niemeyer. His work has always been characterized by the pursuit of architecture rationalization and integration with the environment.

Affonso Eduardo Reidy—Born in Paris, he studied at the National School of Fine Arts, Rio de Janeiro. He was one of the team members who designed the Ministry of Education and Health, led by Lúcio Costa. Among his important projects such as the Museum of Modern Art in Rio de Janeiro—the first work in exposed concrete of the country—is the housing President Mendes de Moraes Residential Complex, or *Pedregulho*, which won the prize at the International Biennial of São Paulo. In this project, there was a social concern to leave open spaces for socializing among residents, besides community service facilities such as health and education. This complex used Le Corbusier's concepts, such as the use of stilts, which in *Pedregulho* were used in different heights to compensate the sloping terrain. Reidy was also responsible for coordinating the urbanistic project of downtown in Rio de Janeiro.

Roberto Burle Marx—Born in São Paulo, Burle Marx studied painting and learned about botany in Germany. Graduated from the National School of Fine Arts, Rio de Janeiro, in 1930, Burle Marx is the most famous landscape architect of Brazil. He designed the first garden for a house designed by Lúcio Costa and Gregori Warchavchik. Costa had an early influence on the work of Burle Marx, and they worked together on several projects, including the Ministry of Education and Health building and Brasília. Another great partner of Burle Marx was Oscar Niemeyer for whom he designed gardens for several projects, such as the Pampulha buildings. Burle Marx worked in several countries in South America and in the United States, South Africa, and France. He was a great supporter of anti-deforestation, always remembering that the balance and survival of the human race depend on nature. In his works, an appreciation of Brazilian native flora is visible, to which he always gave a central role. Burle Marx was also the creator of the Copacabana beach ride, a harmonious blend of Portuguese stone and vegetation.

Postmodernism

Postmodernism came to argue against modernism, aiming to humanize it. It is a questioning of the exaggerated rationalism displayed in some modernist works. Postmodernists fearlessly reintroduced classic elements in their designs, such as ionic columns and pediments, doing often strange mixtures, arranged in an unconventional way, with the only intention to question and bring controversy.

Postmodernism did not have much impact on Brazilian architecture, nor was it very much discussed, because of the strength and tradition of modernism. It has a strong presence in the city of Salvador due to the designs of architect Fernando

Peixoto. He designed several colorful buildings in Salvador, causing much controversy. The emphasis of his projects is the use of colors, using the plastic strength of its facades to highlight his works. The buildings are considered by some to be bold and creative and are criticized by others who argue that the architect was concerned only with the facades, leaving the internal space often claustrophobic by having their windows semi-hidden by decorative panels. However, pleasing or not, the buildings changed the face of a part of town, making it colorful and youthful. On the one hand, it contrasts with the historical side of the city, and on the other hand, the strong and mixed colors remind us of the African roots of the city.

Another prominent architect in Brazilian postmodern architecture was Eolo Maia, born in the historic city of Ouro Preto, Minas Gerais. Maia used and abused verticality to highlight his works, seeking striking symbols of colonial architecture, in which many times he sought inspiration for his projects. Maia mixed color and diversity of materials and shapes on the facades of his buildings making them stand out from the others on site. As Peixoto, Maia causes controversy with his works, which are pleasing to some and harshly criticized by others.

Contemporaneous

After the controversial phase of postmodernism in the 1980s, modernist elements returned to contemporary architecture, with the resumption of the use of technology, as well as of features used in colonial and modernist eras. For instance, tiles are used as decorative elements and *cobogós* are widely used in various types of materials. An example is the *Casa Cobogó*, projected by the architect Marcio Kogan, which also brings back the wooden trusses (latticework) used in former *muxarabiês*. It is a conscious and harmonious style taken from past architectural periods. Today's architects no longer bother to follow a style, each seeking his or her own based on what he or she thinks best from each period. The highpoint was the Mies Van der Rohe Prize for Latin American Architecture and the Pritzker Prize assigned to Paulo Mendes da Rocha, in 2000 and 2006, respectively. Born in 1928, Mendes da Rocha belongs to the latest generation of modernist architects, generating more concern with the rationalization of construction processes. During his career, he has had several winning works in public competitions, has received several awards and has given several conferences throughout Brazil and abroad.

We could say that nowadays the architecture of Brazilian cities is the result of all styles that have marked its history, influenced by cultural, economic, and social values. There are cities such as Salvador, Ouro Preto, Olinda, São Luís, and Rio de Janeiro, where the living history is still preserved. In these cities, we can see the passage of styles that marked different epochs and thereby understand the context and importance of each. On the other hand, there are cities like São Paulo that, despite having a large part of history preserved, are cosmopolitan and may even be said to be different from the rest of the country.

Today, Brazil does not have a prominent name in architecture, as it did in previous times. The country currently has many good architecture professionals who seek in history what they think are the best styles and characteristics and reproduce them in their own personal styles. There is a great concern for the preservation of cultural heritage; because of this, thousands of buildings were listed for preservation as protected, historic buildings. There is a growing concern with sustainability, so current projects aim to have minimal impact on nature. With such diversity of trends, Brazil tends to be a country of updated architecture within the international context while also preserving its history.

REFERENCES

Andrade, Mário. 1993. *A Arte Religiosa no Brasil.* São Paulo: Experimento/Giordano.

Costa, Lúcio. 1952. *Arquitetura Brasileira.* Rio de Janeiro: Departamento de Imprensa Nacional.

Costa, Lúcio. 1995. *Registro de uma Vivência.* São Paulo: Empresa das Artes.

De Fiore, Ottaviano C. 1985. *Architecture and Sculpture in Brazil.* Albuquerque, NM: University of New Mexico.

Fabris, Annateresa. 1993. "Arquitetura Eclética no Brasil: o cenário da modernização." Anais do Museu Paulista Nova Série N. 1.

Fracalossi, Igor. "Manifesto: Acerca da Arquitetura Moderna/Gregori Warchavchik." October 24, 2013. ArchDaily Brasil.

Lemos, Carlos A. C. 2005. "O modernismo Arquitetônico em São Paulo." Paper presented at the III Seminário Docomomo, São Paulo.

Niemeyer, Oscar. 1961. *Minha Experiência em Brasília.* Rio de Janeiro: Itambé Sociedade Anônima.

Niemeyer, Oscar. 2000. *The Curves of Time.* London: Phaidon Press Limited.

Oliveira, Giovanna Ortiz de. 2005. Lucio Costa: *Entrevista,* São Paulo, ano 06, n. 023.03, Vitruvius.

Peter, John. 1994. *The Oral History of Modern Architecture: Interviews with the Greatest Architects of the Twentieth Century.* New York: Abrams.

Santos, F. Paulo. 1981. *Quatro Séculos de Arquitetura.* Rio de Janeiro: IAB.

Smith, Robert C. 1979. *Igrejas, Casas e Móveis, aspectos de Arte Colonial Brasileira.* Pernambuco: Universitária.

Veríssimo, S. Francisco, and William S. M. Bittar. 1999. *500 anos da Casa no Brasil.* Rio de Janeiro: Ediouro.

Suggested Reading

Cavalcanti, Lauro. *When Brazil Was Modern.* Translated by Jon Tolman. New York: Princeton Architectural Press, 2003.

Goodwin, Philip L. *Brazil Builds.* New York: The Museum of Modern Art, 1943.

Lara, Fernando Luiz. *The Rise of Popular Modernist Architecture in Brazil.* Gainesville: University Press of Florida, 2008.

Le Corbusier. *Toward a New Architecture.* Translated by Frederick Etchells. New York: Dover Publications, 1986.

Lemos, Carlos A. C. *Cozinhas.* São Paulo: Perspectiva S.A., 1976.

Lemos, Carlos A. C. *Arquitetura Brasileira.* São Paulo: Melhoramentos, 1979.

Lemos, Carlos A. C. *A Casa Brasileira.* São Paulo: Contexto, 1989.

Segawa, Hugo. *Architecture of Brazil.* São Paulo: Edusp, 2010.

Underwood, David. *Oscar Niemeyer and Brazilian Free-Form Modernism.* New York: George Braziller, Inc., 1994.

Music

Angela Lühning

Michael Iyanaga (Translator)

INTRODUCTION

Imagine a country in which music, present in the street, at home, and at every moment, is a fundamental expression of social life. Imagine a country in which people know an infinite number of lyrics, as well as sing, play, and create new versions of songs, where people adore music. This country is Brazil. Still, it should be noted at the outset that not "everything ends in *samba*" (or, in Portuguese, *tudo termina em samba*), which is a very popular expression throughout Brazil. Indeed, many people around the world associate Brazilian culture with music, happiness, and spontaneity, all of which are represented in the expression *tudo acaba em samba*, or "everything ends in *samba*," in reference to one of Brazil's most well-known musical styles. Thus people point to Rio de Janeiro's samba as the poster image of Brazilian culture. However, this representation quickly becomes a stereotype when focusing on the diversity of Brazilian music over the course of this enormous and incredibly heterogeneous nation's history. Undeniably, music is of great importance in Brazil: it is associated with movement and corporality; is present (or not) in public processions and street parties; and expresses faith, happiness, leisure, or, quite simply, creativity. Still, music says more about the country and its culture; it also speaks about its people. After all, it is often too easy to forget that people are responsible for making music, a nonexistent phenomenon without them. But this is not limited to famous musicians, those who are well known to the media. Music-makers also include "common" people uninterested in fame; individuals who make and create music for other motives will be discussed here.

This section aims to discuss the various types of music that different people have made in Brazil, without limiting the scope to one ethnic group or musical genre, one geographical region, or one historical moment. Rather, the objective is an attempt to present Brazil's musical diversity by way of a contextualized ethnomusicological approach. This constitutes a tremendous challenge, for this ample diversity makes it necessary for me to make choices about what to include here. After all, the country is plural, immense, and diversified, despite the fact that Brazil's culture if so often seen in segmented, stereotyped, and even prejudiced ways. Before discussing the musical diversity, however, it is important first to contextualize the country and its inhabitants in order to then understand the multifarious influences present in Brazil's many musical traditions.

HISTORY AND GEOGRAPHY: SOME BACKGROUND

Brazil—United States?

Despite how distant and foreign Brazil may appear from an American point of view, the United States actually has much in common with its South American neighbor.

Some of the most pronounced similarities include a demographically diverse population, the colonial experience, and postcolonial political processes such as independence, abolition, and the construction of a republic. A few of these aspects are presented here as they relate to Brazil, focusing on points that will be pertinent later to the discussion of musical practices and traditions.

In territorial extension, Brazil is larger than the continental United States and the largest country in South America. Its current population is over 200 million. Since Portuguese colonization began in 1500, Brazil has been involved in a protracted confrontation with the indigenous population, or "Indians," as they have been called since the 16th century. Over time the number of "Indians" in Brazil has decreased significantly and today represents less than 1 percent of the total population. Nevertheless, there are nearly 200 indigenous languages still spoken today. Due to the shortage of manual labor necessary for the exploitation of the country's resources during the colonization process, millions of enslaved Africans were brought to work on plantations to cultivate sugarcane, care for livestock, extract precious minerals, and work as domestic slaves or wage-earning slaves (*escravos de ganho*). The colonization process began in the Northeast and descended along Brazil's coast. Later, when the nation's central regions were finally integrated during the 20th century, the indigenous populations—especially those of the Amazon region—were uprooted and dislocated.

With the arrival of Dom João VI and the royal Portuguese family in 1808, fleeing the turbulent politics of Europe following the French Revolution, Brazil received important cultural installations in the nation's capital such as libraries and academies of fine art, all of which stimulated urban cultural life. But the royal family returned to Portugal in 1821 and declared Brazil's independence in 1822. With this, Dom Pedro I, son of Dom João VI, became the country's emperor. Under the rule of Dom Pedro II, Brazil became a republic in 1889, a year after the abolition of slavery in 1888. The 19th century further marks the start of a series of immigrant waves coming from several different places: Europe (Germany, Poland, Ukraine, Switzerland, Italy, and Spain), the Near East (Syria, Lebanon, and Turkey), and Japan. This resulted in an enormous ethnic diversity in Brazil, though concentrated in different geographical regions.

Within Brazil, there are significant differences among the various regions of the country, as much in relation to biological, climactic, ethnic, and cultural diversity, as in relation to technological and industrial development. The North is situated near the equator, encompassing the largest part of the Amazon rainforest and also the greatest ethnic diversity; the culturally rich Northeast has historically suffered from long periods of drought, which have catalyzed migratory movements to other regions of the country; the Southeast and South have a more temperate climate with extensive agricultural and industrial production, and indeed the Southeast—and particularly São Paulo—has been the country's economic center for quite some time; and the Midwest, a relatively newly settled region, represents the new frontier for agricultural and meat production, and is enwrapped in constant conflict with the forests of the Amazon region. The predominant populations are distinct in each of these regions, resulting in markedly different local cultures, including the music.

The most prominent symbol representing the integration of national territory and culture is the construction of Brasília. Brasília became the nation's capital in 1960, taking the place of Rio de Janeiro, which had itself replaced Brazil's first capital, Salvador (in Bahia), in 1763. Today Brazil is a federative republic with 27 states, employs Portuguese as its official language, and constitutes a majority of Catholics and a growing number of people of Protestant denominations, especially Neo-Pentecostal. According to the latest censuses, roughly 50 percent of Brazil's population is African descendants, with African descendents making up more than 70 percent in some regions.

The Myth of the Three Races

The African descendent population is today concentrated in the same states that until the 19th century served as major disembarkation ports for the transatlantic slave trade: Bahia, Pernambuco, Maranhão, Rio de Janeiro, Rio Grande do Sul, and also Minas Gerais. Fairly recently, the country's urban population has grown to outnumber its rural population, resulting in new configurations relating not only to Brazil's insertion in the globalized world but also to its social and economic inequalities, which continue to be quite prominent. The 20th century has seen the rise of the myth of the three races, that is, the widespread notion—reiterated by innumerable authors—that the Brazilian people are the consequence of the miscegenation and coexistence of the three races: Amerindians, Portuguese, and Africans. Of course the term "race" is questionable. But, more importantly, this concept neglects to consider the asymmetrical power relations of the social roles of these groups, and the history of important ethnic groups such as the Arabs and the Japanese in Brazil. Further, the concept ignores the diversity of African and indigenous ethnic groups. Indeed, to this day, African-descendent Brazilians find themselves at the bottom of the social pyramid, with disproportionately high rates of unemployment and low rates of education, among other inequalities. At the same time, Amerindians find themselves in a similar situation while also continuing to fight for land rights. These issues began to receive significant national attention and be debated starting only in the late 20th century, and finally at the beginning of the 21st century, the political discussions began to lead to affirmative actions for the underprivileged and socially excluded populations.

Music Comes on the Scene

To be more effective in this discussion about Brazilian music(s), it is important to consider two central issues beyond the sociological and political contexts. We have already addressed sources (written, iconographic, or oral) and technologies (along with their subsequent products). Both not only offer information about music but also are a reminder that the preoccupation with diversity observable today was hardly common only a few decades ago. Brazil's ethnic composition allowed for marked cultural diversity. Still, many scholars have yet to take note of this plurality. For quite some time, culture and music were viewed in much more restricted and ideologically distorted ways: music meant either the European songs played in upper-middle-class Brazilian salons or the sacred music played in churches. In fact, the cultural expressions of some of the populations already mentioned, particularly those of black and indigenous groups, were for a long time viewed as incompatible with modern ideas of civilization and progress. These conceptions had been imported from Europe during the late 19th century and circulated widely within elite social circles.

Attention to other cultural forms finally began to appear at the turn of the 20th century as concern grew for defining the Brazilian nation-state. Indeed, the developments in this post-abolition period—from the late 19th century to the first decades of the 20th century—made it impossible to continue to deny the existence of Afro-Brazilians and Amerindians in the country, even if these groups occupied then—and still today occupy—the lowest substrata in Brazilian society. It is important to remember that these groups' music and lyrics (which are often in languages other than Portuguese) could express group identities, address origins, establish notions of belonging, emphasize differences and otherness, and even convey esoteric knowledge. Consequently, musical practices were not always performed near strangers, overseers, or masters. Thus many musical expressions of enslaved and freed blacks, as well as their descendents, were veiled in the limbo of invisibility and "inaudibility," not only because the ruling classes' ideologies prevented them from recognizing the music *as music* but also because often the musical expressions themselves were performed distant from external observers and left few traces.

Sources and Technologies

There are few historical sources (written or iconographic) detailing the musical traditions practiced during the country's colonization process and political transformations. There are even fewer sources produced by the traditions' practitioners themselves, since in general these practices were transmitted orally, without any form of material documentation. But fortunately, we can find third-party narratives and descriptions written by visitors/travelers from other countries and people in administrative positions, in addition to some testimonies by the protagonists themselves, all of which contribute clues to understanding cultural forms at specific moments of Brazilian history. As such, the first publications about the more general aspects of Brazilian music date to the turn from the 19th to the 20th century (Andrade 1972 [1928]; Lopes, Abreu, Ulhôa, and Velloso 2012]).

This same period, the end of the 19th century, marks the arrival of innovative new technologies in Brazil. One such technology was sound recording and reproduction. This technology led not only to ethnographic recordings, including the documentation of music of a few indigenous groups as early as 1911 (by the Berliner Phonogramm-Archiv), but also to commercial recordings and the subsequent widespread dissemination of urban popular music, such as *samba* starting in 1917. These historical recordings are important for they document the people who were themselves responsible for the traditions and musical practices, even if during the recording process a number of external factors interfered. For more information about the history of the recording industry in Brazil, see the website of the Instituto Moreira Sales (http://ims.com.br/ims/explore/artista/humberto-franceschi).

Taking Brazilian Music Seriously

Mario de Andrade (1893–1945) was one of the most important researchers to dedicate himself to comprehending the diversity of Brazilian music. In addition to being a writer, musician, musicologist, art critic, and researcher, he worked in every imaginable field related to music. By the 1930s one of his literary works had been published in the United States, "Amar, verbo instransitivo" (To Love, an Intransitive Verb), and another of his books, *Macunaíma*, which later became a famous film of the same name (1969). Andrade conducted two fieldwork trips in the Brazilian North and Northeast before anyone had ever ventured to do such research. The two research trips, each lasting three months, took place between 1927 and 1928 and allowed him to generate many observations and ideas for future research (Andrade 1976). Andrade was also the mastermind behind another research project which, for the first time, systematically surveyed "traditional music" (*música popular*) of the country's North and Northeast. This undertaking, the Missão de Pesquisa Folclórica, or the Folkloric Research Mission, was conducted in 1938.

Andrade's efforts also propelled the work of a number of classical music composers who began to take interest in the specifically Brazilian characteristics of Brazilian music, distinct from the European classical music which had been their primary inspiration up until then. The movement dealing with these issues began with the Modern Art Week, which took place in 1922 in São Paulo, involving writers, musicians, painters, and thinkers. The book *Macunaíma* expresses masterfully this search for a Brazilian identity and thus became a symbol of the movement.

Concepts and Problems

It is necessary to consider some of the concepts accompanying this period of awareness regarding Brazilian cultural diversity. After all, initially researchers and artists interested in the issue called the various traditions they observed "folklore," a term that appears to have been introduced into Brazil near the end of the 19th century. It was used to designate traditions transmitted orally, especially those of Africans and Amerindians, but also, in the mold of the myth of the three races, those of Portuguese influence. Looking at the term's usage during the period, we can identify a certain dichotomization of characteristics: oral transmission was opposed to writing, repetitive simplicity to erudition and culture, the archaic to the modern, and, finally, the anonymity of everyday people to the concept of individual authorship. The term "folklore" remained popular until the 1970s, with widespread use during the years of the military dictatorship (1964–1985). The military regime highly valued the term, as it was present in various actions designed to integrate cultural activities in the definition of the nation, though the term was used in a very static way. Among the period's authors, there are but few exceptions in which preference is given to terms such as "popular culture" or "popular traditions," since these authors understood culture in a broader way and as something which was always dynamic (Ayala and Ayala 1995). A slow process of conceptual redefinition began only in the 1980s, with the employment of more anthropological and ethnomusicological approaches. In many ways, these perspectives built on discussions originally initiated by Mario de Andrade.

Given this complex history, it is today a great challenge to work with the diversity of Brazilian musical expressions in a broad way and without prejudices, understanding it as an expression of the diversity of social and ethnic groups, from rural and urban areas, connected or not to the market and technologies. Furthermore, it is vital that this diversity be made part of a critical discussion related to the human experience, without conceptual restrictions or barriers. In the same way, the search for origins, exact locales, or precise moments marking the birth of musical practices—projects that were quite common only a few decades ago (with little success)—has been abandoned in favor of more contextual, processual, and interdisciplinary understandings.

ORCHESTRAS, BANDS, AND ENSEMBLES

There have always been a large number of orchestral groups in Brazil, but of quite diverse instrumental formations. Musicologists have for some time given special attention to orchestras (in the Western sense), with a specific focus on their activities in churches and theaters. There is a wealth of documentation about the period known as the "Baroque of Minas Gerais" (*barroco mineiro*), particularly regarding the production of 17th- and 18th-century ecclesiastical music, with ensembles including violins, violas, flutes, oboes, and basso continuo (Duprat 1985). The composition of music for theaters began in the 19th century, with the arrival of the royal family. Indeed, the imperial family held classical music in high regard as an expression of culture, giving special preference to opera (Bezerra, Schwamborn, and Soares 2010). One of the most well-known Brazilian composers of this period was Carlos Gomes, who received much praise for his operas. This also included success in Italy, where he lived for many decades, which subsequently gave his compositions a much more Italian sound than a Brazilian one. Another important classical music composer, active in the 20th century, was Heitor Villa-Lobos. With his nationalist focus, Villa-Lobos sought to infuse his compositions with a decidedly Brazilian character. His oeuvre does not solely comprise orchestral music, including instead a number of different types of ensemble types. It is important to note that the orchestras and composers I have thus far discussed used sheet music for performances and have been a preferred topic of study for researchers due to the ease of access to historical material (see Lopes et al. 2011).

However, in addition to these large-scale classical music orchestras, there were several other types of ensembles. Though generally less well known, they were hardly less important to the formation of Brazilian culture. Beyond the mentioned orchestras, there were also private orchestras made up entirely of slaves who attended to the whims of their masters, who kept an orchestra on their property. Even more numerous were barber bands, which slaves or freed slaves themselves decided to put together. These individuals worked primarily as barbers and in their free time performed in small groups mainly with wind and percussion instruments, but also stringed instruments. It seems that the repertoires these groups performed were quite varied (Tinhorão 1998, pp. 163–186).

Among these instrumental music ensembles must be included *choro* music groups. Choro is a style of urban music which became popular in Rio de Janeiro in the late 19th century. Some of the first choro musicians include pianist Chiquinha Gonzaga (one of the first female Brazilian musicians), the flutist Antônio Callado, and also the pianist Ernesto Nazareth. Notably, all of these artists were also composers (Diniz 1999). The most widely celebrated choro musicians were Pixinguinha and his group Os 8 batutas, a group formed by black musicians in 1919. The group became wildly successful in a short period and traveled to Paris in 1922, where it spent half a year touring and performing to an appreciative public. In many ways, the group's experience was not unlike that of the Fisk Jubilee Singers, the famous U.S. black vocal group, when it toured London and Paris in the late 19th century (Gilroy 2001, pp. 182–192). The instrumentation of a choro group includes flute, clarinet, bandolim (similar to a mandolin), six- and seven-string acoustic guitars, and tambourine. Choro has spread all over Brazil and is quite popular in many social clubs and among groups that continue to play this style.

MUSICAL INSTRUMENTS

Brazil is today home to an infinity of musical instruments, with diverse origins and which are employed in any number of traditions. We can detail their primary characteristics by way of one of the most common classification systems, developed by Hornbostel and Sachs (1914). It divides instruments into four groups according to the vibration of the sound-producing material: chordophones, whose strings vibrate; aerophones, generally, blown with a vibrating column of air; membranophones, whose membranes vibrate; and, finally, idiophones, instruments without strings, columns of air, or a membrane, but whose physical structure permits it to vibrate by way of movements such as scraping, shaking, plucking, striking, and so forth. In Brazil the last two groups (largely of African origin) predominate, although there is also a wide variety of aerophones among the Indians. Moreover, common electronic instruments today sit somewhere between the acoustic and electric.

In the rural areas and in small cities of the inner part of the country, distant from the metropolises of the Brazilian Northeast, we find another type of musical ensemble that has great social and musical importance: fife bands (*bandas de pífano*). These are ensembles composed of two *pífanos*, or fifes, which are handmade transversal flutes, accompanied either by a large membrane drum called a *zabumba* or by a snare drum. More recently, triangles have been added to the groups and the percussive part has become more prominent. The repertoire played by this ensemble, made up of farmers, marketers, or ranchers, depends on the social function and context. After all, the repertoire performed for a religious *novena* (series of nine nights of prayer) or a procession for the city's patron saint is quite different from that used during a private celebration or on civic dates, when the groups play anthems, marches, and two steps. Some bands, such as the band Banda de Caruaru, have reached a significant level of fame and recorded a number of LPs (long-play recordings), introducing thus their musical tradition to other regions of Brazil (Crook 2005).

Today the fife bands' function and repertoire tend to be executed by a different type of equally popular ensemble: philharmonic bands. These are composed of wind instruments (aerophones) and perform a repertoire primarily comprising marches and two steps, in addition to arrangements of other songs. Because they have traditionally performed from sheet music, the philharmonic bands in many ways constitute music schools of Brazil's interior even though they are more generally found in small- and medium-sized cities. They share many similarities with bands found in other contexts; military bands are just one example.

ENTERING THE RING: *BATUQUES, JONGOS, SAMBAS,* AND *UMBIGADAS*

The cultural context of those who performed in late 19th-century barber bands included a vibrancy of other types of musical expressions, such as *batuques, jongos, sambas,* and certainly also a number of Afro-religious traditions. As already noted in the introduction, much of the music performed by the black population in Brazil was relegated to a space between invisibility and rejection, which made it necessary for practitioners to make constant adjustments to the practices themselves and/or negotiate the spaces in which they were performed. *Batuque* was basically a generic term applied to dances led by voice and drumming. *Batuques* often took place on city outskirts, in somewhat distant places, and generally on weekends, when the slave population could get together. Even still, they were susceptible to reprisal by the ruling class and the repression of their musical practices.

In a famous early 19th-century quote, the Count of Arcos, one of Salvador's political authorities, argues that it would be more advantageous to tolerate slave encounters and their ensuing batuques than to prohibit them, as had been advocated by those who not only were bothered by the percussion sounds but also believed the dances to be obscene. The author declares that the batuques guaranteed that the tremendous number of slaves, larger by far than the number of masters, would remain segmented ethnically according to each group's ethnic origin and mother tongue. Therefore, he surmised, not only would the slaves renew close ties within each ethnic group, but they would also renew historical rivalries with other groups. In this way, by dancing and playing in groups distinguished by ethnic origin, the slaves would forget their shared problems and consequently inhibit larger rebellions that would take place were they left without a release valve (i.e., the *batuque*) (Verger 1981, p. 225). It is important to remember that there were a number of slave rebellions during that period—a famous one being the Malê Revolt, led by literate Muslim slaves in 1835—which impelled authorities to crack down on how free black people were to move about the city and congregate in groups, for example, around fountains, where slaves usually got water for their masters' homes.

We know of the existence of *calundus, lundus,* and *batuques* by way of police documents and accounts written by travelers who were witness to dances—generally described as very sensual—which accompanied rhythmically heavy music whose sung melodies sounded strange to European ears (Tinhorão 1998). These public outdoor affairs seem primarily to have been viewed as recreational, though it is quite possible that these also had religious connotations. Due to the general lack of more precise sources, it is impossible to probe these issues any further. But we do know for certain that the musical expressions of people who were still enslaved or already freed, whether recreational or not, were subject to constant criminalization.

The available sources from the 17th and 18th centuries attest to the existence of calundus, characterized as musical encounters of a religious nature. In later centuries, the *calundu* appears to have become the *lundu*, a salon dance, which was a contemporary to the *batuque*, the generic name given to all black festive music and dance events. Sources describe batuques in parts of the country's Northeast and Southeast, generally mentioning also that performers dance in a ring (Tinhorão 1998). As such, the *batuque* appears to have been a precursor of, as well as a contemporary to, the samba. *Samba* was initially also described as a ring dance that included the belly bounce (*umbigada*), a gesture, akin to an invitation to dance, in which the person inside of the ring uses his or her belly to touch the belly of the person who will take his or her place in the ring. Consequently, there are many ways to dance samba along with many different types of ensembles; some sambas were more danced, while later sambas were more sung. This leads to a long—and no doubt endless—controversy regarding *samba*'s origins (Vianna 1995). Similar musical practices coming from other parts of Brazil include the *tambor de crioula* (from Maranhão), the *carimbó* (from Pará), the *côco de roda* (from the Northeast), and the *jongo* (from the Southeast). Each of these shares some characteristics with the other traditions already mentioned but nonetheless with some important differences.

. . . STILL IN THE RING: AFRO-BRAZILIAN RELIGIONS

A subject that has provoked great curiosity while it has at the same time been profoundly misunderstood by Brazilian society and others from elsewhere is that of Afro-Brazilian religious traditions. In the 19th century the apparently ludic encounters (that probably included religious elements) of Africans and their descendents were generically denominated *batuques*. By the 20th century, a much wider range of terms applicable to religious lifestyles radically different from the Catholic Christian conception became the norm. Most unfamiliar to observers of these religious encounters were the distinct ways in which devotees dealt with the divine: they did not pray, listen to, or interpret religious texts, nor did they simply sit stilly in contemplation. Rather they sang and danced, which seemed absurd compared to the austerity and chastity valued in Christianity. To this day, rituals and celebrations continue to be conducted similarly in both Afro-Brazilian religions and related religions of the Caribbean, in countries of the African Diaspora.

Consequently, it is quite impossible to discuss music in these religious contexts without also addressing dance and other spiritual experiences such as mystic trance, which means the incorporation of divinities cultivated by adepts (Rouget 1990). Overwhelmingly, these divinities are African and depending on whether they are historically linked to the cultures of the Yoruba, Fon, or Bantu are called Orixás, Voduns, or Inquices, respectively. However, there are also non-African entities of Brazilian origin. These are called *Caboclos*, *Pretos Velhos*, *Ciganas*, and *Pomba-Giras*. Indeed, "public" rituals (i.e., rituals open to outsiders) as well as many of the most private moments of these religions are conducted with music throughout—from beginning to end—given that the sound communicates dialogically with the dances which are, generally, also performed in a ring, permitting a very specific kind of sensory experience. This link makes it difficult to distinguish who leads whom, whether it is the drums leading the people who—possessed by the religious entities—are in a trance state, or whether it is the movements of these manifested deities that guide the drums (Lühning 1990).

Even though a number of regional differences exist among Afro-Brazilian religious traditions, generally, the musical sounds are constituted by an ensemble of three drums, a percussion instrument (idiophone) of acute timbre—such as an *agogô* (type of metal bell)—to mark the complex rhythmic structures, and songs that are still today sung in African languages. Lying outside European perceptions of time structures, these rhythms have always been of interest to outsiders. Thus, after several attempts to confine the complex rhythmic structures to measures, scholars have finally adopted the concept of "time-lines," a means of understanding rhythmic patterns that was first created by Ghanaian ethnomusicologist J. H. Kwabena Nketia (1974) and later elaborated by his colleague Gerhard Kubik (1994/2010).

AFRICAN COGNITIVE STRUCTURES

Nketia and Kubik developed the idea of the time-line as a graphic system that recognizes the existence of different cognitive structures in African and African American musics, called a "pattern" or "clave." According to the authors, these structures, with internal asymmetric characteristics, would not fit into the European perception of a measure (with its strong and weak beats), nor would it be adequately represented by way of Western notation. Thus, Kubik developed a so-called impact notation, using only two symbols: "x," to signal a beat, and "." to indicate the absence of a beat. Consequently, a pattern of 16 pulses, typical of Rio's samba, would be written: x . x . x x . x . x . x . x x .

The spread of these religious ways of life within Brazil led to the creation of a new religion, an entirely Brazilian religion, called Umbanda. Umbanda was created a little over a century ago and thrives particularly strongly in large cities. Musical repertoires consist of songs (*pontos cantados*), all sung in Portuguese and thus different from the songs performed in other Afro-Brazilian religious traditions which still today use Yoruba, Fon/Ewe, and Kimbundo/Kikongo. Yet Umbanda, born in the country's Southeast, nevertheless shares and coexists with the numerous other traditional Afro-Brazilian religions existing all over the country: in Maranhão, it is Tambor de Mina, in which devotees venerate Voduns; in Pernambuco, it is Xangô and Xambá; in Bahia there are many types of Candomblé, one of the most well-known Afro-Brazilian religions (*Angola, Ijexá, Jeje, Ketu,* and *Caboclo Candomblé*) which vary depending on the ethnic link; in Rio de Janeiro, it is *Macumba*; and in Rio Grande do Sul, it is Batuque. Furthermore, there are Candomblé temples found all over Brazil, each of which was founded by adepts who increase their religious experience. There is no central regulatory structure: the reputations of the religious leaders are built on spiritual force, or *axé*, and the efficacy of their religious work.

Given the previously noted misunderstandings and even unabashed police persecutions against African-derived religions which lasted until the mid-20th century

Brazilian Candomblé practitioners dance to the beat of drums during a ceremony to honor their orixás. (Stephanie Maze/Corbis)

(Lühning 1995/96)—making it so that up until the 1970s people needed to receive official approval from government regulatory authorities to conduct religious ceremonies—many people still do not today openly identify as adepts of Afro-Brazilian religions. In official censuses, for instance, many claim to be Catholic, since this is the religion into which most adepts were baptized. But in recent years this trend has begun to change, in part because the people have come to understand their identification with Afro-Brazilian religions as a political act.

LEAVING THE RING: CAPOEIRA, MARACATU, AND THE BOI

No doubt one of the most well-known Brazilian cultural practices in the world is *capoeira*. Defined variously as a game, dance, fight, and even a sport, capoeira consists of attempting at striking an opponent while executing the fundamental movement known as the *ginga*, a constant oscillation of the body that is designed to fool the opponent prior to hitting him or her. Capoeira is always accompanied by music, which is performed by an ensemble of several instruments, including, most importantly, the berimbau, a musical bow of African origin which became an integral part of capoeira only in the late 19th century (Röhrig-Assunção 2005). In a number of older iconographic sources, the berimbau is generally depicted in the hands of street vendors. The imprecision in the definition of the practice—whether it is a martial art or not—has probably been its best defense. After all, not unlike the African-derived religions already mentioned, capoeira has a long history of vilification and open persecution. Even at the end of the 19th century, soon after the proclamation of the republic, the 1890 penal code allowed for capoeira practitioners to be imprisoned or deported to a correctional colony (Rego 1968).

Like traditions generally, capoeira has never been static, even in relation to the practice of the two styles that became popular during the early 20th century: *capoeira angola* and *capoeira regional*. The former is slower and utilizes movements closer to the ground, while the latter is faster and more acrobatic. The percussion section used in *capoeira angola* comprises three berimbaus, an *atabaque* (conical membrane drum), a tambourine, an *agogô*, and a *reco-reco* (scraped percussion instrument). Conversely, for *capoeira regional* the percussion includes just three instruments: berimbau, tambourine, and atabaque. Today, however, there is an increasing number of capoeira practitioners (*capoeiristas*) who claim to do nothing more than practice capoeira (calling it "contemporary *capoeira*"), thus undoing the distinctions between *angola* and *regional* that were solidified by the mid-20th century. Still, what remains constant is that independent of where in the world capoeira is practiced, the singing continues to be in Portuguese, for Brazil's capoeira conquered the world long ago.

Even when capoeira is not practiced nor are its typical instruments performed, the characteristic movements seem to have influenced other practices such as the steps of *frevo*, a carnival dance from Pernambuco that is always performed with a little umbrella and fast leg flexions. Indeed, the jumping movements alternate between squatting and an outstretched leg position, requiring tremendous physical conditioning. Also from Pernambuco is the *maracatu* tradition, a type of procession that was originally religious—linked to lay black brotherhoods—but is today more closely related to carnival (Albernaz 2011). Maracatus revolve around the coronation of the black king and his queen. Although this was a representation of the corresponding brotherhood during the colonial era, this is today mostly a theatrical gesture.

The theatricality of processions based on a variety of thematic plots is still quite common in the North and Northeast. In addition to the *cavalo-marinho* of Pernambuco, one of the most well-known traditions is the *boi bumbá* or *Bumba meu boi*, a theatrical tale involving a *boi*, or bull, though other animals also appear, such as birds in the *cordão de pássaro* of Pará (Moura 1997) or donkeys in the *burrinha* of Bahia. The varied choreographies are accompanied by various types of musical ensembles, which include not only percussion but also often *rabeca* (a chordophone similar to the violin) and singing (Gramani 2009). Innumerable variations of this tradition exist throughout the country such as in Paraná, where it is known as *boi mamão*. Interestingly, the tradition has been recast in the Amazon region: a large festival is held annually in Parintins, near Manaus, which attracts thousands of people to the small city to participate in the event during which two of the city's *boi* groups compete against one another.

MUSIC AND POETRY

The Northeast, which suffers from such brutal droughts, presents even more surprises: there has traditionally been a strong poetic tradition expressed in a number of musical forms. The most emblematic of these are probably *cantoria* and *embolada*. Both share the spontaneous creation of sung verses in duos, constituting a competition between the singers vis-à-vis each performer's ability and capacity to think, construct meaning, and rhyme. The poetic genres employed are incredibly diverse, no doubt originating partially in medieval European chansonniers, and range from sextains to more complex structures, such as *martelos* (strophes with decasyllabic lines) and *galopes* (strophes of 10 hendecasyllabic verses) (Sautchuk 2011; Travassos 1997). The difference between cantoria and embolada is that the former is accompanied by two violas (10-string guitar), while the latter's accompaniment is provided by two tambourines. Furthermore, the singing of the embolada is not only faster but also more aggressive, frequently employing profane terms and name calling in an attempt to defeat the opponent in the verbal dispute.

These traditions are practiced in open-air markets, public squares, or private events such as birthdays for which the performers are invited to weave vocal duels as a form of entertainment for the party guests. Sung verbal competitions are also common in other parts of Brazil: practiced in the Midwest region is the *Cururu*, which has religious overtones, as performers sing and dance for St. Benedict and the Holy Spirit (*Divino Espírito Santo*). Here the biggest difference from the Northeast is that rather than play a Northeastern viola, practitioners play the *viola-de-cocho*, an instrument made of a single piece of wood whose soundboard has no sound hole.

Northeastern poets sometimes get their inspiration from the poetry published in small pamphlets called Cordel literature (*literatura de cordel*), a name that derives from the fact that they were originally hung on *cordas* (or lines) at open-air markets. However, their creative work is essentially oral, and they have developed a very elaborate artistic technique, facts that seem incompatible with the chronically low education levels and subsequently high illiteracy rates in the Northeast. Moreover, since the poets employ terms and speech patterns often viewed as deviations of normative Portuguese, their poetry has been treated as a testament to learning deficiencies among Northeasterners.

New lines of interpretation, however, point to the manners of speech prevalent in the Northeast and other regions of the Brazilian interior as linked to the nheengatu language, originally a lingua franca used by Jesuits for the conversion of the Brazilian indigenous populations until the late 18th century. This is a mixture of a number of indigenous languages of the era, called *língua geral* (general language) or nheengatu, and it continues not only to be employed today throughout the Amazon region but has also left traces in several other regions. This explains Northeastern preference for nasal sounds and the constant vocalization of word endings, for these are also present in many indigenous languages (Martins 2004/2005).

Beyond these linguistic issues, a notable musical characteristic is present in many Northeastern practices: the systematic use of a specific musical interval, the neutral third, situated between the major and minor third, linked to the use of modal scales. As such, it appears that a locally specific Northeastern musical system exists, and though scholars have not yet explored its geographical extension, the system may actually reach far beyond the Northeast.

DISCOVERING INDIGENOUS MUSICS

The mutual exchanges between the Northeastern and indigenous cultures continue to be—as it has always been—quite frequent. However, it is important to emphasize that there is not, as many erroneously believe, a single indigenous music, even in Brazil. In fact, this would be impossible, considering that there are at present nearly 200 different indigenous languages spoken in Brazil, a number actually smaller than the number of ethnicities, since a number of ethnic groups have lost their languages. And, of course, the existence of a language also indicates a distinct culture that no doubt possesses distinct musical habits and expressions. The majority of Brazilian indigenous groups live in the Amazon region, which extends across nine states and represents more than half of Brazil's territory. Each of these groups possesses individual musical forms of expression.

In general, one cannot separate music from dance, both of which are present in innumerable rituals and celebrations related to life, death, hunting, agriculture, harvesting, and many other aspects of social life. Among the musical instruments found in indigenous cultures, there is a clear predilection for aerophones, which are made from plant materials: flutes, instruments with internal reeds, as well as others played with lip vibration. There are also a large number of idiophones, especially shakers, each with different names, played in many *pajelanças*, or shamanistic rituals, that also include very elaborate chants. Considering the significant number of different indigenous groups, there are still proportionately few recordings (i.e., *in loco* field recordings) portraying indigenous musical diversity. Even so, the recordings that do exist clearly demonstrate the significant variety of the indigenous musical universe (Coelho 2004). As such, there are not only unison songs performed by many voices, with special vocal techniques, sung by men and women, but also songs without instruments for couples, young women, plants and animals, elders, and, ultimately, for everyone (Pamfilio n.d.).

The scope of the indigenous musical aesthetic is so broad that groups have worked with musicians from other cultural universes, such as the singer Sting (Bastos 1996) and the rock group Sepultura. But in the end, the possibility of exchange among these traditions depends largely on the Amerindians themselves, for contact with the groups that reside in the more distant areas is officially mediated by FUNAI (the National Foundation of Indian Support), the governmental agency responsible for indigenous issues in Brazil whose function is to protect these groups' rights and autonomy, preventing potentially harmful contact.

Given the historical invisibility of the indigenous populations, largely decimated over the past several centuries, indigenous issues have come to significant national attention only since 2000, the year of the (much criticized!) official festivities of the quincentennial of the Portuguese "discovery" of Brazil (Tugny and Queiroz 2006). Since then indigenous groups have protested actively, generally for reasons related to land rights, health, and indigenous-language education. This has encouraged many Amerindians to relocate to the large capital cities, whether for legal claims, to publicize their existence as people, to study, and even to attend universities (Macedo 2011). Some of the younger Amerindians have used musical genres different from those of their own indigenous traditions to raise awareness about their situation. Just one example is the group Brô MCs of the Guarani Kaiowá, who use rap as a means of denouncing the constant friction between the Amerindians and the agribusiness farmers, as well as the subsequent genocide to which the Guarani have been subjected.

THE ARRIVAL OF TECHNOLOGY: RADIO PROGRAMS, CINEMA, AND "JAZES"

The Amerindians point us back toward the urban world. After all, as early as the 1930s, well before the Kaiowá rap group, an indigenous group had already caused a stir. The Irmãos Tabajaras (or Tabajara Brothers) were Amerindians from Ceará who, by way of a series of chance occurrences, arrived in Rio de Janeiro, where they sang and played guitar. They achieved commercial success through radio and live show performances, though their popularity was largely built on how exotic they were perceived to be. After touring Latin America they went to the United States, where for decades they did quite well, even recording brilliant versions of classical masterpieces (Nassif 2004). A similar trajectory, which also ended in the United States, was taken by musician, maestro, and composer Moacir Santos. Also a Northeasterner, Santos spent years in Rio de Janeiro before moving in 1967 to the United States, where he became a well-known film score composer whose work was strongly influenced by jazz (Ernest 2010). Other Northeastern musicians, however, remained in Brazil, becoming successful by establishing themselves in Rio. Two illustrious examples are Luiz Gonzaga and Jackson do Pandeiro. Both musicians were responsible for disseminating Northeastern music in Brazil's Southeast, the former with his accordion and the latter with his tambourine, playing many musical genres—*baião*, *coco*, *forró*, and so on—on the radio (Moura and Vicente 2001).

These trajectories elucidate some important facts: for decades many musicians' commercial and professional success required an obligatorily stop in Rio de Janeiro, the home of the country's only recording studios. Even though broadcasting technology existed in a number of states in Brazil, Rio was the only place where the recording industry coexisted with the radio, which had been employed in Brazil starting 1923. This made the city famous and helped consolidate it as a vibrant cultural hub, as it became home to stars and aspiring artists from all over Brazil, feeding its cultural life. In addition, this brought national exposure to musicians who began to build their own careers, particularly radio singers who concomitantly made records. One of the first such singers and no doubt the most well known, even outside of Brazil, was Carmen Miranda. In the 1930s, she began working increasingly in the United States, where she performed on Broadway and in a number of movies (Castro 2005). Her song repertoire consisted primarily of Rio de Janeiro style samba (*samba carioca*). Carmen Miranda was also regarded as the muse of the Good Neighbor Policy, as the Roosevelt administration called its foreign policy toward the countries of Latin America. Though this policy was bound up in uneven power relations, it did lead to a notable dissemination of Brazilian culture in the United States (Moura 1984, 2012).

Cinemas were also important to Rio's cultural milieu, for they were used to show films and house live musical performances. Movie showings largely included exports from Hollywood, whose soundtracks many musicians learned by ear, mentally recording them for arrangements, adaptations, or simply new compositions. This also happened in other major cities, helping thus to propagate these new musical styles. Furthermore, local radio stations had their own instrumental ensembles called "jazes," an adaptation of the word "jazz." Despite the name, these small orchestras, which also included singers, did not play jazz but rather dance music, with a special penchant for arrangements of Caribbean music that arrived via radio, cinema, or contact with U.S. sailors stationed in port cities along the coast. These ensembles comprised winds, banjo, bass, and drum set, as well as the occasional inclusion of other instruments. The jazes were active in Brazil between the 1920s and 1960s, gradually losing their importance in part because they were competing with radio orchestras that also played popular commercial songs.

THE STRENGTH OF THE URBAN MUSIC SCENE: MPB, ROCK, AND REGGAE

The postwar period brought profound changes to Brazilian politics, society, and culture along with new demands, possibilities, and challenges. Although television arrived in Rio in 1950, its exorbitant price limited its accessibility, and it took another 10 years to reach other states and widespread popularity. The 1950s also marked the beginning of the mass production of consumer goods such as refrigerators and cars (the "fusca," as the Volkswagen Beetle was called, is just one example) (Moura 1984, 2012). Brasília was constructed during the 1950s and inaugurated in 1960. Indeed, the 1950s were generally a period of political effervescence in the country. Rio de Janeiro's cultural dominance lasted until the 1960s and 1970s, when other music production and distribution centers began gradually to appear, challenging, thus, Rio de Janeiro's hegemonic position. Furthermore, the names of new artists began to emerge on the national music scene outside of Rio, decentralizing production and distribution (Tinhorão 1997).

Given the sheer number, it would be impossible to mention every artist on the Brazilian popular music scene. Indeed, there are so many completely original musicians with individual sounds. Some musicians have been on the scene for decades; others had only brief careers, and still others are just beginning today; some use more modern technologies, and others use more traditional processes of dissemination. Noticeably, vocal music is especially well represented on the popular music scene in part because audiences have preferred it to instrumental music for many decades. Given these concerns, the names mentioned in the following paragraphs have not been included in accordance with any personal preferences but rather because they are linked to specific movements or historical moments, or because they have great relevance today.

Part of the expansion of Rio's samba included the *bossa nova* movement of the 1950s. This new way of playing samba was characterized not only by its similarities to American jazz but also by its heavy use of offbeat accents. Bossa nova was idealized by middle-class youth of Rio and soon gained international recognition, linked in particular to João Gilberto, Tom Jobim, Vinícius de Moraes, and Baden Powell. An emblematic song of this period is "Garota de Ipanema," or "The Girl from Ipanema," as it was called in English. Even today, many people outside of Brazil see bossa nova as the country's soundtrack, without necessarily realizing that there have been many musical developments in the past several decades. Bossa nova's success all over the world is partially the result of its connection to musicians from the United States, such as Stan Getz and Frank Sinatra, both of whom recorded LPs with their Brazilian colleagues. Indeed, many Brazilian musicians spent short or long periods of time in the United States, living there, as did guitarist João Gilberto, for many years.

ORFEU NEGRO

The film *Orfeu Negro* (*Black Orpheus*) or *Orfeu do Carnaval* (1959) portrays Rio de Janeiro's culture with a bossa nova soundtrack. Based on a play by Vinícius de Moraes, the film was made by French director Marcel Camus. It garnered worldwide success and won the highest award at the 1959 Cannes Film Festival and the 1960 Academy Award for Best Foreign Language Film.

An important phenomenon began in the 1960s: Brazilian popular music festivals. These were televised and broadcast all over Brazil, dialoguing indirectly with a traumatic political moment: the military dictatorship that lasted from 1964 to 1985. The festivals were important showcases for many of the most well-known names of "MPB," at first simply an acronym for Brazilian Popular Music (*Música Popular Brasileira*) but which later came to denominate a whole musical genre. Some of the most consecrated artists include Elis Regina, Milton Nascimento, and Maria Bethania (Napolitano 2002).

With the revelation of these new talents came a wide-reaching artistic movement known as Tropicália, or Tropicalism, which included phases of experimental music, and was led by a number of young singers and composers from Bahia: in particular, Gilberto Gil, Caetano Veloso, Tom Zé, and Gal Costa (Tom Zé 2008). Importantly, Gil and Veloso were exiled to London during the harshest period of the dictatorship. After all, most of the musicians of this era also composed protest music, impressively demonstrating their ability to fool government censorship by way of lyrics that became veritable works of art given their capacity to escape the watchful eyes of the censors. Particularly significant are the lyrics of Chico Buarque's "Construção," "Apesar de você," and "Cálice," as well as Caetano Veloso's "Alegria, alegria" (Oliven 2011).

The 1950s also marked the beginning of the Brazilian rock movement, with Roberto and Erasmos Carlos, and Rita Lee, a movement initially rejected by MPB singers. Tropicália, on the other hand, embraced the Brazilian rock movement, and even more so after the arrival in the early 1970s of another singer from Bahia, Raul Seixas. Subsequently, rock music became popular in a number of different cities, especially during the 1980s, around the time of the country's redemocratization. The bands are emblematic, and many of them are not only still active today, but due to their success can also be heard regularly on the radio. It is noteworthy that all of these bands chose to sing in Portuguese: Paralamas do Sucesso and Legião Urbana, with the singer Renato Russo (from Brasília), Titã and Ira! (from São Paulo), Kid Abelha (with the female singer Paula Toller), and Barão Vermelho (from Rio), with the front man Cazuza. All of these groups possess unique identities and are thus able to maintain legions of fans to this day (Dapieve 1995).

The heavy metal movement also began in the 1980s, represented particularly well by two very influential bands, Angra (from São Paulo) and Sepultura (from Minas

Gerais), even recording with Xavante Indians. Both groups have always preferred to write their lyrics in English and are well received on the international scene. By the 1990s, a number of bands, such as Skank and O Rappa, emerged with a mix of rock and reggae, and helped generally to introduce reggae to a national audience. The reggae scene in São Luís, Maranhão, tends to favor recordings imported from the Caribbean (Silva 1995). Bahia's reggae scene includes a number of composers who write religious lyrics, and even some composers involved in evangelical Christian churches that specifically preach with reggae music. One example is Bola de Neve Church (Falcón 2012; Mota 2012), which has even been present in Los Angeles for the past several years.

THE ROMANTIC AUDIENCE: *SERTANEJO,*
VANERÃO, AND *ARROCHA*

There is, however, also a more romantic audience in Brazil, which is clear with genres such as *sertanejo, vanerão, tchê music*, and *arrocha*, in addition to a number of other similar styles. While each of these represents distinct regions and genres, all have something in common: lyrics packaged into agreeable melodies that have mass appeal. All of these musicians move a considerable and ever-expanding commercial market.

Sertanejo is a stylistic continuation of country music (*música caipira*), the music of the interior regions, especially of the Midwest, consisting of singing in parallel thirds to the sound of the *viola caipira*, a ten-string Brazilian viola, and with a tuning different from the Northeastern viola. Indeed, although there are a fair number of similarities between the singing styles of the Midwest and the Northeast, each region has an entirely different repertoire. However, the so-called sertaneja duos (*duplas ser-tanejas*) dispense all together with the violas and simply become singers accompanied by electric instruments and bands (Santos 2011). These duos are commonly formed by brothers or cousins who learn at a young age how to sing in precise ways with the desired tuning/intonation. More recently, a new generation of solo musicians, who perform *sertanejo universitário* (university sertanejo) in major cities, have become very popular. One of many well-known performers is Luan Santana. While Brazilian sertaneja music certainly has a history and social role similar to those of country music in the United States, there is no direct link between the two.

A similar process of bringing rural traditions into contact with new media demands has occurred in southern Brazil. *Vanerão* is one of the most common genres of the South. It is performed on the harmonica or accordion, an instrument traditionally played in only two regions of Brazil: the South and the Northeast, though their names, sizes, and repertoires are different. In *vanerão* the harmonica is the main instrument of the ensemble, which furthermore includes voice and other instruments. The name "vanerão" appears to be a variation of *havaneira/habanera*, serving as yet another example of the Caribbean's profound influence on Brazilian culture, as we have already seen. It is a genre ideal for pair dancing. In recent decades *vanerão* has influenced the more urban *tchê music*, a mix of music from Rio Grande do Sul and music from other places in Brazil.

Finally, *arrocha*, a recent innovation that is (debatably) from the interior of Bahia, is an offshoot of *brega* (lit. "corny") music (Araújo 1999), with marked sentimental appeal and palatable melodies. Its success is partly justified by its operational and musical logistics, for initially the ensemble included only a single person: a singer who accompanied himself (or herself) on the keyboard, facilitating mobility and small performance spaces. As such, the *arrocha* musician could reach audiences in an intimate way, playing music designed specifically for couples wishing to dance peacefully. Soon after, a number of musicians began to perform *arrocha* for large audiences, and thus expanding, of course, the ensemble size to resemble more traditional bands.

YOUTH AUDIENCES: FUNK, RAP, *PAGODE,*
AND *TECNOBREGA*

There is a preference today, particularly among the youth, for "fashionable" genres that, originating partly in Brazil and partly in the United States, have been appropriated and recreated in the Brazilian context (Lima 2002). The four genres I discuss here are associated with different regions of Brazil, though they have in many cases reached a much broader national audience: funk is markedly popular in Rio de Janeiro; rap in the Southeast; *pagode* in Rio, São Paulo, and Bahia; and *tecnobrega* in Pará. All four genres originate among the disenfranchised suburban populations of large cities and represent socially invisible populations, especially Afro-Brazilians.

Rio's *funk*, also called *funk carioca*, is not, as one might think, a version of recent American funk music. Rather it has a much longer history that is linked to dance parties—known as "bailes Black"—of the 1970s at which soul and other African American musical styles were the main musical attraction for an impressive number of young black youths from the suburban areas of Rio (Vianna 1987). Important names from this first period are Tim Maia and Jorge Ben Jor, both of whom helped popularize many songs. Later, with the influence of the Miami bass style of hip hop, the rhythmic and melodic content of *funk* began to change, leading to a broad range of different styles, ranging from "funk melody" to "funk proibidão" (prohibited funk; Essinger 2005). The latter is a style in which drug dealing is glamorized, markedly different from the more recent "funk ostentação" (ostentatious funk) that presents funk as part of the consumer world and expresses interest in such participation (Palombini 2009). *Funk's* "descent from the hills (*morros*)" ("hills" is a geographical reference to the location of the shantytowns that most in the United States would recognize as *favelas*) is nothing new and is indeed no longer just music of the marginalized and disenfranchised Afro-Brazilian youth of Rio: funk is common at middle-class dance parties and is popular all over the country. It has faced discrimination and has caught on like wildfire among teens and youths. For many, funk represents the dream of ascension through music, whether as a male MC or a female singer, as was the case with Anitta (Larissa de Macedo Machado), who has recently become a national star.

Rap became particularly popular in Brazil during the 1980s, with Thaíde, then the MC Racionais (with Mano Brown), Gabriel o Pensador, Planet Hemp (with Marcelo D2), and MVBill. Rap's link to the Brazilian hip hop movement is stronger for some artists than for others, but generally the hip hop movement—with all of its components—is significantly strong in the whole country. Rappers like Rappin Hood emphasize similarities between rap and some of the local song duel traditions such as *repente* and *embolada* traditions, particularly regarding the relationship between rhymed vocal expression and social critique, though they have no historical relationship. Perhaps rap has attracted a smaller audience than other genres because it is more socially conscious and less danceable. Still, it is no less important, even being used as a teaching tool, especially in public schools with predominantly African descendent populations (Andrade 1999).

Samba developed into a number of different styles. For instance, when Rio's so-called backyard samba, *samba fundo de quintal* (Guimarães 1978; Vianna 1995/1999), became *pagode* in the 1980s, it grew popular in Rio and São Paulo, with stars such as Bezerra da Silva and Zeca Pagodinho, whose lyrics dealt with romance or working-class life in Rio. But Bahia also has a type of *pagode* quite different from the homonymous genre in other states. Initially beginning as an urban variation of Bahian *samba de roda*, pagode soon became a genre in itself. The Bahian pagode movement of the 1990s transformed radically after 2000 when a new generation of musicians began to represent their own suburban neighborhoods by singing about them and addressing their problems. Significantly less socially conscious than rap, pagode lyrics are perhaps most similar to those of Rio's funk though without directly addressing issues of drug-related violence. Bahian *pagode* lyrics often stereotype women in ways that are criticized by some and accepted by others; even the women from these same suburban neighborhoods appear to take little issue with the misogynist lyrics—written by male singers and lyricists—that are accompanied by sexually suggestive choreographies normally danced by women (Leme 2003).

Bahian pagode has had such wide acceptance that for the past several years it has been a major part of Bahia's street carnival, a carnival far better known for processions of *blocos-afro*, *afoxés*, and *trio elétricos*, which is performed along the city's major roads. As such, *pagode* bands perform along with today's *axé music* stars such as Ivete Sangalo, Daniela Mercury, and Claudia Leitte, as well as other exponents of Bahian music such as Carlinhos Brown and the percussion group Timbalada. Since the 1980s, *axé music*, a term that began as a pejorative, has been a genre that brings together all of the sounds of Bahia's carnival, melding the percussive force of the *blocos-afro*—which play a local version of samba fused with reggae called *samba-reggae*—with the electrified instruments of the *trio elétrico* bands, which play songs initially inspired in the aforementioned frevo of Pernambuco (Guerreiro 1999). From Bahia, pagode has now spread to many other states and has garnered a significant amount of success.

CARNIVAL

Today's Brazilian carnival is divided into three different models that compete against each other for national and international appeal: Rio de Janeiro is known for the regal parades of its samba schools, which compete against each other in the Sambódromo; Salvador is known for its open-air carnival, *trio elétricos* (trucks adjusted to carry bands and amplify sound through enormous speakers), and a variety of different-sized percussion groups; and Recife, where carnival generally includes local musical traditions, displaying the dominance of Northeastern acoustic music. (Examples of Bahian carnival are found at http://editora.globo.com/epoca/edic/351/audios_carnaval.htm.)

Finally, *tecnobrega* from Belém (Pará), also a local derivative of the romantic *brega* style, expanded the sound system parties held primarily in the suburban areas of the city. These parties were led by DJs who played a mix of local *carimbó* and Caribbean-influenced musics such as calypso and other rhythms. The most salient characteristic of this genre, beyond its underground, noncommercial scene, is that technology matters not only for sound production and amplification, but also for visual effects. Similar to recent movements in Rio and Salvador, individual artists have been increasingly successful at building careers performing this genre. Gaby Amarantos, for instance, has in recent years become quite well known. These performers have helped definitively expand the diversity of the Brazilian musical market to include the country's North (Amaral 2011).

CONCLUSION

Brazilian music is extremely diverse and goes far beyond yet another stereotype which claims that in Brazil "everything becomes a carnival." Even though we could not address every type of music, we were able to cover at least some of Brazil's musics and in the process notice some important points. Percussion, feared during the colonial period, has become a representative trait of Brazilian music. It is found in a wide variety of ensembles, genres, and contexts. The constant creation of new music genres, whose protagonists tend to be (youth) populations of the country's geographic and social peripheries—such as Afro-descendents or ethnic minorities— evinces the tremendous capacity of Brazilians to embrace and reinvent nearly everything. It is worth noting that males have primarily been responsible for the Brazilian music scene, though women have an increasingly representative role, particularly as singers and musicians. Still, in some genres male predominance is virtually absolute, a fact that no doubt serves as fodder for future sociological studies on gender.

The primacy of sung music is clear. In fact, in recent years there has even been increased activity of the type of vocal-only groups common in the 1950s. However, the contemporary groups vocally imitate musical instruments, including techniques such as beatboxing, in addition to groups such as the Barbatuques, from São Paulo, that work with body percussion. Orality, or perhaps more accurately acoustic culture (Lopes 2011), continues to be a primary means of learning and transmission, with little importance being given to writing. Furthermore, nothing "is sacred," in the sense that there is no tradition that cannot be "toyed with" and changed. One recent example is the transformation of carnival that appears to be taking place in both Rio de Janeiro and Salvador: as people begin to tire of the current mega productions carnival models that have become powerful industries in themselves, they have reinitiated small-scale, participatory street festivals. Another characteristic of music in Brazil is the predominance of musical expressions involving the body, whether in traditional or commercial musics, linking music not only to movement but also to theatricality.

We can therefore conclude as the quintessence of our musical voyage that not only is playful creativity a major springboard but also that Brazilian culture has an undisputed capacity to reinvent itself continually. For this, it borrows and exchanges with musical styles from other countries or with lesser-known Brazilian traditions, even if there are controversies about the impact of the culture industry and the heavy influence of U.S. culture in Brazil (Carvalho 1996). But, returning to Mario de Andrade, we might call this an unabashed "anthropophagical" (i.e., cannibalistic) posture, as he proclaimed in his novel *Macunaíma*, whose homonymous character represents Brazilian culture in a constant process of identity redefinition. As such, we conclude by noting that the variety of musics presented and discussed here is perhaps the best path to understanding the multiplicity and complexity of Brazilian culture, which might therefore explain Brazil's cultural power as well as its ability to act as a constant source of inspiration for musicians and non-musicians, within the country and abroad.

REFERENCES

Albernaz, Lady Selma Ferreira. 2011. "Gender and Musical Performance in Maracatus (PE) and Bumba Bois (MA)." *Vibrant*, 8 (1) [Dossie Music and Anthropology in Brazil. Carlos Sandroni, Hermano Vianna, and Rafael José de Menezes Bastos (eds.)], pp. 323–354. http://www.vibrant.org.br/issues/v8n1/.

Amaral, Paulo Murilo Guerreiro do. 2011. "Tradição futurista e regionalismo global na performance do tecnobrega em Belém do Pará." http://eiap2011.files.wordpress.com/2011/05/paulo-guerreiro-do-amaral-gt-5.pdf.

Andrade, Elaine Nunes. 1999. *Rap e educação, Rap é educação*. São Paulo: Summus.

Andrade, Mario de. 1972 [1928]. *Ensaio sobre a Música brasileira*. São Paulo.

Andrade, Mario de. 1976. *O turista Aprendiz*. Telê Porto Ancona Lopez (Ed.). São Paulo: Livraria Duas Cidades.

Araújo, Samuel. 1999. "Brega, samba, trabalho acústico." Revista OPUS, no. 6. http://www.anppom.com.br/revista/index.php/opus.

Ayala, Marcos, and Maria Ignez Novais Ayala. 1995. *Cultura popular no Brasil*. São Paulo: Ática.

Bastos, Rafael de Menezes. 1996. "Musicalidade e ambientalismo na Redescoberta da Eldorado e do Caraíba: uma Antropologia do Encontro Raoni-Sting." *Revista de Antropologia*, 39 (1): 145–189.

Bezerra, José Augusto, Ingrid Schwamborn, and Maria Elias Soares, eds. 2010. *Haydn, Mozart e Neukomm na Corte Real do Rio de Janeiro (1816–1822)*. Fortaleza: Edições UFC. [Bilingual German-Portuguese edition, including the facsimile edition, translated and annotations, of *Notéicia histórica da vida e das obras de José Haydn*, by por Joaquim Le Breton, Rio de Janeiro, 1820].

Carvalho, José Jorge de. 1996. "Imperialismo cultural: uma questão silenciada." *Revista U.S.P.*, 32: 66–89. http://www.revistas.usp.br/revusp/article/view/26032/27761.

Castro, Ruy. 2005. *Carmen- uma biografia*. São Paulo: Companhia das Letras.

Coelho, Luís Fernando Hering. 2004. "Música Indígena no Mercado: Sobre Demandas, Mensagens e Ruídos no (Des)Encontro Intermusical." *Campos*, 5 (1): 151–166. http://ojs.c3sl.ufpr.br/ojs/index.php/campos/article/viewFile/1640/1382.

Crook, Larry. 2005. *Brazilian Music and the Heartbeat of a Modern Nation*. Santa Barbara, CA: ABC-CLIO.

Dapieve, Arthur. 1995. *BRock. O rock brasileiro dos anos 80*. São Paulo: Editora 34.

Diniz, Edinha. 1999. *Chiquinha Gonzaga, uma história de vida*. Rio de Janeiro: Record.

Ernest, Andrea. 2010. "Mais 'coisas' sobre Moacir Santos ou, os caminhos de um músico brasileiro." PhD dissertation, Universidade Federal da Bahia. http://repositorio.ufba.br/ri/handle/ri/12623.

Essinger, Silvio. 2005. *Batidão: uma história do funk.* Rio de Janeiro and São Paulo: Record.

Falcón, Bárbara. 2012. *O reggae de Cachoeira. Produção Musical em um porto atlântico.* Salvador: Pinaúna.

Gilroy, Paul. 2001. *O Atlântico Negro. Modernidade e dupla consciência.* São Paulo: Editora 34.

Gramani, Daniella Cunha da. 2009. "O aprendizado e a prática da rabeca no Fandango Caiçara." MA thesis, Universidade Federal do Paraná. http://www.sacod.ufpr.br/portal/artes/wp-content/uploads/sites/8/2012/12/Daniella-da-Cunha-Gramani.pdf.

Guerreiro, Goli. 1999. *A trama dos Tambores. A música afro-pop de Salvador.* São Paulo: Editora 34.

Guimarães, Francisco (Vagalume). 1978. *Na roda do samba*, 2nd ed. Rio de Janeiro: FUNARTE (1st ed., Rio de Janeiro: Typ. São Benedicto, 1933).

Hornbostel, Erich M. V. and Curt Sachs. 1914. "Systematik der Musikinstrumente. Ein Versuch." Zeitschrift für Ethnologie, Band 46, Heft 4-5: 553–590.

Kubik, Gerhard. 1994/2010. *Theory of African Music.* Chicago: Chicago University Press.

Leme, Monica. 2003. *Que "tchan" é esse. Indústria e produção musical no Brasil.* Rio de Janeiro: Anna Blume.

Lima, Ari. 2002. "Funkeiros, timbaleiros e pagodeiros: notas sobre juventude e música negra na cidade de Salvador." *Cadernos CEDES*, 22 (57): 77–96. http://www.scielo.br/scielo.php?pid=S0101-32622002000200006&script=sci_arttext.

Lopes, Antonio Herculano, Martha Abreu, Martha Tupinambá de Ulhôa, and Mônica Pimenta Velloso, eds. 2011. *Música e história no longo século XIX.* Rio de Janeiro: Fundação Casa Rui Barbosa.

Lopes, José de Sousa Miguel. 2001. "Cultura acústica e memória em Moçambique: as marcas indeléveis numa antropologia dos sentidos." *SCRIPTA*, Belo Horizonte, 4 (8): 208–228. http://www.ich.pucminas.br/cespuc/Revistas_Scripta/Scripta08/Conteudo/N08_Parte03_art04.pdf.

Lühning, Angela. 1990. "Música: coração do candomblé." *Revista U.S.P.*, no. 7: 115–124. http://www.revistas.usp.br/revusp/article/view/55867/59265.

Lühning, Angela. 1995/1996. "'Acabe com este santo, Pedrito vem aí, . . .' Mito e realidade da perseguição policial ao candomblé baiano entre 1920 e 1942." *Revista U.S.P.*, no. 28: 1994–1220. http://www.usp.br/revistausp/28/14-angela.pdf.

Macedo, Valéria. 2011. "Tracking Guarani Songs: Between Villages, Cities and Worlds." *Vibrant*, 8 (1) [Dossie Music and Anthropology in Brazil. Carlos Sandroni, Hermano Vianna, and Rafael José de Menezes Bastos (eds.)]: pp. 378–412. http://www.vibrant.org.br/issues/v8n1/.

Martins, José de Souza. 2004/2005. "Cultura e educação na roça. Encontro e desencontros." *Revista U.S.P.*, no. 64 (Dossiê Brasil Rural): 29–48. http://www.revistas.usp.br/revusp/article/viewFile/13388/15206.

Mota, Fabrício. 2012. *Guerreir@s do terceiro mundo. Identidade negras na música reggae da Bahia.* Salvador: Pinaúna.

Moura, Carlos Eugênio Marcondes de. 1997. *O teatro que o povo cria.* Belém: Secult.

Moura, Fernando, and Antônio Vicente. 2001. *Jackson do Pandeiro: o rei do ritmo.* São Paulo, Editora 34.

Moura, Gerson. 1984. *Tio Sam chega ao Brasil. A penetração cultural americana.* São Paulo: Brasiliense.

Moura, Gerson. 2012. *Relações exteriores do Brasil, 1939–1950. Mudanças na natureza das relações Brasil-Estados Unidos durante e após a Segunda Guerra Mundial.* Brasília: Fundação Alexandre de Gusmão; Ministério das Relações Exteriores. http://www.funag.gov.br/biblioteca/dmdocuments/Relacoes_Exteriores_do_Brasil.pdf.

Napolitano, Marcos. 2002. *História & Música. História cultural da música popular.* Belo Horizonte: Autêntica. http://www.nre.seed.pr.gov.br/franciscobeltrao/arquivos/File/disciplinas/historia/historia_musica_marcos_napolitano.pdf.

Nassif, Luis. April 1, 2004. "Os incríveis Índios Tabajaras." *Folha de São Paulo*, Sunday. http://www.revivendomusicas.com.br/biografias_detalhes.asp?id=126.

Nketia, J. H. Kwabena. 1974. *The Music of Africa.* New York: W. W. Norton.

Oliven, Ruben. 2011. "The Imaginary of Brazilian Popular Music." *Vibrant*, 8 (1) [Dossie Music and Antropology in Brazil. Carlos Sandroni, Hermano Vianna, and Rafael José de Menezes Bastos (eds.)], pp. 170–207. http://www.vibrant.org.br/issues/v8n1/.

Palombini, Carlos. 2009. "Soul Brasileiro e Funk Carioca." *OPUS*, 15 (1): 37–61. http://www.anppom.com.br/revista/index.php/opus.

Pamfilio, Ricardo, ed. n.d. *Cantando as culturas indígenas.* Salvador: Thydewa. http://thydewa.org/downloads/cantando.pdf.

Rego, Waldeloir. 1968. *A capoeira Angola. Ensaio sócio-etnográfico.* Salvador: Itapoan.

Röhrig-Assunção, Mathias. 2005. *Capoeira. The History of an Afro-Brazilian Martial Art.* London and New York: Routledge. http://www.e-reading.ws/bookreader.php/134571/Assuncao_-_Capoeira_-_The_History_of_Afro-Brazilian_Martial_Art.pdf.

Rouget, Gilbert. 1990. *La musique et la transe.* Paris: Gallimard.

Santos, Elizete Ignácio dos. 2011. "Modernization and Its Discontents: Discourses on the Transformation of caipira into sertanejo Music." *Vibrant*, 8 (1) [Dossie Music and Antropology in Brazil. Carlos Sandroni, Hermano Vianna, and Rafael José de Menezes Bastos (eds.)], pp. 292–322. http://www.vibrant.org.br/issues/v8n1/.

Sautchuk, João Miguel. 2011. "Poetic Improvisation in the Brazilian Northeast." *Vibrant*, 8 (1) [Dossie Music and Antropology in Brazil. Carlos Sandroni, Hermano Vianna, and Rafael José de Menezes Bastos (eds.)], pp. 261–291. http://www.vibrant.org.br/issues/v8n1/.

Silva, Carlos Benedito Rodrigues da. 1995. *Da terra das Primaveras à Ilha do Amor: Reggae, lazer e identidade cultural.* São Luis: EDUFMA.

Tinhorão, José Ramos. 1997. *Música popular. Um tema em debate.* São Paulo: Editora 34.

Tinhorão, José Ramos. 1998. *História social da música Popular brasileira*, São Paulo: Editora 34.

Tom Zé. 2008. *Tropicalista lenta luta.* São Paulo, Publifolha.

Travassos, Elizabeth. 1997. "Notas sobre a cantoria." In Salwa El-Shawan Castelo Branco (ed.), *Portugal e o mundo: o encontro de culturas na música*, pp. 535–547. Lisboa: Dom Quixote.

Tugny, Rosângela Pereira de, and Rubens Caixeta de Queiroz. 2006. *Músicas africanas e indígenas no Brasil.* Belo Horizonte: Editora da UFMG.

Verger, Pierre Fatumbi. 1981. *Notícias da Bahia de 1850.* Salvador: Corrupio.

Vianna, Hermano. 1987. "O mundo funk carioca." PhD dissertation, Universidade Federal do Rio de Janeiro, Museu Nacional (original PhD dissertation version available at http://www.overmundo.com.br/banco/o-baile-funk-carioca-hermano-vianna).

Vianna, Hermano. 1995. *O mistério do samba.* Rio de Janeiro, Editora Zahar.

Vianna, Hermano. 1999. *The Mystery of Samba. Popular Music and National Identity in Brazil.* Chapel Hill: University of North Carolina Press.

Suggested Reading

Beserra, Bernadete. "The Reinvention of Brazil and Other Metamorphoses in the World of Chicago Samba." *Vibrant*, 8 (1) [Dossie Music and Antropology in Brazil. Carlos Sandroni, Hermano Vianna, and Rafael José de Menezes Bastos (eds.)], 2011, pp. 117–145. http://www.vibrant.org.br/issues/v8n1/.

Crook, Larry, and Randal Johnson (eds.). *Black Brazil. Culture, Identity and Social Mobilization.* Los Angeles: UCLA Latin American Center Publications (UCLA Latin American Studies 86), 1999.

Hanchard, Michael. *Orpheus and Power: The Movimento Negro of Rio de Janeiro and São Paulo, Brazil (1945–1988).* Princeton, NJ: Princeton University Press, 1994.

Horn, David, and John Shepherd (eds.). *Bloomsbury Encyclopedia of Popular Music of the World. Genres: Caribbean and Latin America*, Vol. IX. London, New Delhi, New York, and Sydney: Bloomsbury, 2014.

Murphy, John P. *Music in Brazil.* New York and Oxford: Oxford University Press, 2006 (includes CD).

Perrone, Charles A., and Christopher Dunn (eds.). *Brazilian Popular Music and Globalization.* New York and London: Routledge, 2004.

Silva, José Carlos Gomes da. "Sounds of Youth in the Metropolis: The Different Routes of the Hip Hop Movement in the city of São Paulo." *Vibrant*, 8 (1) [Dossie Music and Antropology in Brazil. Carlos Sandroni, Hermano Vianna, and Rafaelde Menezes Bastos (eds.)], 2011, pp. 70–94. http://www.vibrant.org.br/issues/v8n1/.

Souza, Angela Maria de Souza, and Deise Lucy Oliveira Montardo. "Music and Musicalities in the Hip Hop Movement: Gospel Rap." *Vibrant*, 8 (1) [Dossie Music and Anthropology in Brazil. Carlos Sandroni, Hermano Vianna, and Rafael de Menezes Bastos (eds.)], 2011, pp. 7–38. http://www.vibrant.org.br/issues/v8n1/.

Films Cited

Cantando as culturas indígenas. http://www.iteia.org.br/audios/cantando-as-culturas-indigenas.

Library of Congress: Endangered Music Project. *The discoteca collection: Missão de pesquisas folclóricas.* HRT 15018 (1997), also at http://www.folkways.si.edu/the-discoteca-collection-missao-de-pesquisas-folcloricas/world/music/album/smithsonian.

Library of Congress: Endangered Music Project. *L.H. Correa de Azevedo: Music of Ceara and Minas Gerais.* HRT 15019 (1997), também em: http://www.folkways.si.edu/l-h-correa-de-azevedo-music-of-ceara-and-minas-gerais/latin-world/music/album/smithsonian.

Library of Congress: Endangered Music Project. *The Yoruba/Dahomean Collection: Orishas across the Ocean.* HRT, 15020 (1998), also at http://www.folkways.si.edu/the-yoruba/dahomean-collection-orishas-across-the-ocean/world/music/album/smithsonian.

Site of the Instituto Moreira Sales (São Paulo), presenting the Humberto Franceschi collection: http://ims.com.br/ims/explore/artista/humberto-franceschi.

Site with links to documentaries about Brazilian music: http://pedroconsortebr.wordpress.com/2012/10/30/documentarios-sobre-a-musica-brasileira-lista-completa/.

Site with list of Brazilian Indigenous Music CDs: http://www.iande.art.br/musica/musica1.htm.

Food

Elisa Duarte Teixeira

BRAZILIAN CUISINE(S) AND ITS ORIGINS

Brazilian cuisine of nowadays is a colorful and vibrant mosaic of several and diverse cooking traditions from all over the world—Italian, Japanese, German, Middle Eastern, Polish, Jewish, to name a few. But the origins of many traditional dishes of this nation, known for having one of the highest degrees of intermarriage in the world, can be credited to three groups that constituted the basis for this melting pot of races and cultures: the indigenous inhabitants of modern-day Brazil, the Portuguese settlers, and the African slaves.

When the first Portuguese arrived in Brazil, around 1500, they were mainly traders and explorers interested in *pau brasil*, brazilwood used as a red dye pigment for textiles in Europe—hence the name of the country and its people, *brasileiros*. The first settlers, minor nobility from Portugal sent by the Portuguese Crown to colonize and oversee the vast tracts of land called *capitanias hereditárias*, came only after the 1530s and established themselves in the Northeast region. They quickly learned that the climate and soil were appropriate for growing sugarcane, and sugar was a much-valued commodity in Europe back then. So they brought the plant from the Azores and decided they were going to enslave the indigenous peoples—mostly seminomadic tribes, subsisting on gathering, fishing, hunting, and migrant agriculture—to take care of the crops. The attempts failed miserably, leading to a drastic decimation of the local native populations, especially the *Tupis* and *Guaranis*.

Plantation owners then turned to importing slaves from West Africa to do the job. When the first ships bringing starving captured Africans began to arrive in Brazil, by the mid-16th century, the Portuguese diet was already heavily influenced by the indigenous staples—*mandioca*, corn and fish—which they had learned to appreciate out of necessity. Typical Portuguese ingredients, such as olives, olive oil, salt cod, sardines, smoked sausages, salt-cured and smoked pork, tomato, potato, wheat flour, dried fruits, nuts, and wine were too expensive and/or too hard to obtain and to keep in the long-lasting ship voyages.

Because the settlers did not bring women with them until much later, the indigenous women were originally in charge of the settler's meals. They prepared the food using their ancient techniques, namely grilling ingredients in a spit or on the *moquém* (grates made with twigs mounted over a fire); baking ingredients by wrapping them in banana or other leaves and either burying them under fire pits or placing them on the *moquém*; and preserving surplus food by smoking and then grinding the ingredients to a fine powder, that they mixed with several other native ingredients, such as copious amounts of hot peppers of all kinds, and toasted manioc flour.

These concoctions constituted, for many decades, the main source of energy for the settlers, alongside fresh fruits, most of them plentiful year round.

The art of grilling all kinds of meat in a spit or over grates by the fire and consuming them rare is a tradition that survived in the Brazilian barbecue, *churrasco*. Mixing smoked and/or dried ingredients with different kinds of *farinha de mandioca* and corn flours is a practice still found in several preparations all over the country, such as the savory *paçocas* of North and Northeast states, and the *farofa*—a national favorite all over the country.

VERSATILE MANDIOCA

Known in English as cassava, manioc, and yuca, *mandioca* (*Manihot esculenta*), also called *macaxeira* and *aipim* in different regions of Brazil, was a major starchy staple food for the Amerindians when the Portuguese first arrived in Brazil.

Called "the bread of Brazil," this new world tuberous root contains large amounts of carbohydrates, potassium, and good-quality fibers, but also compounds that can combine to form hydrocyanic acid—a poisonous substance that can be neutralized by cooking and a number of other techniques developed by the native peoples. The fresh roots of nonpoisonous varieties sold in markets are elongated and usually 2–4 in. thick and 10–15 in. long. The chalk-white or yellowish pulp is surrounded by a pliable rind about 1/16 in. thick, and a leathery brown outer skin. Both peel off easily and are discarded before cooking. Once cooked, the woody texture of the raw ingredient becomes very soft and creamy.

The plant can be used in its entirety to produce foodstuff for both humans and livestock. More than 4,000 subspecies have been catalogued in Brazil, from which only a few are grown commercially, mostly to produce *farinha de mandioca* of various types and *polvilho* (tapioca starch), also called *goma*—two of the most emblematic and distinctive ingredients of Brazilian cuisine, used to prepare national favorites such as *farofa*, *tapioca/beiju* and *pão de queijo* (see recipe on p. 290).

Brazil is the only producer of *farinha de mandioca* in Latin America, but the root is appreciated in other parts of the Central and South America, in many countries in Africa (Nigeria being a major producer), and in Asia, where India is the largest producer.

With slavery, the African women took the place of the Amerindians in the master's kitchen, and that went on for the following almost 350 years—the amount of time it took Brazil to abolish slavery in the country. More than 3.5 million Africans were captured and sent to the country mainly to work on the sugarcane plantations of the Northeast region, but also in some farming ventures in the Amazon, and in the mines of gold and precious stones of present-day Minas Gerais state region. They were assigned all sorts of tasks, in the sugarcane fields, sugar mills, and elsewhere—including doing all the cooking both for themselves and the masters' and other upper-class families' households.

The women from Africa were skillful cooks. In the beginning they just tried and learned from the indigenous women how to use their native roots, vegetables, fruits, and herbs, while searching for ingredients that looked familiar and were similar to the ones they were used to cooking with back in Africa. Over time, as ingredients from Brazil, Portugal, and Africa were traded back and forth in the innumerous ships that crossed the Atlantic Ocean connecting these three continents, the flavors of the slaves' pans at the *senzala* made their way into the table of the *casa grande* (master's house). The concoctions, mainly soft purées and porridge-like mashes, such as *vatapá*, used plenty of *dendê*, oil extracted from a palm tree nut brought from Africa to Brazil with the slaves, as well as thick, freshly extracted coconut milk, roasted peanuts and cashew nuts, ground dried shrimp, okra, malagueta pepper, and spices such as ginger, coriander, and cumin.

While preparing Portuguese dishes, such as salt cod *frigideiras* (similar to fritattas), African cooks added coconut milk; stews were colored bright yellow or orange with the added *dendê* palm oil, while enhanced with the addition of bananas and yams, fried or boiled. Amerindians' dish *pokeka*, consisting of fish and vegetables baked in banana leaves, ended up being prepared in a clay pot, enhanced by a pungent broth redolent of lime, coconut milk, malagueta hot pepper, cilantro, and *dendê*—and Bahian *moquecas* were born.

As the masters learned to appreciate these somewhat exotic flavors and combinations, new renditions of traditional African, indigenous, and Portuguese dishes began to appear, originating a brand-new, all-Brazilian culinary identity.

Descendants of these slaves comprise today between 30 percent and 50 percent of all Brazilian population, making the country the largest African-descent population outside Africa today. Along these lines, some sources affirm that Brazil also has the largest Japanese population outside Japan, the largest Italian population outside Italy, and the largest Syrian-Lebanese population outside the Middle East. Immigrants from these and other war-torn, impoverished areas of Asia and Europe, such as Russia, Germany, Spain, Switzerland, and Poland, were drawn to the country by the promising new agricultural and pastoral ventures that started to thrive after slavery was abolished in 1888. By 1930, over 4 million immigrants had come to Brazil, especially to the relatively unoccupied South region, and to take the place of the slaves in the coffee plantations and milk farms of São Paulo and Minas Gerais states.

Each one of these immigrating groups gave their contribution to the Brazilian cuisine. Italians brought their pasta, cheeses, polenta, risotto, gelatos, and pizza. The Middle Eastern brought their kibbeh, tabouleh, sfiha, and their elaborate and very sweet pastries—a taste Brazilians had already acquired from the Moors through the Portuguese, especially the nuns, who have had a very strong cooking tradition in Brazil, since the colonial times, of making delicious preserves, cookies, sweets, and confections containing large amounts of sugar and eggs. The Japanese brought to Brazil, among other things, sushi, tempura, yakisoba, soy sauce, tofu, and other soy products. Just to name a few others, Russians brought stroganoff, mayonnaise, and cooked vegetable salad and borsch; Germans brought sauerkraut, strudel, beer, kuchen (called *cuca* in Portuguese), and black forest cake; Poles brought pierogi and stuffed cabbage leaves; Swiss brought fondue; Spaniards brought sangria and paella; and French brought crepes, croissant, and many other classic French dishes. Chinese contributed spring rolls and, indirectly, *pastel*—a much loved Brazilian *salgadinho* (see sidebar "*Salgadinhos*" on p. 305). More recently, following the country's economic development and globalization, sandwiches and all sorts of fast food also made their way into the Brazilian diet, for the good and for the bad. The country's population is now suffering from alarming rates of obesity and other health conditions related to overweight.

All in all, what is considered Brazilian food nowadays comprises a series of regional favorites and a few national dishes that have some typical ingredients and preparation techniques in common, but are as varied as they can be. A visit to one of the innumerous *restaurantes por quilo* (restaurants selling *comida por quilo*), found everywhere in Brazil, not by coincidence, proves this theory right—alongside *moqueca, pirão, feijoada, bolinho de bacalhau* (see sidebar "*Salgadinhos*" on p. 305) and *farofa*, one can find sushi rolls, stroganoff, pasta of all kinds, tabouleh, barbecued cheese, you name it! But all these dishes, as Amerindian-African-Portuguese Internationally Brazilian as they are, syncretize the soul of the Brazilian cuisine: a flavorful intermarriage of several culinary traditions that evolved, together or distinctively, to form what is known today, among Brazilians and abroad, as the authentic Brazilian cuisine, or better said, cuisines.

Croquete de carne: one of the traditional Brazilian "salgadinhos." This dish consists of breaded and deep-fried croquette made of cooked beef chunks that are then ground and mixed with milk-soaked fresh breadcrumbs and/or mashed potato. The traditional shape is oblong, but it can also be formed into balls, as depicted here. (Carlanichiata/Dreamstime.com)

BRAZILIAN STAPLE FOOD AND
EVERYDAY EATING HABITS

What's in the Brazilian Plate?

Brazilians usually have three main meals a day—a light breakfast right after waking up, a substantial lunch between noon and 2:00 p.m., and dinner around 7:00 or 8:00 pm (usually a lighter meal). For any time in between, people love their *salgadinhos*, savory snacks they eat at any time of the day they are not hungry enough to eat a complete meal.

For breakfast, a lot of people have coffee (very strong coffee!), or coffee and milk (called *média* when lighter, or *pingado* when darker), and *pão francês* (similar to a Portuguese roll, but lighter), with butter or *requeijão*. When the bread is toasted on a griddle, it is called *pão na chapa*—a tradition in virtually any Brazilian *padaria*, a bakery that, besides selling freshly baked bread and other pastries, sometimes serve lunch, sandwiches, juices, and other snacks. Some people make their breakfast more substantial by adding cheese and/or ham to it, and fruit to round off the meal. But bacon, sausage, and even eggs are mostly regarded as lunch/dinner (or maybe brunch) items, not breakfast.

A variety of fresh fruit and their juices (*sucos*), sometimes mixed with milk, orange juice, or water in a blender (see sidebar *Lanchonetes: Salgadinhos, Vitaminas, and Sucos* on p. 290), may be consumed as accompaniments or to replace coffee/milk. There are also the *pão de queijo* (a savory cheese roll made with sour manioc starch and *queijo (da) canastra*, typical of Minas Gerais state and the Southeast region in general—see recipe on p. 301), *tapioca/beiju* (manioc starch crepes, very common in the Northern region), and *cuscuz nordestino* that sometimes replace the bread. The latter two are often accompanied by *manteiga de garrafa*, *queijo (de) coalho*, and/or coconut milk.

LANCHONETES: SALGADINHOS, VITAMINAS, AND *SUCOS*

Lanchonetes are a kind of food service establishment spread all over Brazil, and their concept is somewhat unknown to most foreigners. They usually have a very laid-back atmosphere, with metal or plastic tables and chairs, and a large counter surrounding the "kitchen," where most of the action happens. People can sit at the counter and watch their servers/cooks prepare their orders from scratch. The kinds of foods one can find in places like that are mostly sandwiches, *salgadinhos* (see sidebar on p. 305), *vitaminas* and *sucos* (fruit juices).

Being a tropical country with a vast territory, Brazil has a dazzling array of fruits that can be found year-round, either fresh or frozen. Fruit menus with over 40 different fruit offers to choose from are not uncommon. They are usually mixed in a blender with either water or orange juice. Some of them can be mixed with milk and/or have more than one fruit added, and sometimes instant oats—these are called *vitaminas*. Some classic combinations are papaya with orange juice and raw carrot (sometimes with raw beets); avocado and milk; banana with milk and instant oats; guava and milk.

Brazil also has a unique soda, *guaraná*, made with a fruit with the same name, native to the Amazon forest, which is sometimes served with orange slices. Ice-toasted matte tea, *chá mate gelado*, mixed with fruit juices such as lime, grape, and *guaraná*, or milk, is also a popular and nutritious drink Brazilians sip at *lanchonetes* and *padarias*.

For lunch, people usually rely on *arroz com feijão* (rice and beans—see recipes on p. 292) for their starch; a small amount of protein (about 5–8 oz; chicken, pork, and beef are the most popular; fish and shellfish are more common in the cities along the coastline); and at least one cooked vegetable (often braised), such as winter squashes, chayote, zucchini, or broccoli—and salad greens (lettuce, arugula, watercress), with tomatoes and/or other raw or cold additions, such as grated carrots or beets, hearts of palm, cucumber, and sliced onion.

Dinner is usually a lighter fare, often consisting of some soup and bread, especially in the winter, and a sandwich, a salad with some protein on it, slices of pizza or quiche, or some pasta.

Desserts are appreciated both after lunch and dinner, and options can range from fruit compotes, such as banana, milk caramel, fig, papaya, and guava (the paste version, when eaten with cheese, is called *Romeu e Julieta*). There are also more elaborate preparations, such as caramel pudding, coconut flan with preserved plum compote, and passion fruit mousse. Some candies, very popular at birthday parties—like *brigadeiro*, *beijinho* (coconut, egg yolks, and clove), and *cajuzinho* (ground peanuts, sugar, and chocolate)—can also be served to round off a meal, accompanied by a demitasse of very strong coffee, or a cordial.

This is the basic "rule," but there are countless exceptions. The point here is highlighting some basic overall characteristics of Brazilians' eating habits:

- Lunch is the most substantial meal of the day.
- Brazilians tend to complement their main meals with savory snacks and fruit juices, instead of sweet pastries such as muffins and scones.
- Rice and beans are the base of Brazilian diet, and the preferred starch to accompany everyday meals, as opposed to, say, potatoes or bread in the United States.
- Although meat is an important part of the meal, it is usually consumed in smaller amounts.

In a country where hunger has been an issue for so long, the "rice-and-beans" diet may have saved many impoverished people from starvation and even malnutrition, as these two, when eaten together, constitute a high-quality, or complete protein, that is, one containing the 20 essential amino acids. Brazilians eat an average 130–168 g (4.5–6 oz) of rice a day and 127–195 g (4.5–7 oz) of beans a day—the lower their income is, the more they eat these two food items.

Rice and Beans

Brazilian rice, white and usually long grain, is prepared more or less like a pilaf, on the stove (not the oven), but the fat of choice is vegetable oil (a plain, flavorless one, preferably), as you can see in the recipes provided. A well-prepared Brazilian rice is always fluffy, never mushy, and yet tender to the bite, with a mild onion (or onion and garlic) flavor, or unflavored.

Beans are mostly cooked in a pressure cooker with water only. Then, they are sautéed with garlic right before serving and seasoned mainly with salt and sometimes bacon or other smoked pork products, such as *linguiça* (pork sausage). In some regions, cilantro and cumin are commonly used to season the beans, whereas other regions prefer to add chopped green onions only.

BRAZILIAN-STYLE WHITE RICE

2 tbsp vegetable oil
1 tbsp finely chopped onion
1 cup white rice (long grain)
2 cups cold water, approximately (preferably filtered)
1 tsp salt

1. Heat oil in a saucepan and add onion. Fry over medium heat until soft and translucent.
2. Add rice. Sauté, over medium heat, stirring constantly, until grains are whitish and chalky and start to form lumps.
3. Add 1½ cups of the water and the salt. Stir well. Bring to a boil, lower the heat, and simmer, partially covered, until water is almost gone.
4. Taste the rice for doneness. If it is still dry in the center, add remaining water, a little at a time, and continue cooking, covered, until all water has evaporated and the grains are cooked but slightly firm to the bite (cooking will continue after you turn off the heat).
5. Cover the pan, remove from heat, and let stand for about five minutes. Fluff the rice with a fork and serve.

Quick Brazilian-Style Stewed Beans

1 tbsp vegetable oil
1 tbsp diced bacon (optional)
1 garlic clove, finely chopped
1 (16 oz) can unseasoned pinto beans (or small black beans)
1 cup water, approximately (preferably filtered)
Salt to taste
1 tsp finely chopped green onions or chives

1. Place the beans in a colander to drain and rinse under cold running water until there's no more frothing on the top.

2. Heat oil in a small saucepan over medium heat. Add bacon, if using, and fry until golden brown. Add chopped garlic and fry until golden brown. Lower the heat.
3. Add beans and, using the back of a spoon or ladle, smash some of the beans into a paste.
4. Add about 1 cup of filtered water. Bring to a simmer and cook until the broth is thick (heavy cream consistency) and brown colored—you might need to add a little more water.
5. Season to taste with more salt, if needed. Add chopped green onions or chives and serve immediately.

There are many varieties of dried beans in Brazil—the largest consumer and producer of the legume in the world, with approximately 3.5 million tons harvested every year. The most popular are *feijão carioca* or *carioquinha* (similar to pinto beans in color and size), with 85 percent of the market, followed by *feijão preto* (black beans), with 10 percent of the market, more appreciated in Rio de Janeiro state, but mandatory for the preparation of Brazilian national dish *feijoada*. The remaining 5 percent of the sales are specialty beans, such as *jalo, fradinho, rosinha, bolinha, branco, verde, azuki,* and *roxinho.*

Feijoada, *the national dish of Brazil, is a hearty stew made with black beans and several kinds of meat, mainly pork. It is usually served with white rice, sauteed finely shredded collard greens,* vinagrete *(Brazilian chopped tomato and onion salad),* farofa *(toasted manioc flour), and orange segments. (iStockPhoto.com)*

Meat, Poultry, Pork, Fish, and Shellfish

Meat is considered a very highly prized item in the Brazilian eating culture, synonymous with strength, satiety, and energy. The country is a major producer and exporter of beef, being the second in number of heads of cattle, with 212 million. In 2014, the production of beef, pork, and chicken in Brazil was estimated at 25.8 million tons, 75 percent of which was consumed locally.

Beef is the preferred protein for many Brazilians, especially more toward the south of the country. Beef cuts, fresh or cured, have a very important place in the Brazilian plate. The consumption of beef has increased in the past years due to the economic improvement of the country in general, and especially of the low-income populations. In 2003, for example, the consumption of beef per capita was 35.6 kg (78.5 lb)/year, and in 2010 the number jumped to 37.4 kg (82.5 lb).

But due to the high costs of beef, chicken and pork consumption grew even more: from 33.3 kg (73.4 lb) in 2003 to 43.9 (96.8 lb) in 2010, for chicken, and from 12.4 kg (27.3 lb) to 14.1 kg (31 lb) for pork. Brazil is the fourth largest pork producer in the world, and in 2011 the country produced 13.1 million tons of chicken, 9.1 million of which was consumed locally, the rest sent to over 140 countries. Turkey is primarily eaten in Brazil at Christmas and as luncheon meat for sandwiches. Annual consumptions is around 0.6 kg (1.3 lb) per capita.

CHURRASCO AND *RODÍZIO*

Brazilian Barbecue, *churrasco* in Portuguese, became more widely known in the United States and elsewhere in the past decades because of the popularity and ubiquitousness of *churrascaria* chains, such as Fogo de Chão and Texas de Brazil. In this kind of restaurant, also very popular in Brazil, barbecued meat and other products are served in a system called *rodízio* (literally "rotation"). Several beef joints and large cuts of meat, such as *picanha*, the most famous and prized one, pork sausages, chicken legs and hearts, pork ribs and loin, lamb cuts, and even cheese and pineapple, to name a few, are stuck in long metal skewers and roasted slowly in giant coal barbecue pits. In the dining room, there's always a beautifully arranged, bountiful buffet with plenty of cold salads of all kinds, rice, beans, *farofa*, sometimes *feijoada*, and, believe it or not, even sushi and stroganoff!

These restaurants are almost often "per-person" priced, and the clients receive a card (or there will be one waiting for them at the table) with a green and a red side—green means "bring me more" and red means "I've had enough." Servers circulate among the tables offering the barbecued food to clients and cutting slices or taking pieces off the skewered item directly onto the clients' plates until they turn the red face of their card up. Meats are seasoned usually with coarse salt only and served rare or medium rare.

Alcoholic and soft drinks, such as *caipirinha* and *guaraná*, are usually offered at an additional cost. Desserts are either served in the buffet or by ordering from a menu. Some classic offerings are caramel pudding; cream of papaya and vanilla ice cream with cassis cordial; passion fruit mousse; and pineapple with fresh mint.

Even though Brazil has a vast coastline, a plethora of rivers, and incomparable biodiversity when it comes to fish and shellfish varieties, the consumption of this source of protein is relatively small in the country, compared to other places that have less access to these resources. However, it has been increasing steadily in the last couple of years. In 2011, the annual production of fish and shellfish in the country was 1.43 million tons, nearly 629,000 tons of which was produced in farms. The Ministry of Fishing and Aquaculture of Brazil estimated that the average consumption per capita that year was 11.2 kg (24.7 lb), 14.5 percent more than the previous year, which was already 7.9 percent greater than 2009. This means an accumulated 23 percent increase in two years.

Among the emblematic fish and shellfish of Brazil are the *piranhas*, from the Pantanal region, where *pintado* (Brazilian tiger fish), *surubim* (a catfish species), and pacu also come from; the *pirarucu* (arapaima) and *tambaqui* (a pacu species) fish, and the *aviú* (tiny little shrimp), are common in the Amazon region; and shrimp, grouper, oysters, mullet, and crab are more popular toward the South region. Nevertheless, Brazil is still below the worldwide annual mark for per capita consumption, estimated at 18.8 kg (41.5 lb), and merely reaching World Health Organization–recommended consumption of 12 kg (26.5 lb)/year.

Caprine and ovine meat are not widely consumed in Brazil. Lamb accounts for a mere 0.7 kg (1.5 lb) yearly per capita consumption. Almost 57 percent of the herd is in the Northeast region, which also accounts for more than 90 percent of the goat meat and cheese production. The region's preference for this kind of meat is reflected in some regional favorites, such as the *Buchada de bode* (mutton stomach stuffed with the animal's offal—kidney, liver, etc.). Brazil also has some 1.15 million heads of buffalo, the majority of it, 39 percent, in the North region, more specifically in Pará state. The meat is much leaner than beef and the milk has a higher fat and milk solids content, favoring the production of high-quality cheeses.

Brazilians, generally, are not big fans of game and foul, or exotic animals either. There are some small-scale, isolated producers of pheasant, duck, geese, deer, boar, and ostrich in the country, as well as of large reptiles, such as *jacaré*, a type of caiman.

Fruits and Vegetables

Brazil is the world's third-largest producer of fruits and vegetables, behind China and India, with 43 million tons produced per year. But in terms of consumption, the annual per capita average is only 86 kg (190 lb), with the aggravating fact that the vast majority of the population (90%) eats less than 400 g (0.9 lb) of fruits and vegetables a day, below the WHO-recommended amount.

Since the late 1970s, the commercialization of fruits and vegetables, until then centered at open markets and fairs, where consumers could buy directly from producers, has shifted to the supermarkets, now responsible for more than 50 percent of the market (supermarkets used to sell about 12% of the products back in the 1980s). This may have contributed to make fruits and vegetables more expensive, but, on the other hand, more advanced packaging, transportation, and storing techniques allowed supermarkets to offer more regional products to a broader audience, contributing to variety and quality.

Potato, tomato, onion, carrot, pumpkin and winter squashes, cabbage, lettuce, chayote, sweet potato, and bell peppers are, in that order, the 10 most consumed vegetables in Brazil, totaling 27 kg (59.5 lb) a year per capita (2008—POF/IBGE). The most popular fruits are, in this order, banana, orange, watermelon, apple, papaya, other citric fruit, pineapple, mango, grape, and melon. An average Brazilian eats approximately 7.7 kg (16.1 lb) bananas per year (Americans ate an average of 10 lb/ person in 2010). In terms of regions, people in the South of Brazil eat more fruits and vegetables than the rest of the country, followed by the Southeast, Northeast, Midwest, and North regions. People with a higher income eat more fruits and vegetables than low-income individuals.

BRAZILIAN REGIONAL CUISINE

Brazil has a vast and diverse territory, both in terms of geographic conditions, such as climate and vegetation—which affect directly the kinds of ingredients available to each local cuisine—and in terms of cultural and ethnic traditions and eating habits. Thus, to speak of a national cuisine is, often, to speak of several regional cuisines, which are represented by various traditional dishes. Some of these dishes, such as *feijoada* (black beans and meat stew), *churrasco* (Brazilian barbecue—see the *Churrasco* and *Rodízio* sidebar on p. 292), and *moqueca* (Brazilian fish stew), often prepared with ingredients easily available throughout the country and/or abroad, become well known and appreciated outside their *terroir*, or place of origin and, eventually, get elevated to the status of "national dish." Combined, regional and "national" favorites give character and life to the Brazilian culinary identity.

Until very recently, the circulation of markedly regional specialties, especially fresh ingredients, such as fish from the Amazon, and products manufactured on a small scale, such as the Minas Gerais's *queijo (da) canastra* cheese, was limited to their adjacent areas, because of transportation and other distribution problems. *Maniçoba*, for example, is a regional dish of the North region containing some ingredients that can be easily found everywhere else in the country, such as the meats, the same used for *feijoada* (*carne seca*, salt and smoked pork cuts and sausages, etc.). But the distinctive ingredient *maniva* (*mandioca* leaves) that gives it character, although easily found in any region of Brazil, has to go through a very time-consuming (one-week-long) process to get its toxins removed before it can be used in the dish. That makes the dish hard to reproduce outside that region, where you can buy *maniva* ready to use in street fairs and markets.

When it comes to regional cuisine, the vast majority of Brazilian cookbooks and compendiums about Brazilian gastronomy divide the dishes into the country's five geographic regions: North (*Norte*), Northeast (*Nordeste*), Midwest (*Centro-Oeste*), Southeast (*Sudeste*), and South (*Sul*). More recently, though, some Brazilian authors have suggested that the country should be divided into Gastronomic Biomes, that is, regions that share the same climate, fauna and flora. There are six gastronomic biomes, or *terroirs*, in Brazil, corresponding to the terrestrial biomes of the country as defined by IBGE (Brazilian Institute of Geography and Statistics).

The Amazon Biome (*Amazônia*) roughly corresponds to the North region and covers 49.29 percent of the country's territory. The Savannah Biome (*Cerrado*) covers the majority of the Midwest region, except for the Wetlands region (*Pantanal*), and also parts of the North, Northeast, and Southeast regions, totaling 23.92 percent of the country's territory. The Atlantic Biome (*Mata Atlântica*) corresponds to 13.04 percent of Brazil, covering the majority of the South region and about half the Southeast region, as well as the east coastal parts of the Northeast region. The Scrublands Biome (*Caatinga*) represents 9.92 percent, corresponding roughly to the Northeast region, excluding the coastal area. The Grasslands Biome (*Pampa*) can be found only in Rio Grande do Sul state, in the South region, occupying 63 percent of the state's territory, which corresponds to little more than 2 percent of the country's area. Finally, the Wetlands Biome (*Pantanal*) is the smallest in Brazil in terms of area covered, occupying 1.76 percent of the national territory and restricted to parts of Mato Grosso do Sul and Mato Grosso states.

In the North region, the Amazon forest, the largest forest formation of the planet, and the Amazon River, the largest river in the world in terms of water flow, are home to a plethora of fauna and flora, which heavily influence the local cuisine. The Amazon Biome is the Brazilian *terroir* where the cooking traditions of indigenous peoples of Brazil can be most strongly felt.

Major ingredients of the North region include *mandioca* (cassava/manioc/yucca) and several products made with it, especially *farinha de mandioca* of various types (*farinha-d'água, farinha de tapioca, farinha de suruí*, etc.), *tucupi* (broth made with the fermented juice of the grated root cooked with local spices), and *maniva* (ground leaves); *pupunha* (peach palm), both the fruit, the nut, the hearts of palm, and the oil; several fruits and nuts, such as *cupuaçu, açaí, guaraná, bacuri, castanha-do-pará* (Brazil nut), and *cumaru* (tonka beans). Spices and herbs that give character to the region's cuisine are *urucum* (annatto/achiote), *jambu* leaves (known in English by several names, such as para cress, Para/Amazon watercress, toothache plant) and their flower buds (buzz buttons); several hot peppers, such as *cumari-do-pará, murupi*, and *pimenta-de-cheiro*; *chicória-do-pará* (culantro), *alfavaca* (Amazonian/wild sweet basil), and banana leaves, which are used to wrap several foods before baking. Some of the most used fish and shellfish are *pirarucu* (arapaima, usually salt cured), *filhote* (young specimen of *piraíba*, the largest Amazon fish, similar to a catfish, sold as peiche in the United States), *tucunaré* (peacock bass), *tambaqui, aviú* (very tiny shrimp found in the Tocantins River), *sarnambi* (type of clam); endangered species such as *tartaruga* (turtle) and *peixe-boi* (*manatee*) were once part of the diet too. As for meat and poultry, buffalo, duck, and salt-cured beef are used in several dishes. Two regional specialty ingredients worth mentioning are *piracuí* (fish dried on a *moquém* and reduced to a flour) and *queijo do Marajó*, made with buffalo milk.

Some famous dishes of the North region are *Pato no tucupi* (roasted duck stewed in *tucupi* sauce); *Maniçoba* (*maniva* cooked with smoked and salt pork parts and sausages, and sometimes giblets); *Tacacá* (salt shrimp, manioc starch porridge, and *tucupi* soup); *Pirarucu de casaca* (fish baked with seasoned *farinha de mandioca* and plantain); *Tambaqui na brasa* (barbecued fish); *Caldeirada de peixe* (Pará-style fish stew); *Mojica/Mujica* (soup/stew made with either fish or shrimp, usually *aviú*); and *Pirão* (fish stock thickened with *farinha de mandioca* and seasoned with local herbs and hot peppers). Ice cream of local fruits and delicacies such as *Pudim de tapioca* (caramelized tapioca pudding), *Bombom de cupuaçu* (*cupuaçu* and chocolate truffles), and *Beijus* (tapioca flour crepes, usually filled with coconut) are popular desserts.

The Midwest region encompasses two major biomes: the *Cerrado* and the *Wetlands*. The *Cerrado* biome region has basically two seasons: the dry and the rainy, with temperatures ranging from 70oF to 80oF. For many years, it was regarded as poor in terms of flora and fauna, but today, the open savannah-like prairies, with patches of twisted trees and low scrub vegetation, are considered one of the richest biotas of the planet, with over 330,000 identified animal and plant species. Livestock and large crops of soybean, rice, and corn, among others, are now the main economic activity of the region, and that has also left a mark in the region's cuisine.

The Wetland Bioma (*Pantanal*), on the other hand, is the largest continuous flood plan on the planet, bearing a fauna and flora with many species endemic to the region, which heavily influences the local cuisine. Culturally, the region was influenced by Paraguayan-, Bolivian-, and Argentinian-eating habits, as well as the customs of migratory groups coming from Minas Gerais, Goiás, and the South region of Brazil.

Major ingredients of the Midwest region are *mandioca*, rice and corn, for starch; and vegetables such as *guariroba*, also known by many other names (a bitter specimen of hearts of palm), *serralha* (sowthistle), and *dente-de-leão* (dandelion leaves). The *Cerrado* has many endemic fruits, the most popular being *pequi* (souari nut), which has many uses and gives color and flavor to many savory dishes of the region. Other common fruits are *maracujá* (passion fruit), *baru*, also called *cumaru* (tonka beans), *jenipapo*, *mangaba*, *cagaita*, *jatobá*, *cajuzinho-do-cerrado*, *araticum*, *gabiroba*, *mama-cadela*, and delicious *jabuticaba*. Besides *pequi*, seasonings such as *urucum* (annatto/achiote) and *açafrão-da-terra* (turmeric) are used to flavor the regional dishes. The region is also a major producer of honey. From the wetlands come the freshwater fish (e.g., *pintado*, *surubim*, *pacu*, *piraputanga*, and the feared *piranhas*) and reptiles (e.g., the *jacaré*, a species of caiman), as well as game and fowl, such as wild boar and capybara.

Among the most famous dishes of the Midwest region are *Galinhada* (rice cooked with *pequi* and marinated, cut-up free-range chicken); *Arroz com pequi* (rice cooked with *pequi*); *Empadão goiano* (also called *empada goiana*, it is a double-crust savory pie filled with shredded chicken and pork meat, hard-boiled eggs, cheese, *guariroba*, and local seasonings); *Sopa paraguaia* (literally "Paraguayan soup," it is actually a savory cake similar to corn bread, made with corn meal, cheese, and onion); *Caldo de piranha* (piranha cooked in a flavorful broth and then carefully deboned and creamed with its cooking liquid in a blender and seasoned with tomato, bell pepper, and fresh herbs); *Saltenhas* (similar to empanadas, they are filled with shredded chicken); *Escabeche de pacu* (breaded, deep-fried fish in a tomato and onion sauce); *Locro* (short ribs with hominy); *Peixe/Pintado na telha* (fish baked in a clay baking dish similar to a tile); and *Caribéu* (salt-cured beef cooked with *mandioca* and bell pepper). For dessert, *Bolinho/Bolo de arroz* (rice flour and coconut sweet cakes), *Alfenins* (snow-white sugar confections of Arabic origin brought by the Portuguese), *Furrundum* (dessert made with green papaya, *rapadura* sugar, and ginger), and *Pastelim* (mini tarts filled with caramelized milk custard) are very traditional.

The Northeast region corresponds roughly to the *Caatinga* biome, but also includes portions of the coastal *Mata Atlântica* biome on the east, and of *Cerrado* biome, on the west. The backlands have a dry, semiarid climate. The shortage of water that has historically characterized the region, making it difficult to maintain crops or livestock, favored a diet based on starchy, high-fat, and preserved food. On the coastal areas, as well as on the *Cerrado* portion of the region, the proximity to the water brings more diversity to the dishes, brimming with fish and shellfish as well as fresh hot peppers, coconut milk, cilantro, and the African touch of *dendê* palm oil. Fruits of all shapes and flavors, some of which found only in the region, abound and are used to make juices, ice creams, compotes, pastes, and pastries.

The Northeast, the first region of the country to be colonized, is where the indigenous-African-Portuguese fusion cuisine can be more strongly felt. The African inheritance is more evident in Bahia and Pernambuco states, due to slavery, which also fomented the creation of the *cozinha de santo*, a collection of dishes commonly used as religious offerings to the African deities (called *orixás*) of *Candomblé*, an African Brazilian religion. More toward the Maranhão state, Portuguese and Amerindian traditions prevail, and the food is not as spicy as it is on the coast. Catholic and Amerindian influences survive in the *Festas Juninas* festivities (see sidebar on p. 299), celebrated in the month of June all over the country, but more traditionally in the Northeast.

FESTAS JUNINAS

Festas juninas are a mix of the fertility rites of the Amerindians, which coincided with the dry period of the year, the winter (June, hence "juninas," to September), and the days that the Catholic Church celebrated the June saints: Anthony (13), Peter (29), and especially John the Baptist (24)—João, in Portuguese, reasons why they were formerly called "joaninas" festivities.

Some of the elements that characterize the celebration are a large campfire, multicolored flags, and balloons (that were once lit and left to go up in the skies, but the practice caused many fires and is now prohibited), fireworks, and an improvised play/dance called "quadrilha," which involves a mock marriage followed by a coordinated dance of the pairs, supposedly the bride and groom godparents, all dressed in stereotyped countrymen costumes. Dancing to the sound of traditional rhythms such as *baião, forró,* and *boi bumbá,* they perform circle dance evolutions that have French names, inspired by the saloon dances of 19th-century France.

Some of the specialties commonly served at these festivities are corn concoctions, such as *curau* (sweet corn porridge with cinnamon, also called *canjica*), *pamonha* (similar to tamales, but made with freshly grated sweet corn), and *broa* (corn meal puff, sometimes baked wrapped in banana leaves and flavored with peanuts); *quentão* (hot beverage made with *cachaça* and spices); and *pé-de-moleque* (peanut and *rapadura* sugar brittle).

The staple starches of the Northeast region are, once again, *mandioca*, and all its by-products, especially the flours and starches, as well as other roots, such as potato and yams (sweet potato, taro, etc.); corn; rice and beans, such as *feijão-de-corda* (black-eyed pea) and *feijão-andu*; and winter squashes and pumpkins. Common vegetables are okra, *maxixe* (West Indian Gherkin), *caruru* (type of amaranth), *viangreira* (sorrel), and *taioba* (tannia leves). To add flavor to savory dishes, they use *dendê* palm oil, coconut milk, ground peanuts and cashew nuts, dried shrimp, cilantro, several hot peppers, such as malagueta, and *jurubeba*, which is also mixed with *cachaça* to produce a bitter digestif, also used as medicine. Among the most consumed proteins are beef, usually salt-cured by several distinct methods (resulting in several products, such as *jabá*, *carne de sol*, *carne seca* and *carne de sereno*), goat, sheep, chicken, and, on the coastal areas, fish and shellfish, such as shrimp and crab. Among the plethora of fruits found in the region are *caju* (cashew fruit/apple), *cajá*, *cacau* (cocoa), *umbu*, *carambola* (start fruit), *fruta-pão* (bread fruit), *araçá* (strawberry guava), *graviola* (soursop), *pitomba*, *sapoti*, *goiaba* (guava), *acerola* (Barbados cherry), and sugarcane, which is used, in the form of sugar, *rapadura*, or treacle, to make several desserts and regional specialties. Some regional specialties worth mentioning are *queijo (de) coalho* (curd cheese) and *manteiga de garrafa* (similar to clarified butter, used to season food), used in many regional dishes.

Some of the most well-known dishes of the Northeast are *Acarajé* (fritter similar to a falafel but made with black-eyed beans, usually served stuffed with spicy dried shrimp and other concoctions typical of the Northeast region, such as *Vatapá* and *Caruru*—see next); *Vatapá* (porridge made with bread soaked in coconut milk and seasoned with ground cashew and peanuts, ground salt shrimp, ginger, *dendê*, and malagueta hot pepper); *Caruru* (similar to *Vatapá*, but made with okra instead of bread and coconut milk); *Moqueca baiana* (fish steaks, bell pepper, and tomato slices stewed in coconut milk, *dendê*, malagueta, and cilantro broth, usually served with fish *pirão*); *Pirão* (flavorful meat or fish broth thickened with *farinha de mandioca*); *Paçoca de carne* (salt-cured beef ground to a powder and mixed with *farinha de mandioca*); *Bobó de camarão* (shrimp cooked in a coconut and *mandioca* thick sauce); *Buchada de bode* (mutton stomach stuffed with the animal's offal—kidney, liver, etc.); *Caldo de mocotó* (beef calf broth); *Escondidinho* (salt-cured beef "hidden" under *mandioca* purée); *Baião de dois* (rice and black-eyed beans cooked together with *carne seca* and *queijo (de) coalho*); *Xinxim de galinha* (chicken stewed in ground peanuts and/or cashew nuts, dried shrimp, and dedê sauce); *Arroz de cuxá* (rice with sorrel, sesame seeds, dried and fresh shrimp); *Quibeb*e (pumpkin with salt-cured meat and cilantro); and *Sarapatel* (pork offal stew).

For dessert, among many others, *Bolo de rolo* (very thin and large layer of sponge cake covered with guava paste and rolled as a large jellyroll); *Cartola* (caramelized banana and cheese); *Quebra-queixo* (literally, "chin breaker," it is a sticky toasted coconut and caramel sugar candy); *Bolo de aipim* (grated fresh *mandioca* baked with coconut); and *Tapioca* (manioc starch crepes, also called *beiju*, in some regions of Brazil), also served with savory fillings, and fruit ice creams and compotes.

The Southeast region is the most densely populated and economically developed region of Brazil and has two major distinct biomes: the *Mata Atlântica*, on the east coast and at the west side of the state of São Paulo, and the *Cerrado*, which covers a large portion of Minas Gerais state and the central area of São Paulo. The diet of the region is characterized by a massive presence of freshwater and saltwater fish and shellfish, in the coastal areas, as well as fresh and cured pork and beef meats and sausages in the interior region. The fact that Rio was the capital of Brazil for a while, when the Portuguese Crown came to the country to escape the war in Europe, left indelible marks in the eating habits and preferences of the region. Also, due to the economic development of cities such as São Paulo, Rio de Janeiro, and Belo Horizonte, the cuisine of the Southeast region was also heavily influenced by foreign immigrants, especially the Italian, the Japanese, the Lebanese, and the Syrian.

Among the most used proteins are fish and shellfish, such as *garoupa, tainha* (mullet), *siri* (crab), *camarão* (shrimp), and *ostra* (oyster), in the coastal areas; and beef (salt-cured, more to the north of Minas), pork, and chicken, in the *Cerrado* area, which used to be raised in the backyards of family houses, in the country side, until very recently. For vegetables, *palmito* (hearts of palm), *ora-pro-nóbis* (Barbados gooseberry leaves), *almeirão* (flat-leaf chicory), *couve* (collard greens), summer and winter squashes of many kinds, including *chuchu* (chayote) and *baldroega* (purslane), and tomato are commonly used. The fruits of the region include large *abacates* (Florida avocados), banana, guava, jackfruit, *pitanga* (Surinam cherry), *jabuticaba*, and orange.

Together with potato, *mandioca*, and *mandioquinha* (also known as *batata-baroa*), rice and beans are the staple starch—Rio, where *feijoada* was born, prefers the black beans, whereas other states like the brown varieties better; other regional species, such as *feijão-guandú* (pigeon pea), are also used. Some of the regional specialties that are worth mentioning are *farinha de milho*, a flour made using a technique similar to the one used to make *farinha de mandioca* "bijuzada" (flaked), but from corn (*milho*) instead of *mandioca*, and the cheeses: *queijo Minas frescal* (soft cheese similar to farmer's cheese), *requeijão cremoso* (creamy spreadable cheese), *catupiry* (named after the brand, an all-Brazilian cream cheese), and *queijo (da) canastra meia-cura*, a semisoft, sharp cheese produced in the Canastra Mountains, in Minas Gerais state— a quintessential ingredient for the authentic *pão de queijo* (see recipe provided) and *Romeu e Julieta* dessert (cheese with guava paste).

PÃO DE QUEIJO MINEIRO

1/2 cup vegetable oil
1¼ cup whole milk, approximately
1 tbsp salt
1 lb (500 g) *polvilho azedo*
3 eggs, lightly beaten
1 cup shredded *queijo (da) canastra* (or you can try a mix of two parts sharp cheddar cheese and one part mozzarella, or one part mozzarella and two parts good parmesan cheese)

1. Combine oil, milk, and salt in a saucepan. Heat to just below boiling point (watch closely—when the mixture starts to rise, remove immediately from heat and use it in step 3).
2. Place the *polvilho azedo* in a large bowl.
3. Pour boiling mixture all over the *polvilho azedo* and, using a wooden spoon, start stirring the dough.

4. When the dough is cold enough to be kneaded by hand (but still hot), add eggs and cheese and knead until it is very sticky and elastic, about 15 minutes (you will need a spoon or scraper to get it off your hands).
5. Let the dough rest while you preheat the oven to 425oF (it is very important that the oven is at a high temperature when you bake the *pão de queijo*; if they start to get too brown on the bottom before getting golden brown on top, reduce the temperature a little bit).
6. Oil two large baking pans or line them with parchment paper. Oil your hands with vegetable oil and form golf-sized balls with the dough (40–45). Place them 2–3 in. apart in the pan, as they grow considerably when baked. (You can, at this point, freeze the balls and then store them in zip bags to be baked straight out of the freezer, at your convenience—they will take a little longer to bake, though.)
7. Bake until puffed and golden brown (about 15–20 minutes). Serve immediately.

In the *Cerrado* portion of the Southeast region, some of the favorites are the national dish *Feijoada* (black beans—or brown, in Bahia—stewed with salt-cured beef and fresh, salt and smoked pork cuts and sausages, traditionally served with white rice, sautéed collard greens, *farofa*, a tomato salsa called *vinagrete* and orange slices); *Feijão tropeiro* ("mulle driver's beans," cooked beans sautéed in lard, onion, and garlic and thickened with *farinha de mandioca*, served with hard-boiled eggs, pork sausage, and grilled pork chops); *Tutu de feijão* (mashed beans with pork lard or bacon and *farinha de milho* or *mandioca*); *Canja de galinha* (chicken, rice, and vegetable soup); *Virado à paulista* (cooked beans sautéed with lard, onion, and garlic, then mashed and thickened with *farinha de mandioca* and served with rice, fried egg, sausage, sautéed collard greens, and breaded and deep-fried banana); *Bife a rolê* (beef minute steaks rolled up with carrot and bacon and cooked in tomato sauce); *Empadão de frango e palmito* (double crust savory pie filled with shredded chicken and hearts of palm); *Frango com quiabo* (chicken and okra stew); *Lombo à mineira* (braised pork loin, sometimes stuffed, served with white rice, *tutu de feijão*, and collard greens); *Galinha ao molho pardo* (chicken cooked in its own blood, also called *Galinha de cabidela*); *Cuscuz* (flavorful tomato-based broth, usually made with sardines or chicken, thickened with *farinha de milho* and/or *farinha de mandioca*; it can be steamed or cooked as a porridge and served molded and decorated with tomato slices, hard-boiled eggs, peas, etc.); *Vaca atolada* (literally "swamped cow," marinated beef short ribs cooked with *mandioca*); *Torresmo* (Brazilian-style pork cracklings); *Dobradinha com feijão branco* (shredded honeycomb tripe stewed with white beans and tomato); and *Leitoa/Leitão a pururuca* (whole suckling pig roasted until the skin become crisp).

Some of the most well-known fish and shellfish dishes of the Southeast region are *Moqueca capixaba* (Espírito Santo's version of *moqueca*, consisting of fish and/or shellfish cooked in a clay pot with tomato, *urucum*, and cilantro-based sauce); *Torta capixaba* (pie made with seasoned fish, shellfish, and seafood mixed with hearts of palm, olives, *urucum*, and other seasonings, covered with eggs and then baked);

Casquinha de siri (crab meat sautéed with coconut milk, tomato, and cilantro and served au gratin in a crab shell); *Camarão na moranga* (shrimp and *catupiry* stew served on a baked pumpkin); *Azul-marinho* (fish cooked with green bananas, which give it a blue color); *Camarão (ensopado) com chuchu* (shrimp and chayote stew); and *Bacalhoada* (salt cod fish baked with potato, tomato, onion, and black olives). For dessert, fruit compotes and pastes, such as green figs, guava, banana, pumpkin, and quince, sometimes served with *minas* cheese; *Doce de leite* (milk caramel); *Arroz doce* (rice pudding); *Pudim de leite* (caramel cream pudding); and *Manjar de coco com (calda de) ameixa* (also called *manjar branco*, coconut flan with dried plums caramel sauce) are served.

The South region is taken half by the Atlantic Biome and half by the Grasslands. The main economic activity in the area has always been livestock rising. To preserve and transport the meat to other parts of the country, such as the gold mines at Minas Gerais, Goiás, and Mato Grosso states, in the late 17th century, or the coffee plantations of São Paulo state, in the mid-19th century, most of the meat was salt-cured and sun-dried. These techniques are still used today, and many dishes of the region use charque as an ingredient. The cowboys of Rio Grande do Sul, called *gaúchos*, are also credited for creating the Brazilian barbecue, called *churrasco* (see sidebar *Churrasco* and *Rodízio* on p. 303), borrowing from the Amerindian traditions of roasting food on the *moquém*. Another great influence on the South cooking traditions came from the European immigrants—mostly Italian, German, Poles, and Russians—who came to Brazil after the country became independent, in 1822, attracted by the vast empty land available for settlement and the cooler climate. Their wheat- and potato-based diet was soon adapted to the *mandioca* substitutes, although their farming endeavors have introduced some European flair to the Brazilian ingredients list, such as wine, leafy vegetables, and lamb meat, as well as wheat and potato. The African influence is minimal in the cooking traditions of this region, but on the other hand, these are the states where the Spanish influence can be most strongly felt.

Some important ingredients of the South region, besides beef, are rice, potato, sweet potato, corn (sweet and dried—called *canjica*); cool climate fruits, such as grape, peach, and apple; and leaf vegetables such as cabbage and collard greens. Some specialty ingredients worth mentioning are *origone* (packed dried peaches), *pinhão* (large pine nuts from the Candelabra tree, also known as Brazilian and Paraná pine, that are eaten cooked or roasted and used in many preparations), and *charque*, the South region take on salt-cured beef consisting of thin layers of beef, usually fatty, cheaper cuts, that are covered with large amounts of salt (iodized) and left to exude the liquid before being air dried and rolled up to be transported and sold.

Some of the well-known dishes of the South Region are *Churrasco* (Brazilian barbecue—large pieces of meat skewered and roasted over a *fogo-de-chão*, or a bonfire); *Barreado* (beef stew cooked for several hours in a pot sealed to its cover with a flour and water paste and then served with slices of banana over *farinha de mandioca*); *Arroz de carreteiro* ("ox cart men's rice," a mix of white rice and *charque*); *Quebra-bico* (sausage meat sautéed with onion and bell pepper and then mixed with scrambled eggs and *farinha de mandioca*); *Matambre enrolado/recheado* (from the Uruguayan cuisine, lean flank steak stuffed usually with sausage and/or bacon, rolled up like jelly roll and baked or braised); *Roupa Velha* (also called *Xatasca*, pulled *charque* mixed with *farinha de mandioca*); *Porco no rolete* (whole suckling pig stuffed and slowly roasted on a spit over hot coals); *Galeto* (marinated, and barbecued, spatchcocked young chicken), *Cozido campeiro*, and *Puchero* (hearty stews made with ox tail or brisket, sausage, cabbage, collard greens, corn, several root vegetables—potato, mandioca, yams, turnip, carrot, etc.—served with *farinha de mandioca*); *Tainha na telha* (mullet stuffed with *farofa* and baked on a clay dish similar to a terracota roof tile), *Arroz com pêssego/origone* (white rice with peach, fresh or dried—called *origone*); *Cola gaita* (lamb backbone with *mandioca*); *Feijão Pampa* (white beans with shrimp and hard-boiled eggs); *Feijão campeiro* (black beans with *charque*, bone marrow, sausage, and bacon); *Arroz de china pobre* (literally "poor whore rice," white rice cooked with sausage); *Arroz carregado* (*arroz de carreteiro* with vegetables, fresh beef, and sausage); *João trançudo* (tagliatelle with pulled *charque* and onion sauce); and *Charque farroupilha* (stewed *charque* strips served over a *pirão* made with the cooking liquid and *farinha de mandioca*).

Regional sweet concoctions include *Ambrosia* (curdled milk and eggs cooked in sugar syrup, that can also be prepared in the oven—*Ambrosia de forno*); *Cuca de banana* (white cake topped with sliced bananas and a cinnamon crumb); *Doce de abóbora em calda* (pumpkin squares cooked in sugar syrup); and *Sagu com vinho* (tapioca pearls cooked with red wine, grape juice, and cinnamon and served with *crème anglaise*). The people in the South always drink a lot of *chimarrão* (matte tea), either hot or cold (called *tererê*), and sometimes *vinho quente* (mulled wine), to warm up in the colder nights.

SHARING A MEAL

Brazilians really appreciate spending time with their family and friends and easily transform any social gathering into a party, in which food always plays a very important role. National and worldwide holidays such as New Year's Eve, *Carnaval*, Easter, Mother's Day, Father's Day, and Christmas are a good excuse to get together and savor a homemade meal, or any kind of food, at a restaurant or at a friend's house. *Churrasco*, *feijoada*, and *moqueca* are often prepared for large gatherings, but the menu can be any one or, as it is usually the case, a mix of many different cooking traditions. *Caipirinha*, the Brazilian national drink made with *cachaça*, lime (there are versions with many other fruits) and sugar, is the alcoholic beverage of choice, as well as beer and/or *chopp* (Brazilian light tap beer with a tall, creamy head).

At birthday parties, as well as at other reunions, *salgadinhos* (see sidebar Salgadinhos on p. 305) and *docinhos* such as *Brigadeiro* (literally "brigadiers," sweetened condensed milk and chocolate truffles rolled in chocolate sprinkles), *Beijinho* (literally "little kisses," coconut and egg yolk truffles with a clove stuck to the top), *Cajuzinho* (ground peanut or cashew nuts mixed with chocolate and sugar and molded as a cashew apple), and *Olho-de-sogra* (literally "mother-in-law's eyes," dried plums stuffed with coconut) are usually displayed in trays for people to eat at will, over the extended periods of time they spend together, chatting, and, usually, listening to music.

Moqueca, *a Brazilian stew prepared with fish and/or seafood. The flavor profile can be quite different, depending on the coastal region of Brazil. Depicted here is a* moqueca de camarão, *typical of Bahia State: shrimp cooked in coconut milk and dendê oil (red palm oil) served with* caruru, vatapá, farofa, *and rice. (Paul Brighton/Dreamstime.com)*

SALGADINHOS

The Brazilian word *salgadinho* (or *salgado*), which could be loosely rendered as "little savory/salty snacks," encompasses several kinds of bite-sized preparations, usually deep-fried or baked. They are sold in *lanchonetes* and *padarias*, and also traditionally served at birthday parties and other family reunions. Here are some of the most popular ones:

Bolinho de bacalhau: deep-fried quenelle or torpedo-shaped potato and salt cod fritters;

Coxinha: cooked wheat flour dough stuffed with shredded chicken and formed into a mock, small chicken drumstick (hence its name), which is breaded and deep-fried;

Croquete: shredded beef and bread crumbs (or potato) dough, log-shaped, breaded and deep fried;

Empada/Empadinha: double crust tiny little pies made either with mealy or fine, elastic non-fermented dough usually stuffed with chicken, hearts of palm, or shrimp;

Enrolado: baked fermented dough roll, usually rectangular, containing several kinds of fillings, the most common being ham and cheese, pizza (cheese, tomato, and oregano, sometimes olives), and chicken with *catupiry*;

Esfiha: stuffed and baked fermented dough of Middle Eastern origin, it can be found in two versions: *aberta* (round, with an exposed filling) or *fechada* (triangular, with an enclosed filling). The most popular fillings are ground beef, cheese, and *verdura* (usually escarole);

Pão de queijo: originally from the Minas Gerais state, it gained national preference and can be found now in most cities of Brazil. It is a roll made with polvilho, eggs, and cheese. See the recipe on p. 305.

Pastel: a large deep-fried turnover, usually rectangular, square, or half-moon shaped, traditionally served at open street markets accompanied by sugarcane juice (*garapa*). It can have several different fillings, both savory and sweet, such as ground beef with olives and egg, cheese, and hearts of palm;

Quibe: another Middle Eastern–inspired tradition, it is similar to a *kibbeh* but made with ground beef (instead of lamb) and dark bulgur wheat. It can be *frito* (deep-fried) or *assado* (baked and cut into squares).

Risoles: same dough of *coxinha*, but half-moon shaped and containing other fillings besides chicken, such as ground beef, cheese, and shrimp.

Brazilians love to share their food, and they will often offer you to try whatever they are eating. That means they want you to try a bite, and not that they are giving you the whole plate/item. People can get very offended sometimes if you refuse to try their food. Fortunately, most offer you a try before they start eating. But if you, for any reason, don't want to accept it, say *obrigado/a* (thank you) and, if possible, come up with an excuse, such as "Shrimp is not my favorite shell food," or something on the same line. On the other hand, Brazilians also expect you to offer them to try your food (not in all contexts, not all people). As a rule of thumb, if they offer you, offer them back—most times nobody accepts the offer, it is just a gesture of courtesy.

When going out for eating, it is very common too that people share the bills, and sometimes even calculate what each had ordered to have a fairer division, when the choices were too different in terms of price. If they want to pay your part, they can get very offended if you refuse. And if you are planning to treat some friends to a meal, you don't need to say in advance, unless you're planning on ordering some very expensive items. A gratuity of 10 percent is almost always included in the check, and people usually don't tip extra.

A note about timing: like Italians and Spaniards, Brazilians are famous for being late for their appointments of all kinds. When the scheduled time is a more relaxed, social event, delays are even bigger. That means also people are usually not expecting you to be punctual. If you are invited to a party, for example, even if it is a sit-down dinner, you may catch the host in the shower or the hostess finishing up the cooking if you get at their house by the appointed hour. So, it may be nonsense to say that, but "plan on being late," at least half an hour or so. The same is not so true if you're meeting someone at a restaurant where you have a reservation. But even then, show up about 15 minutes late and you will be as punctual as one can be!

REFERENCES

Almeida, S. P. 1998. *Cerrado: Aproveitamento alimentar*. Planaltina: EMBRAPA.

Alves Filho, Ivan, and Roberto di Giovani. 2000. *Cozinha brasileira:* (Com recheio de história). Rio de Janeiro: Revan.

Ang, Eng Tie. 1993. *Delightful Brazilian Cooking*. Seattle, WA: Ambrosia Publications.

Anunciato, Ofélia. 2000. *A Taste of Brazil*. Transl. Julie Martin and First Edition Translations. Cologne: Könemann.

Atala, Alex. 2013. *D. O. M.: Rediscovering Brazilian Ingredients*. London: Phaidon Press.

Bateman, M. 1999. *Café Brazil*. Chicago, IL: Contemporary Books.

Benke, A., and K. L. Duro. 2004. *Cooking the Brazilian Way*. Minneapolis, MN: Learner Publications Company.

Botafogo, D. 1960. *The Art of Brazilian Cookery*. New York: Hipocrene Books.

Castanho, Thiago, and Luciana Bianchi. 2014. *Brazilian Food*. London: Octopus Publishing Ltd. [Original Title: Cozinha de Origem]. Transl. Elisa D. Teixeira.

Cuza, Sandra. 2012. *The Art of Brazilian Cooking*. Gretna: Pelican Publishing Company, Inc.

Fajans, Jane. 2012. *Brazilian Food: Race, Class and Identity in Regional Cuisines*. London/ New York: Berg.

Farah, Fernando. 2011. *The Food and Cooking of Brazil*. Wigston, UK: Anness Publishing Ltd.

Fernandes, Caloca. *Viagem Gastronômica Através do Brasil*. São Paulo: Editoras SENAC/ Sônia Robato, 8th ed.—bilingual. Transl. Doris Hefti.

Fundação Escolar Pan Americana. 1986. *What's Cooking in Rio*. Bilingual edition. Rio de Janeiro: Carioca.

Hamilton, C. 2005. *Brazil: A Culinary Journey*. New York: Hipocrene Books.

Harris, J. 1992. *Tasting Brazil: Regional Recipes and Reminiscences*. New York: Maxwell Macmillan International.

Idone, Christopher. 1995. *Brazil: A Cooks Tour*. New York: Clarkson/Potter Publishers.

Montebello, N. P., and W. M. Coelho Araújo. 2009. *Carne & cia*. Alimentos e Bebidas Series. Vol. 1. 2nd ed. Brasília: Editora SENAC.

Neelemand, Gary, and R. Neeleman. 2007. *A Taste of Brazil*. São Paulo: Marco Editora.

Peterson, J., and D. Peterson. 1995. *Eat Smart in Brazil: How to Decipher the Menu, Know the Market Foods and Embark on a Tasting Adventure*. Madison: Ginko Press, Inc.

Pinto, C. B. 1998. *Brazilian Cooking: Exotic, Tropical Recipes from South America*. London: Quintet Publishing Ltd.

Roberts, Y. C., and R. Roberts. 2009. *The Brazilian Table*. Layton: Gibbs Smith.

Schwartz, Leticia M. 2010. *The Brazilian Kitchen: 100 Classic and Creative Recipes for the Home Cook*. Lenham: Kyle Books.

Sheen, B. 2008. *Foods of Brazil*. From the Series "*A Taste of Culture.*" Farmington Hills: Kidhaven Press/Thomson Gale.

Trevisani, B., N. de Mattos, R. H. de P. Ramos, and T. M. Barbosa. 2004. *Sabores da cozinha brasileira: Amazonica, Baiana, Gaúcha, Mineira, Nordestina, Pantaneira, Paulista*. São Paulo: Melhoramentos.

Leisure and Sports

Erin Flynn McKenna

INTRODUCTION

Sports and leisure activities in Brazil, like music, food, and other customs, are a blending of traditions from the diverse peoples who have come together in Brazil. Brazil's first inhabitants made use of the land's abundant natural resources in their games and competitions. African influences are evident in a variety of athletic and spectator activities, not least of which is *capoeira*, Brazil's iconic martial art. The development of modern athletics in Europe had effects on Brazil, as it did on most of the world. Europeans brought a variety of sports, most notably, soccer. Despite the European origins, soccer as well as other sports took on a character of their own in Brazil. U.S. and Asian influences are also visible in Brazilian recreation activities like basketball and jiu-jitsu.

Over the course of this chapter, these sports and activities, along with many others, will be discussed. This chapter aims to provide an overview, both historical and contemporary, of the sports and leisure activities practiced in Brazil. In some cases, like with soccer and martial arts, an in-depth history is provided, in large part because of these sports' importance to the formation of Brazilian identity.

INDIGENOUS SPORTING TRADITIONS

The indigenous tribes of Brazil have had an assortment of competitions and activities that took advantage of natural resources. For example, the Tikuna tribe in the West Amazon Basin played tug-of-war using a thick vine available in the rainforest. The Paumaris who lived along a southern branch of the Amazon River competed in swimming and canoe competitions in the Amazon River. Tribes in the Mato Grosso region played games using rubber balls made from the rubber trees there. One group, the Paressi, in the southeast of this region played a game with the rubber balls where they could only hit the ball with their heads. If the ball hit the ground, the other team would score a point. Tribes in eastern Brazil also participated in games with rubber balls, or used feathers and cornhusks to make projectiles.

Many of the native games were competitions to prove speed or strength. For example, the Yanomamö people from the northern Brazilian border region engaged in chest-pounding duels to see who could take the hardest beating. Other tribes, such as the Makushi, participated in foot races, where, upon nearing the finish, competitors would be physically restrained by a committee of men. Thus, not only was speed necessary, but also strength to overcome the physical beating. The Shavante of the eastern part of the Amazon had competitions where they had to tread water until exhausted. The Gé-speaking people participated in relay races where they had to carry heavy logs, in some cases as a test of marriage competency.

While some of the indigenous traditions have died out, others continue today and in some cases, variations of the game have been adapted. One such tradition is called *peteca. Peteca* is a hand shuttlecock game originally played by the Tupi Indian tribe around AD 1500. Other indigenous tribes, such as the Taulipang, the Patamona, and the Makushi, played similar games. *Peteca* is a Tupi word that means strike and shuttlecock. The Tupis made their shuttlecocks by tying stones wrapped in leaves inside ears of corn. Modern versions of *peteca* use a leather pad with feathers attached to it. The object of the game is to prevent the shuttlecock from falling to the ground by hitting it among the various team members. Different teams compete with each other to see how many volleys they can accumulate before the shuttlecock hits the ground. *Peteca* is one of a number of Brazil's recreational sports with this volleying objective. Those sports will be discussed later in this chapter. For now, we turn our attention to Brazil's martial arts, other activities with premodern origins.

MARTIAL ARTS

Capoeira

The African slaves arriving in Brazil brought their own recreational and athletic traditions. The slaves were, however, very limited as to what kinds of activities they could engage in. For that reason, different traditions were adapted in Brazil to the context of slavery. One such tradition is *capoeira*. Capoeira is a Brazilian martial art that has elements of dancing, sports, game, and fighting. For some, it is a true Brazilian artistic expression and a philosophy. The origins of capoeira are widely contested. It is understood that African slaves practiced it in Brazil, hence its connection to Africa. However, discrepancy exists over whether capoeira was a means of fighting for runaway slaves or a form of diversion for slaves in captivity. Despite the imprecision of its roots, numerous sources suggest that capoeira appeared in its recognizable Brazilian form during the 1600s, making it essentially Brazilian in nature. This Brazilian essence is crucial to capoeira's becoming a symbol of Brazilian identity.

Despite the pride surrounding capoeira today, it was not always so celebrated. Slave masters were not keen on any slave activities and certainly not one that was seen as aggressive in nature. Then, after the abolition of slavery, capoeira, practiced by Afro-Brazilians, was seen as a threat to elite society. In 1890, Brazil enacted a Penal Code that, for the first time, recognized capoeira by its name for the purpose of outlawing it. Capoeira, however, did not disappear. Rather than abide by the Penal Code, capoeira practitioners, *capoeiristas*, were conscious to disguise the activity, making it less aggressive and more playful, hence its dance-like quality.

Mestre Bimba, creator of capoeira regional one of the two dominant capoeira styles, was instrumental in sanitizing and legitimizing the practice of capoeira. Born Manuel dos Reis Machado, Mestre Bimba started practicing capoeira as a child in the Salvador, Bahia, neighborhood of Liberdade in the first decade of the 20th century. His father had taught him *batuque*, a street dance fight, and he started incorporating some of those elements into his capoiera. During the 1920s, Bimba along with other *mestres* (capoeira instructors) from Bahia started organizing official capoeira events. *Mestres* would train students who would compete with each other. Bimba was very selective in accepting students to train with him. Moreover, he implemented codes of conduct and a formal training that included academic, religious, and military elements. This rigor legitimized the activity.

At that time, Brazilian society was marked by heavy prejudice against Afro-Brazilian culture; thus, creating respectability was key. Bimba's work to gain acceptance paid off. In 1936 he was invited with his students to give a capoeira presentation at the presidential palace. The presentation was an enormous success, and Bimba was awarded the legal right to teach capoeira, then deemed Bahia's "regional fight." In 1953 he was again invited with his students to give a presentation for then president Getulio Vargas, after which capoeira was named Brazil's national fight. This was a great achievement for a sport once disregarded in part for its African roots.

While Bimba's *Capoeira Regional* (Regional Capoeira) was creating a space for its acceptance in Brazilian society, another capoeira style, *Capoeira Angola* (Angolan Capoeira), was claiming its African roots. Vicente Ferreira Pastinha began learning capoeira as a child from an Angolan instructor. Mestre Pastinha opened a capoeira school in 1910. His style was considered more pure because he was opposed to incorporating movements from other martial arts the way Bimba did. Nevertheless, Pastinha also made modifications from the street capoeira, including the elimination of the finger-in-eye movement, which was considered quite brutal. Pastinha received international recognition when he traveled to Senegal to participate in the first World Festival of Black Arts. There, Pastinha and his students were able to introduce this uniquely Brazilian activity to the world.

Both Bimba and Pastinha had profound impacts on the development of capoeira in Brazil and as an internationally recognized martial art. Their two styles of capoeira are the most well known and widely practiced in the capoeira world. The two styles can be distinguished visually and sonically. Capoeira Regional is characterized by its faster-paced music, the more upright posture of the players' bodies, the fast, high kicks, and the aerial acrobatic movements. In addition, the players can be distinguished by their attires. Traditionally *capoeiristas* wore white, so that the best players could proudly display that they had not fallen down and gotten dirty during the game. Following that tradition, Capoeira Regional players train in white and wear a colored belt that signifies their rank.

Capoeira Angola on the other hand is played lower to the ground. Players spend much of the game in a crouched position, with their kicks and spins taking off from there. Their style of play generally requires more strength than speed. Angola players often, but not always, wear black pants and shirts from their training academies. Usually the shirts bear a color that indicates the lineage and affiliation of the player within capoeira.

On an ideological level, these two styles are also different. Capoeira Regional is known for its mixture of elements. Drawing both from African traditions and distinctively Brazilian ones, capoeira regional tends to follow the traditional Brazilian celebration of miscegenation: the mixing of different cultures to produce something new and better. In that way, it claims a particularly Brazilian character. Capoeira Angola stresses the African elements and origins of the art much more.

Despite these differences, these two styles also have some similarities. They both use Afro-Brazilian instruments, like the *berimbau*, a stringed bow with a gourd at the bottom, for the music accompanying the fights. They incorporate songs that discuss slavery and Afro-Brazilian religious beliefs. Both styles use a somewhat similar basic *ginga* step, a rocking side-to-side movement while raising one arm. The *ginga* step is a neutral position that players take during the game.

There are other variations beyond these two styles, for example *Capoeira Contemporânea* (contemporary capoeira), which attempts to combine the two main styles into something new. Aspiring *capoeiristas* in Brazil today practice a particular style based on which academy they enroll in. While some informal practicing occurs in Brazil, it is more and more common for players to enroll in academies and go through a more official training. Through their academies, they earn opportunities to engage in all levels of competition from local to international.

Players also have opportunities to make a living off of their capoeira expertise. Globally, capoeira has grown tremendously in recognition and popularity since the 1970s. Due to capoeira's international acclaim, an increasing number of tourists are drawn to Brazil and particularly Bahia to watch or practice capoeira. Thus, *capoeiristas* are able to do performances for tourists or to teach them in order to make money. In Salvador, Bahia, the cradle of capoeira, it is hard to miss the *capoeiristas* performing in central plazas or tourist venues. In addition, tourists enroll in capoeira academies. The Brazilian government has recognized capoeira's international significance by naming it to its list of intangible cultural patrimony. Nowadays Brazilian capoeira *mestres* travel the world to divulge their art. Capoeira classes are popular in all major cities in practically every continent.

Brazilian Jiu-Jitsu

While not as iconically Brazilian as capoeira, Brazilian jiu-jitsu is another martial art with foreign origins but distinctly Brazilian qualities. Brazilian jiu-jitsu originated with Mitsuyo Maeda, a Japanese fighter who moved to Brazil in 1914. Throughout his career, Maeda was too small to participate in sumo, so he specialized in judo, teaching and fighting all over the world. While in Brazil he met Gastão Gracie, a Brazilian who helped Maeda establish a Japanese community in Belém. In appreciation, Maeda offered to teach jiu-jitsu and judo to Gastão's son Carlos. In 1925, Carlos Gracie established his own jiu-jitsu academy in Rio de Janeiro. With his brothers' help, they established jiu-jitsu academies all over Brazil.

Hélio, one of the Gracie brothers, was a small, sickly child. He studied Carlos's moves from the sidelines, rather than on the floor. However, he was given the opportunity to teach in his brother's absence. Because of his frail physique, Hélio relied on techniques that favored agility over strength. These techniques were different from the Japanese ones, influencing the particular style of Brazilian jiu-jitsu.

Brazilian jiu-jitsu made its way into international circuits after several of the Gracies left Brazil for the United States. Hélio's son, Rorion, was one of the founders of the Ultimate Fighting Championship where people from different martial arts styles would fight each other. Rorion's younger brother, Royce, won the first competition, launching Brazilian jiu-jitsu into international consciousness and legitimating it in the Mixed Martial Arts scene.

Jiu-jitsu is a grappling martial art, more like wrestling than other styles that rely on striking and standing up right. Jiu-jitsu is low to the ground. It differs from judo insofar as it allows different moves like leg-locks. Brazilian jiu-jitsu has elements of street fighting, and fighters often compete against opponents in other martial arts. In some of the most prominent fights, Brazilian jiu-jitsu players have been victorious, giving Brazilian jiu-jitsu fame for being the toughest martial art in contemporary times.

Like capoeira, there is an interest in jiu-jitsu's cultural roots. While undeniably Japanese in origin, jiu-jitsu also has certain elements that are distinctly Brazilian. For example, it incorporates the Brazilian concept of *malícia* or *malandragem,* also common in capoeira, which refers to using trickery or guile to get ahead. Brazilian literary heroes are often portrayed as possessing this quality. In Brazilian jiu-jitsu, the quality is manifested insofar as smaller players are often able to defeat bigger ones, using their shrewdness and speed. There is no comparable idea in Japanese judo or jiu-jitsu. Another distinctly Brazilian quality is the fact that jiu-jitsu instructors in Brazil are called by their first names or nicknames rather than a more formal title. This is characteristic of Brazil on all levels, even in the presidency.

On a political level, questions about jiu-jitsu's roots were important in establishing legitimacy both in the face of capoeira's sordid popularity and during World War II. Now capoeira is greatly revered; however, during the first few decades of the 20th century, people were skeptical of its African roots. Therefore, jiu-jitsu had an advantage by associating itself with Japan rather than Africa. Nevertheless, during the 1930s and 1940s, the Japanese were seen as the enemy; thus, differences, rather than similarities between Brazilian jiu-jitsu and Japanese judo were emphasized. After the war, when relations between Japan and the rest of the world normalized, there was less of a concern with differentiating them, but by then, the Brazilian jiu-jitsu had diverged quite significantly from the Japanese judo that it was notably distinct. Moreover, with the arrival in Brazil of different groups of Asian immigrants with their own varieties of martial arts, the Gracies' academies absorbed aspects of those styles as well. In this way, Brazilian jiu-jitsu, like many of Brazil's athletic and leisure activities, is a blend of elements.

MODERN SPORTS

European Influence

By the time modern sports were developing in Europe, the colonial period in Brazil was ending. Economic rather than political or military ties were dictating the degree of European influence in the Americas. In Brazil, that influence was largely British. For that reason, one can trace many organized sports traditions in Brazil to British origins.

The first European sport to appear in Brazil was horse racing. Beginning around 1810, Englishmen were racing horses in Brazil. In 1849, the first official Racing Club was established. The rules, competition, and organization of the races were based on British and French models. Horse racing largely served as entertainment for the upper classes in Brazil. Although it started in Rio, by the end of the century it was increasingly more organized and it expanded to other urban centers such as São Paulo, Salvador, and Recife.

Rowing is another sport brought by the British that became popular during the mid-19th century. As beaches became a place for social gatherings, rowing competitions became popular and, over time, better organized. By the end of the century, there were 15 boat racing clubs in Brazil and boat racing surpassed horse racing in popularity. In addition to boat clubs, there were gymnasiums and other sports clubs being built. Swimming, shooting, cycling, and fencing were all commonly practiced. Sports were seen as a way to maintain social control and embrace modernity.

Soccer

Brazil has so thoroughly come to claim soccer as its own; it is hard to imagine it as having origins elsewhere. Soccer (in Brazil known as *futebol*) was first played in Brazil, usually near the wharfs and ports of coastal cities, by British sailors around the 1860s. Charles Miller is credited with introducing soccer to Brazilians at large in 1894. Brazilian-born, the son of British parents, Miller studied in England as a young man. When he returned to Brazil, he brought a rulebook and several soccer balls. He started to organize several teams in São Paulo. Soon, rowing clubs were also sponsoring soccer teams.

At first, the majority of players were Brits living in Brazil. However, Mackenzie University organized a team of majority Brazilian players. After that, clubs quickly started forming throughout Brazil. By the second decade of the 20th century, there were leagues in Rio Grande do Sul, Minas Gerais, Bahia, Pernambuco, and Paraná, as well as teams in Amazonia. Soccer's popularity grew astronomically.

The style of play remained very British and European influenced throughout the early part of the 20th century. While Europe was forming professional leagues, there were still a number of Europeans in Brazil who were playing in the Brazilian leagues. Brazilians who played were primarily white elites. In 1923, a club sponsored by Portuguese immigrants started fielding Afro-Brazilian and mulatto players. When the team, Vasco de Gama, won the Rio championship, it paved the way for more non-white players to play for the sporting clubs.

SPORT AND THE TRIUMPH OF THE UNDERCLASS

When organized sports started in Brazil, they were largely for elites. Both horse racing and rowing were activities that were inaccessible to the majority of the Brazilians. Thus, soccer, a sport that required little equipment, offered an opportunity to the many. Moreover, with soccer's universally applied rules, talent, rather than money or personal connections, determined the winner, allowing the underclass a chance to overcome persistent nepotism and unfairly applied laws governing their society. Like soccer, capoeira represents a victory of the underclass. Capoeira is long associated with resistance: slaves' resistance to masters and *capoeiristas'* resistance to authority for enacting the 1890 Penal Code. Thus, Brazil's most renowned sports share a history of the underclass claiming a space to participate.

By the 1950s, soccer had been claimed by the popular classes, who were excelling and giving Brazilian soccer a very national characteristic. While the European style was more physically aggressive and calculated, the Brazilian style was agile, quick, and improvised. An underlying philosophy in Brazil is to figure out a way to get by, a *jeitinho*. The ideal of using a *jeitinho* also applied to soccer. There was always a creative way to play or to win and that's what Brazilian players strove to find. As in capoeira, *ginga* is an important element of soccer playing in Brazil. One needs to know how to move gracefully his or her way around the field. It is the *ginga* that is responsible for the so-called jogo bonito (beautiful game) that Brazilians play.

The spectators also involved themselves in the event through the music, dance, and cheering that accompanied the soccer games. This energy from the fans combined with the creative energy on the field to turn soccer into a sort of party. This spectator participation is what causes soccer in Brazil to be such a major collective activity. Even when not physically in attendance at the game, there is a deep connection between the fans and teams that explains why, when the Brazilian team loses, the country plunges into depression as happened on several historic World Cup occasions.

The World Cup championship was established by the International Federation of Football Associations (FIFA) in 1930 because of the turn toward professional competition in Europe. Up until that point, the Olympics served as the major international soccer competition. However, because the Olympics were only for amateur athletes, many of the top soccer players were excluded from play. Thus, the World Cup offered the opportunity for the world's best players from each country to compete with each other.

The first several World Cup tournaments did not have qualification matches. Rather, teams were invited to play by the FIFA organizers. The first World Cup was held in Uruguay, and neighboring Brazil was invited to participate. Four and eight years later, Brazil also participated because other South American countries were unable to send teams to the tournaments in Europe. During this first decade, Brazil went from highly inexperienced to making it to the semifinals in 1938.

Due to World War II, the World Cup was suspended until 1950 when Brazil was asked to host the event. In preparation, the legendary Maracanã stadium was built in Rio de Janeiro. The stadium is ideal for Brazilian-style soccer. Because it is so large, it does not provide the spectator with the best perspective for viewing the game; however, because of its capacity, huge numbers of fans can attend and create the necessary ambience for the Brazilian game.

Fans showed up in droves at the Maracanã to attend the 1950 World Cup final where Brazil was facing its neighbor and rival, Uruguay. Heavily favored to win on its home turf, all Brazil needed was a tie to claim the cup. Brazil was leading 1–0 before the Uruguayan team scored early in the second half. Then, with 11 minutes remaining, a Uruguayan forward scored a goal to go up 2–1. Brazil did not recover and, to the astonishment of everyone, was tragically defeated. The game that was supposed to crown Brazil as the best soccer team in the world left many Brazilians in despair. Mourning the loss, some Brazilians reportedly took their lives; others vowed to boycott the Maracanã forever.

In 1954, Brazil qualified for the World Cup but lost in the quarterfinals to Hungary. It was not until 1958 that Brazil had the opportunity to make up for the bitter defeat eight years earlier when it made it to the World Cup final. The 1958 team

Brazil's national soccer team poses for a photo before the World Cup final against Sweden in Stockholm on June 29, 1958. Brazil defeated Sweden 5-2 to win its first World Cup. Soccer legend Pelé (pictured bottom row, third from left) was playing in his first World Cup tournament. (AP Photo)

was well coached and well organized and had a batch of new players, including the 17-year-old Pelé and another exceptional player, Mané Garrincha. Despite entering the tournament with a reputation for falling apart in important matches, the Brazilians secured a 5–2 victory against Sweden to take the trophy.

In 1962, Brazil again won the World Cup despite losing Pelé to injury in the second game of the tournament. In 1966, Brazil was going for an unprecedented third title. In anticipation of the historic event, Brazilian fans made the trip to England to witness the moment firsthand. In a tragedy echoing 1950, the fans were again left to weep in the streets. The Brazilian team was admittedly less talented than the previous two tournaments, but no one expected the team to succumb in the first round. However, Brazil bounced back, and in 1970, behind Pelé's mature leadership, the Brazilian team became the first to claim three World Cup titles. At this point, soccer and Brazil became synonymous.

By 1974, Brazil was in the midst of its military dictatorship. Moreover, the game was beginning to change. Increasingly, players were being lured overseas to play professionally for more lucrative contracts than what Brazilian clubs could offer. Despite government efforts to invest in sports, Pelé and other members of the national team decided against playing in the World Cup that year. Brazil made it to the semifinals but lost to the Netherlands 2–0. In 1978, Brazil was eliminated in the second round. Throughout the 1980s, the Brazilian squad suffered. Despite having talented individual players, they could not come together as a team to defeat the Europeans. Moreover, because of overseas influence on individual players contracted in Europe, the Brazilian style was being overtaken and mixed with a European style of play that left much to be desired on the field.

It was not until 1994 that the Brazilian team was able to successfully combine its mixed European and Brazilian playing styles to win the World Cup. Talented goal-scorers on offense combined with a disciplined and skilled defense to claim the title. After losing to France in the 1998 final, Brazil came back in 2002 to claim a fifth World Cup victory. During the next two tournaments Brazil lost during the quarter-final rounds, much to the dismay of fans looking to celebrate an unprecedented sixth victory. In the 2014 World Cup, fans were hoping that the Brazilian National Team would make amends for 1950 and claim the title on home soil. Although the team raised hopes by making it to the semifinals, fans were crushed when the team lost 7–1 in a resounding defeat to Germany who went on to secure the title.

Brazil's World Cup matches are treated as quasi-holidays in Brazil. In some cities, the businesses shut down during games so that employees can watch the games. Often game-viewing parties are organized at private homes or bars and restaurants that set up big screen television. Gigantic screens are also put up in public plazas and on the beach. Cheering for the national team is a collective effort. Scores of people don their green and yellow clothing on game days to show their support. During games, fog horns, firecrackers, and collective cheers erupt all around the country when Brazil scores a goal. Should Brazil lose, a collective depression descends on the country. For all their support to the Brazilian team, Brazilian fans are also deeply motivated by cheering against their rivals, especially Argentina. The only thing that compares to the lauding of a Brazilian victory is the cheering for an Argentina defeat.

Although the national game clearly harnesses the most collective energy in Brazil, Brazilians are also deeply passionate about their local teams. The clubs compete regionally (by state) and nationally for titles. There are different divisions according to the level of each club; the best teams play in both the state and national competitions. State competitions run from January to April. The national competition starts in May and ends in December. Based on the outcome of the national tournament, clubs are assigned to higher or lower categories. There are over 400 registered clubs. This makes for intense intercity rivalries. Because the Brazilian league does not provide the same remuneration as European clubs, top Brazilian players do not play in Brazil. For that reason, European teams are also closely followed so as to keep up with the Brazilian players there.

Internationally Competitive Sports

While soccer's popularity is hard to compete against, Brazilians practice many other professional sports and engage in international competitions. Brazil's beaches provide ample space for playing beach volleyball. The International Volleyball Federation recognized Brazil as an ideal site for beach volleyball when it invited Rio de Janeiro to host the first official Beach Volleyball World Championships in 1987. When beach volleyball was made an official Olympic sport in the 1996 games, two Brazilian duos claimed the gold and silver medals in the women's competition. Brazil has also been very successful in indoor volleyball. Since 2013, both the Brazilian National Men's and Women's teams have been ranked as number one in the

International Federation of Volleyball (FIVB). Among other championship titles, the men secured the gold in the 1992 and 2004 Olympics and the women in the 2008 and 2012 Olympics.

Because 19th- and early 20th-century foreign influences in Brazil were largely British and French rather than American, unlike other countries in Latin America, Brazil did not develop a strong baseball tradition. Nevertheless, there is one notable exception: São Paulo, with its large community of Japanese, has become a bastion for baseball. Another U.S. sport, basketball, does, however, have increasing popularity in Brazil. A number of NBA players have come from Brazil. One of the best female players ever to play the game, Hortência Marcari played for the Brazilian team for almost two decades. In the 1987 Pan-Am games, the Brazilian men's team staged a surprising upset over the United States to claim the gold medal.

Other popular sports in which Brazil makes a strong international showing include gymnastics, auto racing, track and field, swimming, and tennis. Daiane Garcia dos Santos is one of Brazil's best gymnasts of all time having competed in the 2004 and 2008 Olympics. In the floor routine, she introduced two challenging skills, which are referred to as the "Dos Santos" in her honor. Felipe Massa is a current Brazilian Formula One auto racer with 11 wins in his 193 Grand Prix races. In track and field, Brazilian athletes João Carlos de Oliveira (João do Pulo), a long and triple jumper, and Yane Marques, a pentathlete, both brought home bronze medals from the Olympics. Marques is a current competitor and earned her medal in 2012. De Oliveira, who competed and earned bronze medals in the triple jump in both 1976 and 1980, also set the world record for the triple in 1975, which he held until 1985. In swimming, the Brazilian team took home two medals, in each of the past two Olympics: a silver and a bronze in 2012 and a gold and bronze in 2008. César Cielo Filho is the most successful Brazilian swimmer of all time with three of those four medals. Tennis players Maria Bueno and Gustavo Kuerten are discussed in the Spotlight on Brazilian Athletes.

Extreme sports are becoming increasingly popular in Brazil, as in the rest of the world. Such sports include ultramarathon cycling, land-yachting, all-terrain auto racing skateboarding, and inline skating. Brazil has hosted a number of international extreme sports competitions. One of its most successful athletes in extreme sports is Fabiola da Silva, who has won eight world championships in inline skating.

Women and Sports

Although soccer is popular among women, especially for adult women spectators, at a younger age women are more likely to participate in swimming, volleyball, and basketball. Women's participation in organized sports was limited to being a spectator throughout most of the 19th century. Elite women would attend the horse races and rowing events, and it would give them a chance to show off the latest fashions they acquired. However, by the turn of the 20th century, there were more opportunities for women as participants in sports. They include horse racing, rowing, swimming, pole vaulting, fencing, tennis, track and field, volleyball, basketball, gymnastics, archery, and cycling.

In the 1930s, Brazilian female athletes had gained prominent international recognition, with athletes competing in the Olympics in swimming and fencing. The 1960s and 1970s brought further change to the development of female athletes, and women have been far more involved in international competitions. Both the women's volleyball and basketball are consistently ranked among the top in the world.

While progress has been made in terms of women gaining respect as athletes in Brazil, there are still barriers to their full participation. Female athletes can be suspected of being homosexuals or considered unfeminine. Women rarely hold high-level positions in sports administration or policy. Their participation in sports has also brought consequences, such as the commercialization of the female body. Despite this, the achievements that have been made in athletics pave the way to greater equality in other aspects of society.

SPOTLIGHT ON BRAZILIAN ATHLETES

Emerson Fittipaldi, auto racing (1946–), began racing when he was 17 years. He became the youngest racer to win the World Driver's Championship with five Grand Prix victories in 1972. In 1989 after countless victories, Emerson set his sights on the elusive Indianapolis 500. Leading most of the race, an opponent caught him in the last couple of laps. As they attempted to lap other contestants, they collided and only Emerson stayed in the race to claim victory. He won a second Indy 500 in 1993.

Ayrton Senna, auto racing (1960–1994), competed for eight years in go-cart racing before turning to auto racing in 1981. After trying different teams and competitions, Ayrton found success with the Honda Marlboro McLaren, claiming the Formula One championship in 1989 by winning 8 out of 16 Grand Prix races. He took numerous victories over the next five years until he suffered a deadly accident in a 1994 race.

Éder Jofre, boxing (1936–), started boxing at the age of three, when he was taught by his father. After competing in the 1956 Melbourne Olympics, he turned professional. During his professional career, where he fought as a bantamweight and featherweight, he lost only two fights. In 1960 he won the National Boxing Association's bantamweight title and then clinched the world title. He held this title until 1965. After two losses to Mashika Harada, Jofre retired, only to return to the ring in 1969, when he went on another winning streak in the featherweight class. He won the featherweight championship in 1973 and in 1974 was stripped of the title for not defending it. He finally retired for good after a victory in 1976.

Royce Gracie, martial arts (1966–), son of Hélio, was trained as a child in jiu-jitsu by various members of his family. He began competing in tournaments at the age of eight and earned his black belt by 18. In 1993, he entered the first Ultimate Fighting Championship successfully showcasing the Brazilian sport by winning the single elimination tournament against his seven opponents. After winning the second and fourth UFC titles as well, he was inducted into the UFC Hall of Fame in 2003.

Hortência Marcari, basketball (1959–), began playing basketball when she was 13, and when she made it to the national team in 1979, she quickly became a star. A 5′8″ shooting guard, Hortência contributed accurate outside shooting, speed, and intensity to her teams. She was recruited by the University of Tennessee, but family necessity dictated that she remain in Brazil earning money rather than an education for basketball. She dominated the game in Brazil: in one championship game she scored 120 points! Hortência played on the Pan-Am silver medal team in 1987, scoring 30 points. In the 1991 Pan-Am games, Brazil, with Hortência, defeated the United States and went on to win gold. She came out of retirement in 1996 to play on the Olympic squad that earned a silver medal. In 2002, Hortência was inducted into the Women's Basketball Hall of Fame.

Oscar Schmidt, basketball (1958–), made a name for himself during the Inter-club Basketball World Championships as a clutch scorer with limitless range. Oscar played on several European teams as well as the Brazilian national team. In 1987, Oscar and the Brazilian squad defeated the United States to win the gold medal. Oscar scored 46 points. Despite being drafted by the New Jersey Nets in 1984, Oscar turned down the offer, preferring to stay in Europe, playing as a star. Oscar played professional basketball for 26 years, logging 49,703 points. He played in five different Olympic Games and led the games in scoring three times.

Maria Bueno, tennis (1939–), started to play tennis when she was six. She never took private lessons or had a personal coach, but studied the game's top players on her own. When she was 14, she became Brazil's female tennis champion, and in 1958 she claimed her first major title. In 1959 she won both Wimbledon and the U.S. National Championship and repeated at Wimbledon in 1960. In a time when women's tennis was characterized by long volleys, Maria put forth her own more exciting style: charging the net after hitting it over and hitting shots out of reach of her opponents. She battled back from serious illness to claim the 1963, 1964, and 1966 U.S. championship titles and the 1964 Wimbledon. Upon retirement in 1971 she remained active in charity tennis events.

Gustavo Kuerten (*Guga*), *tennis* (1976–), erupted onto the tennis scene in 1997 when he won the French Open despite being ranked No. 66. In 2000, he hit his pique, winning five major tournaments, including his second French Open, and went on to claim the No. 1 spot in tennis by beating both Pete Sampras and Andre Agassi in the Tennis Masters Cup. His tennis achievements catapulted him into the hearts of Brazilians whose interest in tennis grew astronomically because of Guga's success. He used his fame and tennis earnings to encourage the Bank of Brazil to start a development program for junior tennis players, the first of its kind in Brazil. Affected by the condition of his brother who was born with cerebral palsy, Guga also started a foundation that raises money to cure childhood illnesses. In 2012, Guga was inducted into the Tennis Hall of Fame.

Edson Arantes do Nascimento (*Pelé*), *soccer* (1940–), was born to a family of humble resources in the town of Três Corações, Minas Gerais. His father was a superb professional soccer player whose career ended early due to injury. Edson spent all of his free time playing soccer with grapefruits or sock balls when no actual balls were available. His teammates gave him the nickname Pelé when he was eight. Pelé's first organized team soccer experience was for the Bauru Athletic Club (BAC) youth team when he was 13. He took this opportunity as possibly offering him a shot at professional soccer. At that time Pelé was an undisciplined, raw talent on the soccer field. His coach at BAC helped to discipline him. In 1958, Pelé was selected for the national team that clinched World Cup victory. Pelé scored six goals over the course of the tournament. After the 1958 World Cup, Pelé was contracted to play for the Santos team of São Paulo. This was Brazil's most respected club team. Pelé was on the 1962 World Cup team, but sat most of it out because of an injury he sustained early in the tournament. In 1969, while playing for Santos, Pelé scored his 1,000th goal, a feat never before achieved. After getting knocked around pretty significantly in the 1966 World Cup, Pelé was reluctant to play again in 1970. In the end he decided to play, leading the team to its third World Cup victory. Pelé continued his career with Santos until 1974, when he retired from Brazilian club ball. In 1975, Pelé signed a contract with the New York Cosmos of the North American Soccer League, in part to generate support for soccer among U.S. sports fans. In 1978, the Cosmos won the league championship. He then retired from soccer, but did not withdraw completely. He served as the minister of sport and as an international ambassador for soccer. His fame and skill are still unprecedented today, and he sets the standard for aspiring soccer stars.

Manuel Francisco dos Santos (Garrincha), soccer (1933–1983), was born and grew up in Pau Grande, a poor village outside of Rio de Janeiro. As a child he contracted polio, which left one of his legs crooked. Because of this, he was nicknamed Garrincha, a little bird. Despite this abnormality in his growth, he was determined to play soccer. He was completely dedicated to his sport, often choosing to play rather than go to school. This allowed his body to get stronger. Moreover, because of the shape of his legs, he developed a unique dribbling style and he was a natural winger or outside forward. When he was 14, Garrincha joined his local team in Pau Grande. At the age of 20, he was selected for Botafogo. There, he blossomed with better teammates surrounding him, and he contributed to his team's three Brazilian championships between 1957 and 1962. He also played for the victorious World Cup teams of 1958 and 1962.

Marta Vieira da Silva, soccer (1986–), became a fan of soccer at a young age and dreamed of playing professionally. She usually played with boys or men much older than her, who were rough and sometimes verbally abusive. Nevertheless, her skills earned her spot on the field. At 12 she was permitted to play on an all-male regional team. After leading her team to back-to-back championships, she was banned from playing. She then went to Rio to play for a women's team. After playing in several international competitions, she was offered a spot on a Swedish professional team. Marta is the first Brazilian female player to play professionally in Europe and the first female athlete inducted into the Brazilian sports hall of fame. She won a silver medal at the 2004 and 2008 Olympics. In 2009, she began playing for the Los Angeles Sol of the Women's Professional Soccer League.

Roberto Carlos da Silva Rocha, soccer (1973–), started playing on his father's men's team when he was eight. He signed his first professional contract with the Brazilian Palmeiras club at the age of 20. After several strong seasons, he was picked up by Internazionale Milano and then Real Madrid, where he played for 11 years, winning three Spanish Super Cups and three UEFA titles. Despite playing defense he scored 47 goals during his Real Madrid tenure. He was a specialist in free kicks. Roberto Carlos played on several Brazilian national teams, including that which won the World Cup in 2002.

Given Brazil's penchant for soccer, the number of players who have achieved greatness throughout the years is significant, and stories of their achievements could fill a whole book. Other players worth mentioning include Ronaldo Luiz Nazario de Lima (1976–), Ronaldo (Ronaldinho) de Assis Moreira (1980–), Arthur (Zico) Antunes Coimbra (1953–), and Ricardo (Kaká) Izecson dos Santos Leite (1982–). All of these players, and many more, had successful careers with club teams in Brazil and in Europe and represented Brazil on numerous occasions in the World Cup and other tournaments.

RECREATIONAL SPORTS/GAMES

In addition to the recreational play of the organized sports mentioned previously, there are a number of other recreational sports and games that are popular in Brazil. Due to Brazil's concentration of habitants in the coastal areas, a significant number of games are played on beaches or in water. For example, Futebol de Areia (Sand Soccer) is an adaptation of soccer to the beach. Two teams of 12 players face off in the sand on a court the size of an indoor soccer field with two goals. The rules and object of the game are the same as soccer and the game consists of two 30-minute halves.

Frescoball is another example of a Brazilian beach sport. Frescoball was invented in Rio de Janeiro and is a sort of beach tennis in which the two players face off to volley a rubber ball with wooden paddles attempting to keep the ball in the air. The name "frescoball" refers to the *fresco* or "refreshing" waves of the ocean where it is played. Both Futebol de Areia and frescoball have been banned in some beach areas of Brazil because of the tendency to interfere with others' enjoyment of the beach. Beach volleyball is also very popular in coastal cities such as Rio de Janeiro.

Another informal sport in Brazil includes Punhobol or Faustebol commonly played in southern Brazil. This game is a combination of volleyball and soccer. It is played by teams with five players on each side. It can be played in grass or sand with a volleyball net (2 meters high) and either a soccer or volleyball. Players can touch the ball only once either by kicking or by punching the ball. A point is scored by hitting the ball into the net, hitting the ball out of bounds or when the opposing team takes more than three attempts to get the ball back over the net. The team that scores the most points in 30 minutes of play is declared the winner.

Other informal team sports include Biribol, a water volleyball game originating in São Paulo state. This game is played in swimming pools and requires a net. The rules and objective are the same as volleyball and the winner is the best of three sets. Tamboreu is another volleyball-like game, only instead of volleying, the ball is thrown and caught between the two teams. When one team misses a catch, bats the ball, or throws the ball into the net, the opposing team scores a point. The best of three sets wins and to win a set, a team needs to earn 10 points.

Marathon backwards is an example of an individual recreational competition. It is played in the city of São Vicente Ferrer in the state of Pernambuco. Marathon backwards is a foot race in which participants run the course backward while looking straight ahead the entire time. Anyone who faces backward is disqualified. The length of the race is determined by organizers.

In addition to these games, there are others uniquely played by children. A number relay game has children divided into two teams racing each other to transport objects between their respective circles. Tampa is a game using a bottle cap, where a small obstacle course is set up and each child flicks his or her bottle cap along the course to see who can finish first. If the bottle cap goes outside the track or hits another player's cap, the child loses a turn. Jogo de taco, most likely of foreign origins, is a game played with four players: a batter and catcher on each team. Two cans are placed between the batter and catcher on each team. One catcher throws a ball aiming at the opposite team's can. The opposite batter has the opportunity to hit the ball. If he does, the two batters run back and forth between the cans as many times as they can until the original catcher recovers the ball. A point is scored for each time the batter touches the opponent's can. The catcher and batter on a single team switch places whenever the catcher hits the can on his throw.

LEISURE

Leisure and free time in Brazil follow a pattern similar to the rest of the world. Modern, industrialized society paved the way to an expansion of the leisure class beyond those with extreme wealth. Industrialization became prominent in Brazil in the early 20th century. With industrialization, wages typically increased, allowing people to spend money on leisure time and goods. However, Brazil was and remains a country with vast inequalities of wealth. While some people have the opportunity to spend money and time on leisure pursuits, others merely work to make ends meet. Thus, leisure practices are largely a reflection of social class. In addition, notable differences in terms of access to different leisure activities exist between those living in urban and rural settings.

Keeping these differences in mind, one can note some trends regarding leisure in Brazil. Leisure time in Brazil is largely social. As discussed in the soccer section, being a spectator is a social activity. Even watching television is social. While most Brazilian families own their own television sets, joining with friends and neighbors in a private home or public place to watch an episode of a telenovela is not uncommon. Listening to the radio or music or reading newspapers and magazines is far less common than watching television and other social activities.

Because of the social nature of leisure time in Brazil, much of leisure takes place in public spheres. People gather in restaurants, coffee shops, or snack bars during the day. Open air venues are popular throughout Brazil, particularly in coastal areas and places with a warm climate throughout the year. In urban areas there are theaters, cinemas, museums, zoos, parks, plazas, and botanical gardens. Live sporting events attract large crowds. Concerts often take place in public venues, offering entertainment free of charge. In the rural areas, options are much more limited for engaging in these types of activities. There, people may experience temporary diversion when a festival is going on or if there is a traveling show or event.

In the afternoons and evenings, it is common for Brazilian adults to socialize in bars known as *botecos*. In Brazilian bars, beers are typically served in half-liter bottles, perfect for sharing. Thus, rather than ordering a personal beer, a beer is ordered and poured into smaller glasses for everyone. When that runs out, another bottle is ordered. Brazilian beer is typically served, extremely cold (*bem geladinha*), and dividing the beer into smaller servings and sharing is a way to ensure that it is always consumed at the coldest temperature possible. In addition to serving beers, *botecos* often serve appetizers. Again, it is common to share among friends when out at the bar. With food, the social time can go on and on.

For a more active evening experience, going out to dance is a popular activity. In large Brazilian cities, there are a wide variety of venues to dance. Some places are dedicated to specific Brazilian genres of music like *forró*, funk, or samba, while other places are more eclectic or international in their music selection. As in any city, there is a range of venues, some more costly than others; thus, financial resources play a part in one's nighttime entertainment options. In some instances, rather than going out, people host their own parties. In rural areas where there are fewer options for nightlife, festivals or private parties provide opportunities to dance.

Of course spending time at beaches is one of the primary leisure activities for Brazil's coastal inhabitants. Beach culture started to gain popularity in Brazil around the mid-19th century when concerns for sanitation and hygiene arose. Thus, bathing in the ocean began as a health, rather than social, activity. Nevertheless, as people congregated to take care of their hygiene, it quickly turned social. Furthermore, as modern ideas about bodily aesthetics took hold in Brazil, the beach became a place to perfect and show off the human body.

Today's beach is still a place to work out and show off bodies. Brazilian beachwear is known for being particularly revealing. Women of all ages and sizes don bikinis and men tend to sport briefs or lycra shorts rather than the longer surf shorts. Beaches in Rio are equipped with gymnasiums where people can work out on pull-up bars or sit-up benches. There is also a running path along the sand to run, rollerblade, bike, or walk. In addition to the recreational sports mentioned previously, other beach activities include surfing, snorkeling, windsurfing, paddle boarding, and stand-up paddle for those who prefer a more active leisure experience while at the beach.

The beach is also a place to relax and spend the day. Brazil's beaches generally have an assortment of venders offering a wide range of foods and beverages. The items range from popsicles and beer to grilled meat and cheese on skewers. There are also people renting chairs, offering to give henna tattoos or surfing lessons.

Renowned Brazilian anthropologist, Roberto DaMatta has claimed that the beach is Brazil's great democratic space, where people, regardless of race, class, gender, and sexual orientation, all come together, swimming-suit clad and stripped of status markers. However, this is increasingly contested as people consider that even in a bathing suit, a number of distinctions can still be made. In fact, Rio's beaches are reportedly quite segregated as people from different social groups congregate together excluding others. Sometimes, the upper classes disparage the lower classes, calling them *farofeiros*, a term that implies being impolite due to eating the messy, flour-based food, *farofa*. They also complain about them bringing their food to the beach and their different behavior and customs. Transportation policies can make it very difficult for those residing outside of the city to get to certain beaches, revealing structural inequalities. The beach has also become a place of violence and other criminal activities, the threat of which already makes for an unpleasant experience. While beaches in Brazil are public, making it a place where people can come together, there are still inequalities and conflict that should not be ignored.

Travel and vacationing in Brazil is another leisure activity that is on the rise. In the past decade, a growing middle class in Brazil has allowed for an increase in automobile ownership and made domestic flights more accessible. In fact, some airlines even offer airline tickets on credit so that people can take their vacation and pay for it in installments. While travel is still a major obstacle for the majority of Brazilians, people still look for ways to take vacations. One of those is by visiting nearby areas in the interior of Brazil or on the coast that are not hard to get to. For example, families living in the city of Salvador may spend weekends at a second home on the nearby island of Itaparica. Going to the island allows the family to escape from the routine of daily life, while avoiding expensive treks to a faraway destination. Another frequent vacation for North Easterners is during the St. John festivities in June. A harvest celebration, coastal residents in the Northeast frequently take trips into interior towns and cities to celebrate. For those who can afford a bigger journey, Rio de Janeiro is the top domestic destination with over 5 million visitors a year. Brazilians traveling internationally visit neighboring Buenos Aires as well as Miami, New York, and various places in Western Europe, and occasionally other places around the world. Over the past decade, the number of Brazilians travelling to the U.S. has increased significantly. According to the U.S. International Trade Commission, arrivals to the U.S. from Brazil increased by 292% between 2004 and 2011, and Brazilians' expenditures totaled $8.5 billion in 2011.

Because Brazil is such a huge country with great diversity and socioeconomic disparity, there is a tremendous range of sports and leisure pursuits that exist there. Creativity combines with historical influences to produce the sports, games, and activities that Brazilians engage in, and undoubtedly new options will continue to emerge. This chapter attempts to give an overview of the different sport and leisure activities currently pursued, while providing more details about Brazil's popular and emblematic activities.

REFERENCES

Buckley Jr., James. 2007. *Pelé.* New York: DK Publishing.

Christensen, Karen, Allen Guttmann, and Gertrud Pfister, eds. 2001. *The International Encyclopedia of Women and Sports*, pp. 176–178. New York: Macmillan Reference USA.

Green, Thomas A., and Joseph R. Svinth, eds. 2010. *Martial Arts of the World: An Encyclopedia of History and Innovation.* Santa Barbara, CA: ABC-CLIO.

Lever, Janet. 1983. *Soccer Madness: Brazil's Passion for the World's Most Popular Sport.* Chicago, IL: University of Chicago Press.

Levinson, David, and Karen Christensen, eds. 1996. *Encyclopedia of World Sport: From Ancient Times to the Present.* Santa Barbara, CA: ABC-CLIO.

Lisi, Clemente Angelo. 2007. *A History of the World Cup: 1930–2006.* Lanham, MD: The Scarecrow Press, Inc.

Röhrig Assunção, Matthias. 2005. *Capoeira: The History of an Afro-Brazilian Martial Art.* New York: Routledge.

Sagert, Kelly Boyer. 2009. *Encyclopedia of Extreme Sports.* Westport, CT: Greenwood Press.

Talmon-Chvaicer, Maya. 2008. *The Hidden History of Capoeira: A Collision of Cultures in the Brazilian Battle Dance.* Austin: University of Texas Press.

Toledo Camargo, Vera Regina. 2001. "Brazil." In Doris Corbett, John Cheffers, and Eileen Crowley Sullivan (eds.), *Unique Games and Sports around the World: A Reference Guide.* pp. 55–60 Westport, CT: Greenwood Press.

Van Mele, Veerle, and Roland Renson. 1992. *Traditional Games in South America.* Schorndorf, Germany: Hofmann GmbH and Co.

Walder, Marc. 2008. *Essential Brazilian Jiu Jitsu: An Illustrated Guide to the Fighting Art.* Champaign, IL: Human Kinetics.

Cinema

Carla Alves da Silva and Simone Cavalcante da Silva

EARLY FILMS

The history of Brazilian film starts as early as the invention of cinema itself. In 1898, the Italian immigrant Affonso Segretto filmed his arrival on the ship *Brésil* in the Guanabara Bay in Rio de Janeiro, not long after the Lumière Brothers had revealed their invention of the moving picture camera to the world. From that early beginning, Brazilian film production has not only been in synchronicity with important artistic developments of the world cinema industry, but has also established trends of its own. Brazilian cinema has combined international and national aesthetics and themes, and told stories that reflect both the nation's high and popular culture.

According to the late Brazilian film critic Paulo Emilio Gomes, the first cinematographer arrived in Brazil in 1896. Nevertheless, the first recordings for exhibition happened only in 1898. Historically, the invention of the cinematographer was an important sign that the Industrial Revolution was expanding into the entertainment industry. Brazil's economy was still agriculture based at the time, and slavery and the monarchy had ended only a few years before, in 1888 and 1889, respectively. The scarcity of electricity available in the country at the time delayed the booming of film exhibition, which happened only in the first decade of the 20th century. The first films produced in Brazil were short, followed by longer feature films. The first narratives were predominantly based on notorious crimes people knew from the local newspapers. A good example is Francisco Marzullo's *Os Estranguladores* (*The Strangler*, 1906). Drama, melodrama, and comedy also became popular genres, as well as films inspired by Brazilian theater, mainly the "Teatro de Revista," a burlesque type of production trendy at the time. It was common for actors and singers to stand behind the screen, say lines, and sing in synchronicity with the silent movies being projected. Those early years of the new century became known as the "Belle Époque" of Brazilian cinema. An increasing production of silent movies and the expansion of the systems of distribution and exhibition marked the period. In 1911, however, the arrival of Hollywood productions slowed down the booming of Brazilian cinema. Brazilian audiences were seduced by the novelties from abroad, including the first "talkies" (movies with recorded sound) in the 1920s. Brazil would only start producing talkies in the 1930s.

The 1930s was the decade when the first two great Brazilian directors became known: Mario Peixoto (1908–1998) and Humberto Mauro (1897–1983). Both Peixoto's film *Limite* (1931) and Mauro's *Ganga Bruta* (1933) were mostly disregarded by the audiences at the time of their releases, but were later considered masterpieces of Brazilian cinema. *Limite*, the only film ever completed by Mario Peixoto, was a silent experimental feature that told the story of two women and a man on a rowboat. The symbolic narrative of *Limite* has gained the admiration of great names of cinema like French film critic Georges Sadoul and American filmmaker Orson Welles, and it has influenced contemporary names of Brazilian cinema like Walter Salles Jr., who described the feature as "a film of transcendent poetry and boundless imagination" (Rohter 2010). Surprisingly, Mario Peixoto was only 18 years old when he created his masterpiece. Humberto Mauro's drama *Ganga Bruta* tells the story of a cheated groom who kills his bride on their wedding night and escapes to another town where he becomes part of a love triangle. Some critics described the film as "the worst film of all time" then, and its lack of popularity among viewers resulted in enormous financial losses to Cinédia, one of the pioneer film studios of Brazil in the early 20th century, founded by the former journalist Adhemar Gonzaga. The film was restored in 1952 for the first Brazilian Cinema Retrospective, and received raving reviews of intellectuals of the time, including future *Cinema Novo* director Glauber Rocha, who described the feature as "one of the 20 best films of all time" (Rocha 2003).

Brazil's artisanal film industry of the early 20th century was not able to follow the fast development of the cinema industry in Europe and in the United States. Soon an increasing number of films started to be imported from abroad. In the second decade of the 20th century, an average of only six films was produced domestically, mostly in the states of Rio de Janeiro and São Paulo. Films of that time started to include versions of classic Brazilian novels and historical facts, such as the symbolic participation of Brazil in World War I. The industry developed in quality and number after 1925, and more than 10 films started to be produced not only in Rio de Janeiro and São Paulo but also in the states of Minas Gerais, Pernambuco, and Rio Grande Sul.

Humor was also an important component of Brazilian feature film in the first decade of the 20th century. The studio Cinédia launched the genre known as *Chanchadas*, which mixed elements of slapstick, Brazilian carnival, burlesque theater, and Hollywood musicals. The *Chanchadas* attracted a great audience and reached its most successful era with the opening of the Atlântida studio in Rio de Janeiro in the 1940s. Some of the great names of the *Chanchadas* were the comedians Oscarito and Grande Othelo, and the singer and actress Carmen Miranda, who later moved to the United States to costar films in Hollywood. She became known as the "Brazilian Bombshell" and eventually turned into a pop culture icon.

The *Chanchadas* also received a great deal of criticism for the lack of depth in the stories and its failure to portray the country beyond Rio de Janeiro. In 1949, the Italian producer Franco Zampari and the industrialist Francisco Matarazzo Sobrinho opened the Vera Cruz studios, producing over 40 films by 1959. The 1940s was a decade of cultural effervescence in the city of São Paulo: with the opening of the Museu de Arte Moderna (Museum of Modern Art), the Museu de Arte de São Paulo (the São Paulo Art Museum), and the Teatro Brasileiro de Comédia (Brazilian Comedy Theater). A new interest for more professional Brazilian films followed the movement; therefore, Vera Cruz was created in an attempt to bring more depth and variety to national cinema, and to elevate Brazilian film production to international standards. The studio was inspired by the Cinecittá in Italy and intended to operate as a more sophisticated professional company, similar to the Hollywood studios. Brazilian films became more technical and expensive, and professionals and equipment were brought from abroad.

By the end of the 1950s, Atlântida Cinematográfica was still producing *Chanchadas* and films that followed on the Hollywood model of filmmaking, securing a good number of spectators in the movie theaters. Companhia Cinematográfica Vera Cruz, despite some important features like the epic *O Cangaceiro* by Lima Barreto, which won a special mention for the sound track and as "Best Adventure Film" at the Cannes Film Festival in 1953, closed in 1954. *O Cangaceiro* led Vera Cruz to international recognition. It was screened in 80 countries and was eventually sold to Columbia Pictures. Its international success, however, could not prevent the closing of Vera Cruz studios by the end of the 1950s. The high costs of its productions led the company to bankruptcy. In 1962, another Brazilian film won international recognition: *O Pagador de Promessas* (*The Given Word*), directed by a former Vera Cruz actor, Anselmo Duarte. The film won the *Palme D'Or* in Cannes and was nominated for the Academy Award of Best Foreign Film. The prestige brought to Brazilian cinema by *O Cangaceiro* and *O Pagador de Promessas* helped to keep the interest of international cinephiles in a very influential film movement that started in Brazil: the *Cinema Novo*, also known as the New Cinema.

CINEMA NOVO

Influenced by Italian Neorealism, the French New Wave, and Auteur style, *Cinema Novo* arose out of the dissatisfaction of a group of Brazilian filmmakers with the alienation instilled by the *chanchadas* and the strong presence of U.S. productions in Brazilian movie theaters. *Cinema Novo* filmmakers turned away from the Hollywood model presented by the *chanchadas* and focused on themes related to the Brazilian sociopolitical sphere, such as conflicts in rural areas, problems in the large cities, as well as film adaptations of Brazilian novels. Belonging to a preparatory phase of the movement, the films *Rio 40 Graus* (*Rio 100 Degrees*) (1955) and *Rio Zona Norte* (1957) by director Nelson Pereira dos Santos introduced Brazilian moviegoers to major concepts of the *Cinema Novo* aesthetics: the use of ordinary people as actors and low-budget productions. By the early 1960s, the movement had already established a new concept in filmmaking that placed film productions as a vehicle to express Brazilian's struggles and dissatisfaction within those turbulent times. In its first phase (1960–1964), the *Cinema Novo* movement presented a more experimental style of filmmaking, which combined slow rhythm camera movements and harsh images and sounds to major themes that revolved around the rural Northeastern part of Brazil, namely violence, religious alienation, and economic exploitation. Important films and directors from this phase are *Vidas Secas* (*Barren Lives*) (Brazil, 1963) by Nelson Pereira dos Santos and *Deus e o Diabo na Terra do Sol* (*Black God, White Devil*) (Brazil, 1964) by Glauber Rocha. In 1965, Rocha, one of the most distinguished filmmakers of the *Cinema Novo* movement, coined the term "aesthetics of hunger" to refer to the political filmmaking produced in Brazil in that period. In his manifesto, Rocha connects the idea of "hunger" to a source of critical power expressed through a state of philosophical weakness and impotence brought by the economic and political condition of society. Regardless of the social criticism of these films, their crude images and high intellectual content did not draw in enough spectators to the movie theaters, eventually building a gap between audience and Brazilian cinema.

The new repressive regime instituted in Brazil by the military coup d'etat of 1964 and the enforcement of censorship marked by the Fifth Institutional Act in 1968 meant that films could no longer be openly political. Influenced by the Tropicalist movement, *Cinema Novo* filmmakers started to rely on the use of allegories and metaphors in their films as a strategy to trick censorship and have theirs films produced and exhibited. Some of the most representative films and directors of the last two phases of the Cinema Novo movement are *Terra em Transe* (*Entranced Earth*, Brazil, 1967) by Glauber Rocha, *Como era Gostoso o Meu Francês* (*How Tasty was my Little Frenchman*, Brazil, 1971) by Nelson Pereira dos Santos, and *Macunaíma* (*Macunaima*, Brazil, 1969) by Joaquim Pedro de Andrade, a film based on a literary adaptation of the 1928 novel of the same name by Brazilian writer Mario de Andrade.

Parallel with the last phase of *Cinema Novo* (1968–1972), a young group of filmmakers from an area in the city of São Paulo known as "Boca Do Lixo" (Mouth of Garbage) started to produce low-budget, experimental, and extremely provocative films that questioned not only the established military regime but also the ideology behind the *Cinema Novo* movement. These movies, referred to as *marginal* or *udigrudi* (an appropriation or mocking of the English word "underground"), rejected the dominant Hollywood model of filmmaking and challenged the "aesthetics of hunger" in favor of an "aesthetics of garbage"—a critique of the exploitation of Third World countries by the dominant international capitalist powers. Some of the most representative films of the *marginal cinema* include the ultra-low-budget horror movies by the legendary José Mojica Morins, best known as "Coffin Joe," *O Bandido da Luz Vermelha* (*The Red Light Bandit*, 1968) by Rogério Sganzerla and *Matou a Família e Foi ao Cinema* (*Killed the Family and Went to the Movies, 1969*) by Julio Bressane.

CINEMA IN THE LATE 20TH CENTURY

Between the 1970s and 1980s, various decrees and government programs promoted the national cinema and provided financial support to the Brazilian film industry. Despite the military dictatorship, the 1970s and 1980s can be considered the golden age for the Brazilian film industry since as many as 100 films were produced a year with the support of EMBRAFILME—a government film agency created in 1969. One of the biggest box-office hits of Brazilian cinema was *Dona Flor e seus dois Maridos* (*Dona Flor and Her Two Husbands*, 1976) by Bruno Barreto that brought 12 million spectators to the movie theaters. The government's financial support to the national film industry through EMBRAFILME succeeded during the 1970s; however, the increase in film production costs and the massive presence of the U.S. film caused the domestic market to decline in the 1980s. The way found to bring back more public to watch national productions was to invest in a new genre: the *pornochanchada*.

Inspired by the Brazilian *chanchadas* of the 1950s and influenced by the Italian sex comedies of the 1970s, the *pornochanchadas* (or Brazilian sex comedies) became the major national film genre produced from the 1970s until the 1980s. The use of elements such as excess of sexuality to the comic plots alongside the low-budget costs created a trademark for the genre and brought Brazilians back to the movie theaters. Censorship tended to be more tolerant toward *porno-*

chanchadas since the genre focused less on political themes—when compared to *Cinema Novo* films—thus not threatening the military government. While these soft-core comedies of the 1970s and 1980s did not make a significant contribution to national cinema, they surely reflected on the changing of the social roles at a key period in Brazilian history and supported the domestic film market for most of the 1980s.

The decrease of censorship in the 1980s led Brazilian cinema to produce some of its most notable works such as *Memórias do Cárcere* (*Memoirs of Prison*, 1981) by Nelson Pereira dos Santos, *Pixote, a Lei do Mais Fraco* (*Pixote*, 1982) by Argentine-born director Hector Babenco, and *Eu sei que vou te amar* (*Love Me Forever or Never*) by Arnaldo Jabor, a hit at the Berlin festival in 1986 with nearly 5 million spectators in Brazil alone. Whereas the 1980s was a fertile period for Brazilian cinema, the economic challenges the country was facing with the arrival of redemocratization, in addition to the continuous strong presence of Hollywood films in the national market, brought on a heavy burden to national film production.

In the early 1990s, because of the neoliberal economic policies implemented by President Fernando Collor de Mello (1990–1992), the Brazilian film industry experienced a profound crisis. The closing of EMBRAFILME and the suspension of the laws that stimulated the audiovisual production led the national film industry to a scarcely 0.6 percent of all movies exhibited in Brazil. Around 1993, the demand for more audiovisual products prompted the creation of the Audiovisual Law that supported private investment in the film industry by allowing Brazil's businesses the right to receive income tax exemption.

After a five-year period of few productions and almost no audience for national productions, a direct result of the radical cut in governmental fiscal incentives and public funding, a Brazilian film finally hit the big charts. *Carlota Joaquina—Princesa do Brazil* (*Carlota Joaquina, Princess of Brazil,* Brazil, 1995) by Carla Camurati, a historical comedy about the arrival of the Portuguese court to Brazil in 1808, won the public acceptance and made Brazilians reconsider the idea of seeing themselves on the big screen. *Carlota Joaquina—Princesa do Brazil* was the first national production to break the 1 million admissions barrier in the 1990s, confirming a new era in Brazilian cinema known as *Retomada*. Along with Camurati's film, *O Quatrilho* (Brazil, 1995) by Fabio Barreto also made history for its nomination to the Academy Award for Best Foreign Film, a feat achieved only in 1962 with *O Pagador de Promessas*. Between 1998 and 2002, Brazilian film production regained speed and a significant increase in audience, leading to a certain amount of stability. In 1998, another Brazilian film found international fame; *Central do Brasil* (*Central Station*) directed by Walter Salles received two Academy Awards nominations (Best Foreign Film and Best Actress), followed by *Cidade de Deus* (*City of God*, Brazil, 2002) by Fernando Meirelles with three Academy Awards nominations including one for best director. The huge success of *City of God*, which experienced a major worldwide distribution, confirmed the presence of Brazilian cinema in a global context. Another important milestone in the early 21st century was the creation of the state agency ANCINE (Agência Nacional de Cinema), which works to regulate, inspect, and protect the audiovisual industry in the country.

Directors Fernando Meirelles and Kátia Lund, crew members, and actors from the internationally acclaimed film City of God. *The film was the first acting experience of many of the cast members, who were* favela *residents before the launching of the film.* (*AFP/Getty Images*)

POST-RETOMADA FILMS

Brazilian film production *post-retomada* (2003 onward) is marked by the intensification of influence of Globo Filmes (film producer and part of the Brazilian media conglomerate Rede Globo) into the production, distribution, and exhibition of films, which may lead to the establishment of a mainstream industrial cinema within the national cinema. On another note, the *post-retomada* productions are seen as more stylistically hybrid, based on real-life drama, and integrate documental aspects in their productions. Current Brazilian filmmaking is slowly becoming of commercial demand, as well as part of a transnational tendency, as Brazilian directors such as Fernando Meirelles from *City of God* has directed U.K./German film *The Constant Gardener* in 2005 and the U.S. production *Blindness* in 2008, an adaptation of a novel by acclaimed Portuguese writer José Saramago. In 2010, José Padilha's *Tropa de Elite 2 Elite Squad: The Enemy Within* became the largest blockbuster of Brazilian cinema, with 11 million viewers. Padilha's success got him an invitation to direct the 21st version of *Robocop* a couple of years later. Brazil's audiovisual industry, which includes cinema, grew an average of 9 percent per year in Brazil between 2008 and 2011. In 2011, the audiovisual sector increased to become 0.46 percent of Brazilian economy and the world's 10th-largest market for film in number of tickets sold, according to data from ANCINE (Alcântara 2014). The market share for Brazilian films for all the tickets sold at movie theaters in Brazil is 14.3 percent. Currently, state policies of incentive for coproductions have been fostering partnerships between Brazil and a number of countries, including Germany, Spain, France, Spain, Portugal, India, Venezuela, Argentina, and Chile. Those partnerships resulted in 82 coproductions between 2005 and 2013.

REFERENCES

Alcântara, Rosângela. 2014. "Mercardo Audiovisual Brasileiro: Políticas Públicas, Avanços e Perspectivas." Lecture. Brasil Cinemundi, Belo Horizonte.

ANCINE-OCA. 2014. "Dados do Mercado, Filmes e Bilheterias." Accessed May 21, 2015. http://oca.ancine.gov.br/filmes_bilheterias.htm.

Dennison, Stephanie, and Lisa Shaw. *Popular Cinema in Brazil*. 2004. Manchester: Manchester University Press.

Gomes, Paulo Emílio Salles. 1980. *Cinema, Trajetória No Subdesenvolvimento/[Paulo Emílio]; Prefácio De Zulmira Ribeiro Tavares*. Rio De Janeiro: Paz E Terra: EMBRAFILME.

Johnson, Randal, and Robert Stam. 1995. *Brazilian Cinema*. New York: Columbia University Press.

King, John. 1990. *Magical Reels: A History of Cinema in Latin America*. London: Verso.

Labaki, Amir. 1998. *The Films from Brazil: From the Given Word to Central Station*. São Paulo: Publifolha.

Martin, Michael T. 1997. *New Latin American Cinema*. Detroit: Wayne State University Press.

McClennen, Sophia A. 2011. "From the Aesthetics of Hunger to the Cosmetics of Hunger in Brazilian Cinema: Meirelles' *City of God*." *Symploke*, 19 (1): 95–106. *Project MUSE*. Web. May 7, 2015. https://muse.jhu.edu/.

Nagib, Lúcia. 2003. *The New Brazilian Cinema*. London: I.B. Tauris in Association with the Centre for Brazilian Studies, University of Oxford.

Nagib, Lúcia. 2007. *Brazil on Screen: Cinema Novo, New Cinema, Utopia*. London: I.B. Tauris.

Rêgo, Cacilda, and Carolina Rocha. 2011. *New Trends in Argentine and Brazilian Cinema*. Bristol, UK: Intellect.

Rocha, Glauber. 2003. "Humberto Mauro e a Situação Histórica." *Revisão Crítica do Cinema Brasileiro*. São Paulo: Cosac Naify.

Rohter, Larry. November 9, 2010. "Brazil's Best, Restored and Ready for a 21st-Century Audience." *The New York Times*. Web. May 21, 2015.

Shaw, Lisa, and Stephanie Dennison. 2007. *Brazilian National Cinema*. London: Routledge.

Stam, Robert. 1997. *Tropical Multiculturalism: A Comparative History of Race in Brazilian Cinema and Culture*. Durham: Duke University Press.

Xavier, Ismail. 1997. *Allegories of Underdevelopment: Aesthetics and Politics in Modern Brazilian Cinema*. Minneapolis: University of Minnesota.

Films Cited

Os Estranguladores. Dir. Francisco Marzullo. 1906.

Limite. Dir. Mário Peixoto. 1931.

Rough Diamond (*Ganga Bruta*). Dir. Humberto Mauro. 1933.

O Cangaceiro. Dir. Lima Barreto. Vera Cruz, 1953.

Rio 100 Degrees F (*Rio 40 Graus*). Dir. Nelson Pereira dos Santos. Columbia Pictures do Brasil, 1955.

Rio Zona Norte (*Rio Zona Norte*). Dir. Nelson Pereira dos Santos. Livio Bruni Produções Cinematográficas, 1957.

The Given Word (*O Pagador De Promessas*). Dir. Anselmo Duarte. 1962.

Barren Lives (*Vidas Secas*). Dir. Nelson Pereira dos Santos. 1963.

Black God, While Devil (*Deus e o Diabo na Terra do Sol*). Dir. Glauber Rocha. Copacabana Filmes, 1964.

Entranced Earth (*Terra em Transe*). Dir. Glauber Rocha. DiFilm, 1967.

The Red Light Bandit (*O bandido da luz vermelha*). Dir. Rogério Sganzerla. Sagres Filmes, 1968. VHS.

Kill the Family and Went to the Movies (*Matou a família e foi ao cinema*). Dir. Julio Bressame. M.A. Filmes, 1969.

Macunaima (*Macunaíma*). Dir. Joaquim Pedro de Andrade. DiFilm, 1969.

How Tasty Was My Little Frenchman (*Como era gostoso o meu francês*). Dir. Nelson Pereira dos Santos. New Yorker Films, 1971.

Dona Flor and Her Two Husbands (*Dona Flor e seus dois maridos*). Dir. Bruno Barreto. Embrafilme, 1976.

Pixote (*Pixote: a lei do mais fraco*). Dir. Hector Babenco. Embrafilme, 1981.

Memorias do Cárcere (*Memórias do Cárcere*). Dir. Nelson Pereira dos Santos. Embrafilme, 1984.

Love Me Forever or Never (*Eu sei que vou te amar*). Dir. Arnaldo Jabour. Embrafilme, 1986.

Carlota Joaquina: Princesa do Brasil (*Carlota Joaquina: Princesa do Brasil*). Dir. Carla Camurati. Europa Carat, 1995. VHS.

Quatrilho (*O Quatrilho*). Dir. Fabio Barreto. Eurocine, 1995.

Central Station (*Central do Brasil*). Dir. Walter Salles. Buena Vista International, 1998.

City of God (*Cidade de Deus*). Dir. Fernando Meirelles. Miramax, 2002.

The Constant Gardener. Dir. Fernando Meirelles. Focus Films, 2005.

Blindness. Dir. Fernando Meirelles. Miramax, 2008.

Elit Squad 2 (*Tropa de Elite 2*). Dir. José Padilha. Globo Filmes, 2010.

Robocop. Dir. José Padilha. Columbia Pictures, 2014.

Contemporary Issues

Rodrigo R. Coutinho

INTRODUCTION

Brazil is known for being a self-centered country. However, this has not always been the case. Politically, Brazil was, for example, a founding member of the League of Nations and a founding member of the United Nations. It was the only Latin American nation to participate in World War I, and actively participate in World War II with the Allied Forces. It was also the only Latin American country to send expeditionary forces to Europe. Economically, Brazil has been a major exporter of commodities—rubber and coffee, for example, for a long time.

However, Brazil has been, all in all, and during most part of the 20th century, more of a closed economy than an open one, especially during the years of Import Substitution Strategy, and the period of military dictatorship (1964–1985). Since the openness of its economy, and especially with the government of Fernando Collor de Mello (1990–1992), Brazil has become more involved and active in the international scene, playing a significant role in debates, organizations, and the world economy in general.

Part of Brazil's increased participation in the international arena comes naturally, since the country is home to most of the Amazon forest and all its biodiversity. This "simple fact" makes of Brazil a steakholder in most debates that involve the subject, and in some cases even a natural leader. It is also the case in clean energy debates, in large part due to the abundance of water in the country. Brazil is home to about 12 percent of the world's surface water resources. Moreover, most of its energy comes from renewable sources. The country built its first hydropower facility in the world in 1901. In other cases, it is due to historic reasons, as, for example, in discussions about racism. This is because Brazil, as the United States, has a very large population of Afro-Brazilians. The two countries have the largest Afro-descendant populations outside Africa. Finally yet importantly, Brazil's participation in the international arena occurs as it gains momentum as a large economy (currently world's seventh largest), and increasingly participates in international associations, such as Mercosur and Unasur, and BRICS (Brazil, Russia, India, China, and South Africa). The country is also active with the International Monetary Fund, the World Bank, and the World Trade Organization, whose current director-general is the Brazilian Roberto Azevêdo. It is also worth mentioning the United Nations, in which Brazil holds the responsibility for the opening speech of the General Assembly, which happens every September in New York, and participates in all of its specialized agencies, including peacekeeping efforts, such as that in Haiti (MINUSTAH—The United Nations Stabilization Mission in Haiti), which is led by Brazil.

This chapter on contemporary issues is going to discuss a selection of the many possible contemporary topics, some more broadly, others in more detail. They are abortion and family planning, affirmative action, the role of the Catholic Church, the spread of drugs in the country, its preparation to host big events, its position in Latin America, the configurations of nationalism, the manifestations of racism, and television as a civil society group. My concluding discussion is whether the country is finally developing now, or it remains forever the "country of the future."

ABORTION AND FAMILY PLANNING

Abortion is considered a crime in Brazil. It is allowed only when pregnancy puts the life of the mother at risk, when pregnancy is the result of rape, and/or in cases of fetal anencephaly, the latter which have been authorized only since 2012. The government provides the procedures at public hospitals, under the *Sistema Único de Saúde* (Unified Health System). The Brazilian government also tolerates abortion when the procedure happens in countries where it is legal.

Despite being illegal, abortion cases are frequent in Brazil. There are plenty of illegal, specialized clinics. People from higher social classes have access to good clinics, and procedures are usually safe, though expensive. However, those from lower social classes turn to methods not always safe or non-specialized clinics. Nevertheless, unwanted pregnancy is much more frequent in lower social classes, for whom access to contraceptive methods is limited, because of both cost and lack of information. For that reason, pro-choice groups in Brazil aim at legalizing abortion in the country, the main argument being that the matter is of public health interest, as illegal abortion procedures are among the main causes for maternal deaths, and due to the violation of women's rights.

Against those groups stands the Catholic Church, which, according to the last census in 2010, composes around 65 percent of the Brazilian population. Twenty-two percent of the remaining ones are Protestants, and 13 percent have either another religion or none. For the Catholic Church, life begins with fertilization: "human life must be respected and protected absolutely from the moment of conception" (excerpt from the Catechism of the Catholic Church, Fifth Commandment). During the military government (1964–1985), which somehow depended and counted on the Catholic Church to "guarantee" order, or support the regime within the country, the issue was never clearly discussed. Neither was contraception. However, with the emergence of feminism in the 1970s and an increasing will to discuss family planning, the Church, on the other hand, has assumed a rather interesting position: it is clearly against abortion, but does not adopt a clear stand concerning family planning and contraception. That position has become clearer because of the growing influence of other religious groups in Brazil (in 1940 95 percent of Brazilians were Roman Catholics, in 1991 83 percent, dropping to 65 percent in 2010) since the 1980s, which happened largely due to the economic situation of the country, marked by high inflation, debt, and default (1987). Hardships are a common motivation for changing religions.

Since the establishment of democracy, the rise in the debate on women's rights, as well as of STDs (sexually transmitted diseases), the issue of abortion has gained more momentum, and an increasing number of people are in favor of it. The case of Uruguay, for example, where abortion was legalized in 2012, gained great support from some society groups in Brazil, and a lot of attention in the media. At the same time, during political campaigns, candidates tend to avoid the subject, lest losing popular support. The last poll on the topic, made by Vox Populi in 2010, indicated that 82 percent of Brazilians believe the current legislation to be appropriate, while 14 percent believe abortion should be decriminalized. 4 percent did not have a position on the subject, and, when asked, tend to have a position against the procedure, except in specific situations, as the current legislation states.

The subject has traditionally been a taboo in the country, but this has been changing, especially as procedures performed poorly have resulted in growing deaths (estimates suggest around 10,000 per year) and hospitalizations of mothers. Research indicates around 1 million abortions are performed illegally in Brazil every year, which result in around 250,000 hospitalizations. In the meantime, publicity campaigns of contraceptive methods spread in the country, in all media, including in the popular soap operas. The Catholic Church says nothing about them. They have contributed significantly for reducing the fertility rates in the country, from 4.5 in the late 1970s to 1.9 in 2010. If Brazil will ever change its legislation, though, it is hard to tell.

AFFIRMATIVE ACTION

Of the population, 47.5 percent declared themselves to be white, 7.5 percent to be black, 43.5 percent to be multiracial, or *pardos*, 1.0 percent to be Asian, and 0.5 percent to be indigenous in the last Brazilian census (2010). It is important to distinguish the different conceptualizations of being black in Brazil and in the United States, where blacks are all those with an African descent, following the concept of the one-drop rule, which defines as "colored" persons of any African or Native American ancestry. In Brazil, on the other hand, people consider themselves to be black when they are dark skinned. People with "one-drop" consider themselves multiracial, *mulatos* or *pardos.*

The discussion of whether to implement affirmative action in Brazil is very complex because the idea of race, the quality of public schools, and opportunities for college loans there are different from those in the United States. According to *Encyclopedia Britannica* (art. "Race"), "Features of African Americans [in the US] vary from light skins, blue or gray eyes, and blond hair to dark skins, black eyes, and crinkly hair and include every range and combination of characteristics in between. American custom has long classified any person with known African ancestry as black, a social mandate often called the 'one-drop rule.'"

It was in the administration of President Lyndon Johnson (1963–1969) that affirmative action was implemented in the United States. The idea was to improve opportunities for African Americans, as eradicating discrimination was one of the targets of civil rights legislation. The landmark Civil Rights Act of 1964 and an executive order in 1965 were the federal government's first initiatives to institute affirmative action policies. Aptitude tests and similar discriminatory practices against African Americans were no longer allowed in federally funded businesses. The Office of Federal Contract Compliance and the Equal Employment Opportunity Commission (EEOC) were responsible to supervise affirmative action programs. Subsequently, affirmative action was broadened to cover women, Native Americans, Hispanics, and other minorities later benefited from affirmative action, a practice that colleges, universities, and state and federal agencies adopted as well.

In the United States, affirmative action led to policies of inclusion such as the use of quotas based on race, which by the late 1970s began being questioned as being "reverse discrimination." In 1978, the U.S. Supreme Court outlawed quota programs for being unfair to white students, but permitted higher education institutions to consider race when admitting students. Restrictions from the U.S. Supreme Court on race-based affirmative actions increased in 1989. Nowadays the U.S. Supreme Court considers affirmative action constitutional, but prohibits race to be the major factor in college admission decisions.

Affirmative action is still applied in the United States, even though to a lesser extent than it used to be. Despite being controversial, there have been good results. It has helped to improve opportunities for African Americans, which is the focus in this country. Other factors have helped it to be successful. First of all, due to the one drop rule and the subsequent segregation that occurred in the United States (as in South Africa), it is reasonably easy to distinguish who is white and who is black (one-drop rule) there. Therefore, quotas based on race may be applied more easily. Furthermore, in general, public elementary, middle, and high schools in the United States provide a better educational experience for students to enter colleges or universities than the Brazilian ones do. There are also all kinds of loans available for students to finance their college education (including federal ones), as well as housing, books, and additional costs, not to mention subsidized health insurance, in the United States. There are, therefore, ways for African American students to prepare for higher education, as well as to finance it at low interest rates and long payback periods, which allows for a de facto inclusion.

In Brazil, on the other hand, the situation is much more difficult. First, it is very difficult to distinguish who is white and who is black. In the last census (2010), for example, in which people declared spontaneously their own race, Brazilians identified themselves in almost 200 different races, ranging between white and black, indigenous and others. Second, basic (elementary, middle, and high schools) public education in Brazil, although available to most likely everyone, is generally of poor quality. Although higher education may be public and practically free of charge, financing is not widely available, and books and transportation are not provided for students. Financing is not widely available for private higher education institutions either. Public (and free of charge) health is available, but of a poor quality; private health is excellent, but insurance is expensive, and not subsidized for students.

Out of 100 students who enroll in elementary and middle schools, 90 finish it; out of these 90, 75 enroll in high school and only 57 graduate; and of these 57, only 14 enroll in higher education. At the end, only 7 manage to receive an undergraduate degree (*Jornal Valor Econômico*, August 27, 2014, p. A2). Surprisingly, 12 percent of the population had undergraduate degrees at the end of 2012, according to IBGE, the Brazilian Institute of Geography and Statistics (in 2012 PNAD, or National Household Sample Survey).

Affirmative actions have been discussed in Brazil for decades now, but they were first adopted (based on skin color) in the country in 2003 by the State University of Mato Grosso do Sul (UEMS). Candidates would have to submit a photograph to prove certain "African" characteristics so they could enroll under the quota system. Other institutions followed, and the debate gained momentum, generating a lot of controversy. Some claimed reverse racism, others contested it due to the earlier-mentioned difficulty in defining ethnicity with basis on color. A particular case generated a lot of controversy across the country in 2007. Two monozygotic twins, called Alex and Alan, applied for the University of Brasília, which adopted the system of quotas in 2004. Alex was accepted under the quota system and Alan was not.

Controversy apart, the debate over affirmative action matured. In 2005, the federal government established (Law 11,096/2005) the Prouni (Programa Universidade para Todos), a system of social quotas that provides part-time or full-time scholarships for students whose families make up to three minimum wages per person. The focus is students from public schools, and the selection criteria is based on both income and grades from the Exame Nacional de Ensino Médio, an exam taken by all students at the end of high school in Brazil. In 2012, the federal government approved the Lei de Cotas, its *Quotas Law* (Law 12,711/2012), which destines 50 percent of all positions in federal institutions of higher education for students who complete all of their high school studies in public schools. Half of those spots, however, are directed to people whose family income is lower than one and a half minimum wage per capita. Race is also taken into consideration, per state, depending on the racial stratification of each state (e.g., there are more Afro-descendants in northern states, and, allegedly, there are more whites in southern states). Since 2013, all federal institutions have adhered to the law.

As a result, from 1997 to 2013, according to the Ministério da Educação (Ministry of Education), the percentage of Afro-Brazilians enrolled in higher education or having a college degree went from 1.8 to 8.8; and the percentage of *pardos* (racially mixed) enrolled in or having a college degree went from 2.2 percent to 11 percent. The results are impressive, though not yet satisfactory. At the end of 2012, according to Instituto Nacional de Estudos e Pesquisas Educacionais Anísio Teixeira (Anisio Teixeira National Institute for Educational Studies and Research), 35 percent of people enrolled in Brazilian universities were either black or *pardos*, and 62 percent white. As mentioned earlier, they represent, respectively, 51 percent and 48 percent of the population. The gap is still there, but the numbers have improved.

Despite such significant changes, college dropout rates are still higher among less privileged social classes. Common reasons include transportation costs or cost of books and material, and students' need to work and provide for their families. Moreover, their grades at the end of high school are still lower than the ones of those obtained by people who come from private basic schools. Grade averages are similar among those who complete their studies, though.

Affirmative action was a late but pertinent discussion in Brazil, not to happen solely on similar basis to those adopted in the United States, but considering its peculiarities. Very controversial in its early days, it has matured and generated proven results. More has to be done to eliminate social and racial discrepancies in higher education. However, maybe even more attention is necessary at the basis of the educational system in Brazil—especially in public high schools—not only to close the remaining gaps, but also, and most importantly, to increase the quality (and reach) of education as a whole in Brazil, at all levels. After all, good schools are made of good teachers and professors, but also of good students.

CATHOLIC CHURCH

Brazil has the largest Catholic population in the world, with a congregation of around 120 million. It is followed by Mexico (93 million), the Philippines (80 million), the United States (75 million), and Italy (53 million), with less than half as many Catholics as Brazil. However, Brazil is ranked only 36th in representation of Catholics in its society, with 65 percent, behind countries such as Portugal (88 percent), Italy (88 percent), Poland (87 percent), Mexico (83 percent), Philippines (81 percent), Venezuela (79 percent), Argentina (77 percent), Colombia (75 percent), and Spain (72 percent).

As previously mentioned, Brazil lost many congregates in the late 1970s, 1980s, and 1990s, especially due to increasing economic difficulties faced by the country and efforts made by other churches to attract new members. A somewhat similar case that can be used to illustrate that is Haiti's, a country with 8.5 million people, 80 percent of them Catholic. After the earthquake of 2010, Haiti has received the visit of several different priests from different churches, a move that has been widely criticized internationally. Nevertheless, times of weaknesses and difficulties happen to be a good opportunity for converting congregates into different religions. The loss of congregates, which happened not only in Brazil but also in most of Latin America, was one of the reasons, if not the main reason, why Jorge Mario Bergoglio was elected Pope Francis in 2013, after finishing second to Joseph Aloisius Ratzinger (Pope Benedict XVI) in 2005. As expected, the pope's first trip abroad was to Rio de Janeiro, to attend the World Youth Day 2013, in July 2013.

Brazil, like most other Latin American countries, has had strong ties with the Catholic Church since the period of colonization. Such relationship might be best represented through the figure of the Spanish Jesuit Missionary José de Anchieta (1534–1597), one of the founders of the cities of São Paulo (1554) and Rio de Janeiro (1565). He was involved in the conversion of the indigenous population into the Catholic faith, one of the reasons why he is commonly known as the "Apostle of Brazil." Some scholars considered him the father of Brazilian literature. Pope Francis canonized Anchieta in 2014.

The power of the Catholic Church during the Portuguese colonization is still evident. The country has thousands of Catholic churches, several Catholic universities, and numerous cities and streets named after Catholic saints or important Catholic figures. Baptism and marriage under the blessing of the Catholic Church is not only a tradition but also almost a norm in the country. Most Brazilians know how to pray the "Hail Mary" (*Ave Maria*) and the "Lord's Prayer" (*Pater Noster*), and many wear crucifixes and have saint statues and images in their homes. The Catholic Church has always had great power in Brazil, and has always had close ties with and great influence on the government.

On the other hand, its relationship with the public has been changing substantially in a curious way recently. Traditionally, the Catholic Church has been very conservative in Brazil. Historically, it has always had great influence on people's behavior. Both aspects have started to change, especially due to the growing influence of other churches, and the rise in social movements requiring transformations. Brazilian society has changed and so has the Catholic Church in Brazil. Divorce and the struggle for same-sex marriage are examples of these changes. There are still sectors of the Catholic Church that are more conservative. With Pope Francis in the Vatican, however, it has become more open for debate, or at least it has come to accept different opinions and points of view more freely.

DRUGS

Drugs have been a major topic of discussion in Brazil for a long time. Three drugs will be discussed separately in this section: cannabis, cocaine, and opiates, as each has its own "logic." All numeric data reproduced here come from the 2014 World Drug Report from the United Nations.

Cannabis or Marijuana

Although production of cannabis resin, or "hashish," is concentrated in North Africa, the Middle East, and South-West Asia, cultivation and production of cannabis is widespread in the world, ranging from personal cultivation to large-scale operations. Consumers of cannabis ranged from 125 to 227 million people in 2012; that is, from 2.7 percent to 4.9 percent of the global population aged 15 to 64 years. Numbers of consumers in West and Central Africa, North America, and Western and Central Europe are considerably higher than the global average. Brazil is the 14th in per capita consumption of cannabis (8.8 percent), behind countries such as Israel (8.9 percent), Canada (12.2 percent), and the United States (14.8 percent). However, in Latin America and the Caribbean, the proportion of total treatment admission for cannabis use in general hospitals has increased from 23 percent to 40 percent between 2003 and 2012, although it is perceived to be the least harmful illicit drug.

Legalizing production, distribution, and consumption of marijuana, such as Uruguay and the state of Colorado in the United States have done, is a controversial issue. Less so, however, is whether to decriminalize its use or not. While the former raises questions such as the potential increase in consumption given potentially lower prices, drug tourism, or even cross-border leakage, the latter brings to the table a discussion involving incarceration (or not), and treatment of addiction, which as mentioned earlier, has been increasing.

What is surprising, though, in Brazil, which in general tends to be very conservative, is that while cocaine and opiates are "taboos," cannabis is more "open to discussion," though estimates suggest that 75 percent of the population is against legalizing its use, as Uruguay did. Since 2006 (Law 11.343/2006), possession of marijuana for personal use no longer leads to detention or imprisonment, but to socio-educational sentences. Cultivating it for one's own consumption, however, does, as it is considered trafficking. A sentence goes from 5 to 10 years in prison. This law was a step forward, in a country where a quarter of inmates were convicted for drug trafficking.

In addition to that, the discussion gains momentum in Brazil as the former president, Fernando Henrique Cardoso, has acted, since 2010, as the Chair of the Global Commission on Drug Policy. Other former presidents who are also commissioners are Mexico's Ernesto Zedillo, Colombia's César Gaviria, Portugal's Jorge Sampaio, and Chile's Ricardo Lagos. The Commission also includes international names such as Peruvian author and journalist Mario Vargas Llosa, former U.S. secretary of state George Shultz, former secretary general of the United Nations Kofi Annan, and former EU High Representative for the Common Foreign and Security Policy Javier Solana. The purpose of this commission was "to bring to the international level an informed, science-based discussion about humane and effective ways to reduce the harm caused by drugs to people and societies" (*Global Commission on Drug Policy*).

Cocaine

Cocaine is mainly produced by three countries: Peru, Colombia, and Bolivia, respectively. Colombia used to produce more than Peru until 2011, but cultivation decreased by 25 percent in Colombia from 2011 to 2012, due to both eradication of cultivation and aerial spraying by the Colombian government in 2012. Estimates that year suggested that there were between 14 and 21 million users of cocaine globally (0.4 percent of the world's population). Its consumption is concentrated in the Americas (especially the United States, the largest market), Western and Central Europe (the second largest market, especially France, Germany, and the United Kingdom), and Oceania. The prevalence of cocaine use is 1.8 percent in North America, 1.2 percent in South America, 1.5 percent in Oceania, and 1.0 percent in Western and Central Europe, down from 1.3 percent in 2010, a significant and quick drop most likely motivated by the financial crisis.

The main routes of cocaine are through Brazil (especially through the Amazon forest, which has borders with all three major producers—Bolivia, Peru, and Colombia—and is a region of difficult surveillance) and Central America to reach the United States, and through Africa to reach Europe, which, many argue, has helped to increase consumption of cocaine in both Brazil and Africa. Estimates suggest that there are roughly 3.4 million users of cocaine in South America, half of them in Brazil. However, the incidence of cocaine use among college students was estimated to be at 3 percent in 2009 (most recent data available).

In the case of Brazil (and that may be true for all of South America), what is most worrying is the spread of cocaine in the form of crack—the freebase form of cocaine that can be smoked. It is said to be the most addictive form of cocaine and is relatively cheap. Crack is estimated to be available in 95 percent of the 5,570 municipalities across all 27 states in Brazil. Some large cities, like São Paulo or Rio de Janeiro have areas known as *cracklands*, where people go to buy and smoke crack, and where some people "live" for days or even weeks.

Prostitution and crime are some of the consequences of addiction, the constant "need for more." Other corollaries include the increase in violence and the spread of diseases, particularly sexually transmitted ones. Last, inevitably there is a growth in the imprisoned population, as well as in the health costs for the country, where health is a public (and free of charge) service for all. In short, the drug traffic, particularly regarding the use of cocaine, and especially crack, has created a major public health crisis, whose dimension keeps increasing. Real solutions to the problem by the government—such as effectively closing the borders, or dealing openly with the issue through education and/or treatment—are yet to be seen.

Recently, *Verdades Secretas*, a prime-time soap opera that aired on Rede Globo, a very popular TV network in Brazil, portrayed a former top model who became addicted to drugs, in an attempt to bring attention to the subject. It was a noble initiative, in a society addicted to television and soap operas, in particular (a topic from this chapter), but it might not be enough yet to bring such a controversial debate to serious discussion in a conservative society like the Brazilian one. After all, this is not the first time that the Globo Network approaches the theme of drugs in its soaps. A similar story was told in *Passione* in 2010, in which one of the main characters was a former athlete from high society who turned into a crack addict. Although it did lead to national debates at the time, no concrete solutions came out of it.

Opiates

Opiates (e.g., heroin and morphine) are produced mainly in Afghanistan (with 80 percent of global opium production) and Myanmar. Although Colombia and Mexico cultivate illicit opium poppy, they represented less than 3 percent of global seizures of heroin in 2012. Its main markets are in Western and Central Europe, but the United States is becoming an increasing buyer as heroin substitutes synthetic opioids, due to price and increased availability.

The UN estimates that around 1 million people inject drugs in Latin America and the Caribbean (0.33 percent of the population), a figure that is slightly above the world average (0.27 percent); 46 percent of users are concentrated in Russia, China, and the United States. Opiates top the list of drugs that cause diseases and deaths (especially due to overdose). It is estimated that 13 percent of the people who inject drugs (e.g., heroin) are expected to live with HIV, and more than half of those who inject drugs are anticipated to live with Hepatitis C. Opiates do not seem to be, yet, of great concern (or alarm) for countries in Latin America and the Caribbean, including Brazil. Nevertheless, dependency and the spread of diseases, which may occur because of the use of injected opiates, must be addressed properly immediately.

Portugal decriminalized the use of drugs, including opiates such as heroin, in 2001, with the purpose of reducing the spread of diseases and treating dependency, but results are yet to be seen.

HOSTING BIG EVENTS

Prior to the 2014 World Cup in Brazil, many predicted—especially the international media—a chaos during the event. They anticipated the continuation of the political protests that had been occurring in the country, doubted Brazil's logistical efficiency (airports, ports, roads, etc.), and mistrusted its ability to finish all the constructions it was supposed to complete before the event. Brazil spent more than $11 billion to prepare for the World Cup, which took place in 12 stadiums in 12 different cities. It spent about $4 billion in stadiums, 80 percent of which was financed by the local Development Bank, a fact that was the target of considerable criticism, given the country's need for investment in key areas such as health, education, security, and infrastructure.

Brazil, five times world champion of soccer, and known as the "soccer country," wanted to host the event. However, it should have prepared itself accordingly for such an important task. On the eve of the World Cup, numerous projects initiated in order to facilitate the event—such as airports and urban mobility projects—had not been completed. Some of the stadiums themselves were only ready at the very last minute. Nevertheless, the World Cup ended and, it must be said, it was a huge success. The stadiums were crowded with record public, only surpassed by the U.S. World Cup in 1994, and the previous World Cup in Brazil in 1950. Airports had delays of the order of 7.6 percent, which was below the EU average of 8.4 percent in 2013. Almost 1 million foreign tourists visited the country, and 95 percent of them indicated interest in returning to Brazil someday. Brazilians' hospitality gained positive marks worldwide, as did, according to a research from Datafolha (one of the country's most respected polling institutes), the organization of the event, the level of comfort, feeling of security, and the transportation to the stadiums, as well as the quality of air transportation and of the tourist attractions.

The structure that received compliments from Brazil's visitors and its own nationals, it must be said, was created specifically for the World Cup. The security, specific urban mobility schemes, and interventions on traffic were not typical of customary sports events in Brazil. Moreover, it is also good to mention that numerous projects that should have been completed prior to the event were not, like airports (e.g., Rio de Janeiro's airport), and works of urban mobility such as subways. The completion of these projects is of utmost importance, and many of them are already in progress. These imperfections in no way diminish the success of the event, though.

Brazil has about $364 billion in international reserves. It is the seventh-largest economy in the world with a GDP of $2.2 trillion, and its investments in education and health outweigh dozens of times per year the investments the country made in the World Cup. Therefore, it is wrong to blame the World Cup for the country's pseudo lower investments in health and education. Investments in the World Cup were in addition to those, and apparently responsible for a country the size of Brazil, more than six times bigger, for example, than South Africa, which hosted the 2010 World Cup. What one should have questioned, therefore, was not whether to cancel the World Cup in order to further investments in health, education, and safety, as some did, but instead, the quality and fairness of all these investments. Brazil's investments in the World Cup were most likely overpriced, which is perhaps the case with its regular investments in health, education, and security.

One might say that Brazil has done well to host the 2014 World Cup. There have been many criticisms concerning the future use of the stadiums, though. However, among the 12, at least 7 will be (and already are being) used: Porto Alegre, Curitiba, Rio de Janeiro, São Paulo, Belo Horizonte, Salvador, and Recife. Brasília will also likely be used, since there are supporters of large clubs across the country living there. Fortaleza is also home to large soccer teams, though not in the first division of the Brazilian National Championship. Natal, Manaus, and Cuiabá may be less used,— but they were included for political reasons, as they are located in major centers of tourist attraction: the Pantanal (Cuiabá), the Amazon Forest (Manaus), and the Sand Dunes (Natal). Therefore, contrary to what happened in South Africa, where soccer is not a popular sport, the Brazilian stadiums will most likely be used quite often. Many of them needed to be reformed, so as to avoid accidents, such as the one that occurred in Salvador, in 2007, in which 7 people (of the more than 60,000 spectators) died after part of the stands fell.

The World Cup in Brazil was not only a success in itself, but it also helped to raise the self-esteem of Brazilians and lifted, indeed, Brazil to a new level in terms of international image. This was not only because of the visibility that the competition always gives— and it did give the country and Brazilians not only before, but mainly during the event— but also because the country will surely continue to reap benefits from it in terms of tourism, as it did during the event. Something similar happened during the World Youth Day in 2013, in Rio de Janeiro.

Bring on the 2016 Olympic Games to Rio de Janeiro!

LATIN AMERICA

Right in the first chapter of his book *Forgotten Continent: The Battle for Latin America's Soul*, Michael Reid brings into discussion the term "Latin America," and to whom it applies. He considers "Latin America" to include the Spanish-speaking countries from Mexico to Argentina, and Brazil, and occasionally considers Haiti to be part of it. Some people view Latin America as a somehow "cohesive" and integrated part of the globe, while many others consider Latin America as a "pot" in which every country looks pretty much the same. In business, for example, some say, "well, to do this business in Brazil, send Mr. X; he has worked in Chile already, and knows Latin America." The question is, "is that true"?

Reid (p. 22) argues that there is a clear cultural difference between the Andean countries and those in the southern cone. For him, the indigenous populations of Bolivia and Ecuador contrast to Colombia's *mestizo* population. Peru is in between them. The mark of Brazil, Venezuela, and Cuba, on the other hand, is their significant black and mulatto population. Reid also notes differences within Central America: "Relatively equitable landholding, European migration and a strong democratic tradition mark Costa Rica out from the others. Guatemala, with a large indigenous population, suffers from a racist and backward political elite and an over-mighty army, but is showing timid signs of democratic progress" (p. 26).

However, still according to Reid, other authors have argued that "the region constitutes a distinct civilization," emphasizing the similarities among the Latin American countries, as well as the differences between them and other countries in other areas of the world (p. 29). The region might not be as cohesive as some believe it to be. Some of the Andean countries (Peru, Bolivia, and Ecuador) have a huge indigenous population, and the indigenous culture is very present there. The same occurs with Mexico. Brazil, on the other hand, like the United States, has decimated its indigenous population and has a huge Afro-Brazilian population, which is also true for some Caribbean countries such as Cuba, Haiti, and Jamaica, which the author does not include in his definition of Latin America, but other authors would. In both Argentina and Uruguay, on the other hand, like in Costa Rica, according to the author, there are mostly white, European-like populations. In addition to that, while in countries such as Brazil, Argentina, and Chile people are very nationalistic, in others, such as Bolivia and Peru, social grievances are common, especially in the case of Bolivia.

Integration among Latin American countries does not seem to occur in practice. In Mercosur, for example, which includes Brazil, Paraguay, Argentina, and Uruguay, disagreements are a norm more than an exception, especially between Brazil and Argentina, the bigger countries. For example, when Brazil devalued its currency early in 1999, to the surprise of Argentina, many of the Argentinean industries, such as automotive and auto parts and appliances, were affected, and Argentina retaliated. The same disputes among the two are currently going on in regard to establishing free trade agreements (e.g., with the European Union).

Bolivia still argues with Chile over access to the Pacific Ocean, lost during the War of the Pacific (1879–1883). Chile often does not get along well neither with Argentina, nor with Peru. In Peru, for example, there is a great concern over Chilean companies acquiring counterparts in Peru, which is way beyond the endless discussion about the "paternity" over *ceviche* and *pisco*. (Ceviche is a national Peruvian dish, made out of fish. Pisco is a Peruvian brandy.) When Evo Morales took office in Bolivia, he nationalized many assets, including some of Petrobras, the Brazilian national oil company, to the disagreement of the Brazilian government. Brazil and Paraguay are in constant discussion over participation in the Itaipu dam, especially now, with ongoing debates over increasing its capacity. Another dispute is regarding the road from Brazil to the Pacific, which crosses Bolivia. This is a road that Bolivia has always wanted, and that Brazil is financing. Its construction has been blocked by some indigenous groups. The same is true with the dams in Peruvian Amazon, which Brazil wants to finance and build, and are also being blocked by some groups in Peru. Another important fact that shows the current lack of union in the region is that each Latin American country currently negotiates trade agreements individually with the United States and other nations, upsetting its peers. Chile and Colombia have done that. Even Uruguay, which should not do so because of the Mercosur, is apparently willing to do so. The trade agreement of the Americas (Free Trade Agreement of the Americas) has not been viable due to disagreements among its members.

It has also been difficult for Latin American countries to work together against drug trafficking and use. Colombia, Peru, and Bolivia have not been capable of reaching an agreement over drug combating. Colombia signed an agreement of cooperation over this issue with the United States, to the very discontentment of Brazil, for example. Colombia and Venezuela are in constant discussions over whether protecting or not protecting some guerrilla groups, such as the FARC (the Revolutionary Armed Forces of Colombia). Recent not well-explained assassinations have increased tension between the two countries in the past five years, which are only now being contended. It is also worth mentioning the constant disagreements involving the Dominican Republic and Haiti, who share the island of Hispaniola.

These examples are evidence that it makes sense to mistrust the idea of Latin America as a cohesive, integrated part of the globe. Are there commonalities among Latin American countries? Yes, there are, but there might be more differences than most people think.

NATIONALISM IN BRAZIL

Nationalism varies in the region of Latin America according to countries and sub-regions. For example, in countries like Peru, Ecuador, and Bolivia, where there are indigenous populations with substantial representation, there is some sense of nationalism, but it is not very stable. In these countries, there is a permanent risk of social grievances between the indigenous groups and the "whites," the ones with Hispanic background. That is the case in Peru, for example, where segmentation between "whites" and people with indigenous background is notable (e.g., Lima and the Asia District, a vacation destination during the summer), and that is the case more so in Bolivia, where there are even "movements" of independence. In Argentina and Chile, the sense of nationalism is more cohesive, despite the existence of some Amerindian representation in them.

Brazil, though, is in a very specific position in the region. Besides the presence of some indigenous population (less than 1 percent, according to the last census, 2010), the great majority of it, as was the case in the United States, was decimated during colonization. There was a massive inflow of Africans as slaves, as well as of Western Europeans. Besides the Portuguese colonizers, Italians, Germans, Spaniards, Japanese, Hungarians, Ukrainians, Syrians, Lebanese, and Jews also made Brazil their home. More recently, the country has also welcomed immigrants from other South American countries such as Paraguay, Bolivia, Uruguay, and Argentina, and from China, Korea, and Russia. The inflow of Africans and the Portuguese has happened since Brazil's early days; other groups have immigrated there especially since the end of the 19th century and in the early 20th century.

An important occurrence in Brazil was the incentive for miscegenation, under the idea of "whitening" the population. The last census (2010) can illustrate the effects of that racial project. 47.5 percent of the population declared themselves to be white, 7.5 percent to be black, 43.5 percent to be multiracial, or *pardos*, 1.0 percent to be Asian, and 0.5 percent to be indigenous. It is important to distinguish the notions of being black in Brazil and in the United State. In the latter, as pointed out earlier, blacks are all those of African descent, following the concept of the one-drop rule. In Brazil, on the other hand, historically people consider themselves to be black when they are clearly dark skinned. People with "one-drop" consider themselves multiracial, *mulatos* or *pardos*. If Brazil were to follow the American racial criteria, for example, one could argue that Nilo Pecanha was Brazil's first "Afro-Brazilian" president in 1909, 100 years before Barack Obama took office in the United States.

The way nationalism manifests itself in Brazil is very clear: through people declaring themselves as Brazilians, and not by usually clustering themselves ethnically, as Asian-Brazilians, Afro-Brazilians, Italo-Brazilians, and the like. Although academically and politically there are now groups that attempt to follow the U.S. hyphenated model to indicate ethnic identity, in general terms citizenship remains a more powerful identity mark. Putting it simply, they are all Brazilians. A significant number of children of immigrants, for example, do not speak the language of their families' origin, but Portuguese only, despite the fact that a good number of them still hold passports from their family's countries of origin. Nationalism is also facilitated by the fact that Portuguese is a cohesive official language (unique in the region), and it is celebrated through a variety of national symbols (anthem, flag, etc.), and kept alive through Brazilians' passion for national sports teams, such as its famous *Seleção Brasileira*, its national soccer team.

National symbols indeed unite the population and the country as a whole. Everybody knows how to sing the national anthem, can easily identify the national flag, and follows the games of the national soccer team (in addition to their own local ones). There is also a tendency for many people to follow the same news program on television and the same soap opera. Most books read are written in Portuguese and by Brazilian writers. Brazilian music is very popular, although foreign music is common on Brazilian radios. Most Brazilians, unfortunately, cannot understand a foreign language.

RACISM

There is no way to deny the fact that there is racism in Brazilian society, as there is in other societies. However, to understand racism against Afro-Brazilians in Brazil similarly by comparing it to the way it happens (and happened) in other places, such as the United States and South Africa, is a misunderstanding. First, it is important to remember that both in the United States and in South Africa there was a legal segmentation between whites and blacks under the law—the Jim Crow laws. In the United States, it reversed in the 1960s with the signing of the Civil Rights Act of 1964 by President Lyndon B. Johnson. In South Africa, it lasted until 1994, when the apartheid regime officially ended.

In Brazil, after the abolition, which happened in 1888, there has been no more official separation between blacks and whites. In Europe, just to mention a few examples, there is racism against Turks in Germany; against Arabs in France, Spain, and England; and against Africans in France, England, and Russia. Haitians are discriminated against in the Dominican Republic, and so are indigenous populations in countries across Latin America. Racism, thus, manifests itself against different races from different races. When it is discussed in Brazil, it is usually discussed in the way people usually discuss it in the United States and South Africa: racism between whites and blacks.

Although existing in Brazilian society, racial prejudice is not as intense as the social prejudice in the country, for instance. Some might argue that there is a correlation between being black and poor and being white and rich. True. History attests to that. After all, after abolition most former slaves did not have any money, land, means of production in general, or privileges from the government as many whites did. Except for a few privileged ones, they did not have access to education either (illiteracy in Brazil was over 65 percent in 1900; in 2012 it was under 9 percent). They had, therefore, to work their way up in society. Although it has been a slow process, and despite all adversity, Afro-Brazilians have been achieving social mobility, and their acceptance among the more privileged groups has been happening naturally.

If, on one hand, the incarcerated people in the United States are largely black (almost 40 percent of all inmates, while they represent around 13 percent of the U.S. population, thus having an incarceration rate six times higher than that of whites), in Brazil they might be a majority because they are poorer, and thus uneducated (less than 1 percent of inmates have higher education, while they represent more than 11 percent of the population—education and social class are highly correlated in Brazil), but not exactly because of their skin color (54 percent are either black or of mixed race, while they together represent 51 percent of the total population). There is even a belief and saying in Brazil that states influential or powerful people do not go to jail.

Some people (foreigners) argue (actually against the facts) that they see no black faces on Brazilian television, or that there are no black doctors or lawyers, but soccer players or samba singers. The truth is that there are fewer of them because they did not have the proper education and opportunities that the other races did, again for the historical reasons mentioned earlier. Once Afro-Brazilians manage to have access to proper education, they might be represented more proportionally. A similar argument is valid for the situation of women (which applies to all countries), who still are not proportionally represented in all segments of society in Brazil.

In short, prejudice, be it for reasons of skin color, ethnicity, religion, origin, and the like, is always objectionable and reprehensible, but it still happens. Brazil is no exception to that. However, one has to understand that the logic, or rationale, of racism in the United States or South Africa does not apply to Brazil. Racism in Brazil, in essence, is largely connected to the existing and historical prejudice among the social classes, which is not to say that it is not a significant part of the daily lives of Afro-Brazilians.

TELEVISION AS A CIVIL SOCIETY GROUP IN BRAZIL

For many people, it is surprising the extent and/or power that television has on Brazilian society, and the influence it has on the minds of the Brazilian people. It is widely known, for example, that in 1989, Fernando Collor de Mello was elected the youngest (at 40 years old) and first president of Brazil after the military dictatorship that lasted from 1964 to1984 with massive support from Rede Globo. The then still candidate Collor was barely known in most areas of the country then. With support from television, he was able to gain public notoriety and win the elections, after a famous debate hosted by the television network, and which was widely covered the next day on the most watched news program in the country, Jornal Nacional, an 8:00 p.m. news show. The influence of TV Globo in the 1989 Brazilian elections was the theme of the 1993 documentary *Beyond Citizen Kane*, by Simon Hartog, and of the biography of Roberto Marinho, the journalist and owner of TV Globo, written by Pedro Bial. It is also widely known, for instance, that Fátima Bernardes, one of the two anchors of the 8:00 p.m. news who recently left the program after 15 years, is considered to be one of the most influential celebrities in the country.

Brazil has more television sets per household (almost 100 percent) than refrigerators, microwave or landlines, maybe now losing only to cell phones. Despite its decreasing market share, TV Globo still holds around 35 percent of the audience in Brazil. The network's 9:00 p.m. soap opera is probably the second most viewed show on Brazilian television, after its own 8:00 p.m. news. There is certainly plenty of poor quality TV shows aired by Brazilian TV networks. What some people do not know, however, is the extent of television's power to educate people in Brazil. In the late 1980s and early 1990s, television advertisements were responsible for a drastic decline in the rates of breast cancer due to a very popular propaganda in which celebrities performed live self-exams. HIV/AIDS numbers have also dropped dramatically in Brazil not only due to the death of Cazuza in 1990, one of the country's most popular singers who fought the disease publicly and openly, but also due to massive television advertisements that promoted the use of condoms and educated the population on how to prevent contagion. Smoking was also reduced dramatically not only due to the efforts that placed shocking images on cigarette packs, and the country's very strict regulation toward it, but also due to frequent and recurring television advertisements. Now Brazilians smoke less than Americans as a percentage of the population (14 percent of 200 million Brazilians, or 28 million, and 16 percent of 310 million Americans, or 50 million, as of 2012), with less consumption per capita (1.2 kg/person in Brazil and 1.3 kg/person in the United States in 2012).

In addition, soap operas, especially the 9:00 p.m. ones, are a "forum" of debate for several issues. Brazilians were able to discuss and better understand issues such as racism, prejudice against minorities (persons with disabilities, homosexuals, or LGBTs), family planning violence, diseases such as cancer and HIV/AIDS, alcoholism, violence against women, child labor, drug addiction (cocaine, crack, marijuana), and trafficking. Many in Brazil argue that television has been responsible for the drastic decline in fertility rates observed there in the past few decades—from 4.4 in 1980, when 36 percent of households received a soap opera signal, to 1.9 in 2010, when 90 percent of the households were receiving a soap opera signal, according to National Geographic. Television shows have also been vehicles for obtaining further public support for issues the country has to improve, such as combating corruption, improving its infrastructure (e.g., providing better infrastructure and services for people on wheelchairs), and advancing its education and health systems.

Brazilian soap operas usually have 180 episodes and are broadcasted from Monday through Saturday. In every one of them there is at least one character whose story portrays a specific issue that is relevant for Brazilian society: a beautiful woman who, after an accident, ended up in a wheelchair without access to public spaces and/or services in Rio de Janeiro, someone who is addicted to alcohol or drugs, or a businessman who corrupts government officials. These characters and issues are discussed daily among Brazilians (and many remain on their minds for years or decades, such as the characters Odete Roitman or Heleninha Roitman, from *Vale Tudo*, which aired from 1988 to 1989). This is particularly true among the less privileged social classes. Brazilian television networks also have several educational programs per se. There is a very famous one called Globo Rural, which educates farmers on how to produce crops and/or raise cattle, swine, and chicken. Another one, Globo Educação (an Educational Program from Globo TV network), educates children (distant learning) and raises issues about education in Brazil.

Brazilian television shows are usually not as opinionated as they are in some countries (e.g., Fox News in the United States). Many of them are usually considered more informative. After the military dictatorship, and until recently, it was not common for TV networks to take sides on elections (except for the earlier-mentioned case) or support political parties. Brazilian TV programming is free of censorship, which is something very important for Brazilians, especially because, during the military period, censorship was an everyday reality, and is actually very critical of the government and of the problems of the country. Moreover, because of its popularity and the power that some TV networks have, this means of communication deeply penetrates Brazilian society, in a way that books and/or the printed press currently do not.

BRAZIL—DEVELOPING OR FOREVER THE "COUNTRY OF THE FUTURE"?

According to Paul Krugman, Maurice Obstfeld, and Marc Melitz, in *International Economics: Theory and Policy*, although several characteristics distinguish the developed from the developing countries, five are particularly important for macroeconomic analysis. This does not mean, however, that these characteristics are all necessarily present and are present only in these economies. They are as follows:

1. Financial markets in developing countries are limited in size and are subjected to rigid official control. Usually, the governments of these countries leave the interest rate a little below the level that would equate supply and demand for loans, rationing them.

2. The direct involvement of government in the economy goes beyond financial markets. Governments usually hold significant shares of companies involved in the economy, and government spending is significantly high in relation to the GDP.

3. The government finances a large part of its spending by printing money, which results in higher average inflation rates and, sometimes, in the indexation of wages, contracts and loans, and other prices in the general price index.

4. Exchange rates are established by the government rather than determined in the foreign exchange market. International loans are restricted and the government may allow residents to buy foreign currency only for specific purposes.

5. Natural resources or agricultural commodities account for a large share of exports of many developing countries.

Given these characteristics, let us consider the following, in the case of Brazil:

In regard to financial markets, although the Brazilian stock exchange (BM&FBovespa) is among the largest in the world, and debt negotiation focus changes from short term to long term, there is a great control of the state in the credit market and the level of savings is still very low (around 14 percent of the GDP).

Although credits are growing, they are still very expensive, keeping investments low (around 18 percent of the GDP), thus slowing growth. Besides, the credit available for investors at lower rates is through the local development bank, which favors specific sectors chosen by the government.

Government's involvement in the economy is still quite large. Many industrial sectors are still heavily regulated; a good number of companies continue to be state owned, either partially or totally. In several sectors, investments remain a state monopoly or are restricted to foreign capital, which reduces investments in the economy. Moreover, government spending is not only high in relation to the GDP (19 percent of GDP in 2013) but it is growing faster than the economy.

Although under control since the mid-1990s, inflation in Brazil is still higher than that of most developed countries; it has risen considerably recently. However, several contracts, including wages, are indexed to inflation indices (in the case of the minimum wage, with a guaranteed increase over inflation). Prices such as that of energy, fuel, and transportation are kept lower than they should in order to control inflation.

Exchange rates are still under close supervision by the Brazilian government, under a floating exchange rate regime. The Central Bank constantly intervenes by purchasing or selling dollars in the market in order to control the direction or the speed of the exchange rate moves. International loans are not restricted but are more accessible to larger corporations, the ones with shares traded on foreign stock exchanges or with presence abroad; the purchase of foreign currency by residents is not restricted but is somehow controlled through taxes.

Finally, in regard to trade, Brazilian exports are still concentrated primarily in commodities—iron ore, crude steel, soybeans, sugar, alcohol, orange, cattle, and so on. Brazil also exports manufactured goods, such as cars, airplanes, motorcycles, and electronics, but not enough yet to counterbalance its imports; thus, the country is having increasing deficits in its trade balance. It should, therefore, focus more on stimulating the export of manufactured goods, with higher value added, as most developed countries do.

Another characteristic of developing countries, and which is present in Brazil, is the heavy dependence on foreign capital to finance investments and development—which is reflected in the high debt/GDP ratio. That tends to decrease the country's credibility (increasing its country risk premium), increase even more its interest rates and interest payments, and attract more speculative investments rather than foreign direct investments. That trend is mostly due to the fiscal debt and the low levels of savings in the economy.

Therefore, it is fair to say that Brazil is still "condemned" to be considered a developing country for years to come, for all the characteristics mentioned that continue present in the economy. However, there are signs of improvement—a more mature financial system and financial markets, a lesser involvement in the economy, through privatizations, concessions, and public–private partnerships, a more independent Central Bank, and a trend to export more value-added goods, among others. Yet, for the time being, one might say that it is still "the country of the future," as it is commonly known to be, due to the title of Stefan Zweig's 1941 book *Brazil: Land of the Future*.

REFERENCES

Almeida, Paulo Roberto de. 2009. "Lula's Foreign Policy: Regional and Global Strategies." In Joseph L. Love and Werner Baer (eds.), *Brazil under Lula: Economy, Politics, and Society under the Worker-President*, pp. 167–183. New York: Palgrave Macmillan.

Baer, Werner. 2008. *The Brazilian Economy: Growth and Development*, 6th ed. London: Lynne Rienner.

Bastide, Roger. 1978. *The African Religions of Brazil.* Baltimore, MD: The John Hopkins University Press.

Beyond Citizen Kane. Dir. Simon Hartog. Channel 4, 1993.

Bial, Pedro. 2004. *Roberto Marinho.* Rio de Janeiro: Zahar. —— The author cites these works in his text but did not include them in the bibliography.

Busch, Alexander. 2010. *Brasil, País do Presente: o Poder Econômico do "Gigante Verde." (Brazil, A Country of the Present: The Economic Power of the "Green Giant").* São Paulo: Cultrix.

Cardoso, Fernando H. 2010. *Xadrez Internacional e Social Democracia.* São Paulo: Paz e Terra.

Edwards, Sebastian. 2010. *Left behind. Latin America and the False Promise of Populism.* Chicago, IL: University of Chicago Press.

Encyclopedia Britannica. 2015. Race. www.britannica.com.

Freyre, Gilberto. 1987. *The Masters and the Slaves: A Study in the Development of Brazilian Civilization.* Trans. Samuel Putnam. Oakland: University of California Press.

The Global Commission on Drug Policy. 2015. http://www.globalcommissionondrugs.org/about/.

Halperín Donghi, Tulio. 1993. *The Contemporary History of Latin America.* Ed. and Trans. J. C. Chasteen. Durham, NC, and London: Duke University Press.

Hayek, F. A. 2007. *The Road to Serfdom.* Bruce Caldwell (ed.). Chicago, IL: The University of Chicago Press.

Holanda, Sérgio Buarque de. 2012. *Roots of Brazil.* Trans. G. Harvey Summ. Notre Dame, IN: University of Notre Dame Press.

Instituto Brasileiro de Geografia e Estatística (IBGE). 2013. http://www.ibge.gov.br.

Jornal Valor Econômico, August 27, 2014, p. A2.

Krugman, Paul, Maurice Obstfeld, Marc Melitz. 2014. *International Economics. Theory and Practice.* Upper Saddle River, NJ: Prentice Hall.

Magnoli, D. 2009. *Uma gota de sangue: história do pensamento racial.* São Paulo: Contexto.

Montaner, Carlos Alberto. 2003. *Twisted Roots: Latin America's Living Past.* New York: Algora Publishing.

National Geographic. 2014. How Brazil's Fertility Fell. http://ngm.nationalgeographic.com/2011/09/girl-power/fertility-graphic.

Passione. Writ. Sílvio de Abreu. Rede Globo de Televisão. May 5, 2010, to January 14, 2011. Television.

Reid, Michael. 2014a. *Brazil: The Troubled Rise of a Global Power.* New Haven, CT: Yale University Press.

Reid, Michael. 2014b. *Forgotten Continent: The Battle for Latin America's Soul.* New Haven, CT: Yale University Press.

Ribeiro, Darcy. 2000. *The Brazilian People: The Formation and Meaning of Brazil.* Gainesville: University of Florida/Center for Latin American Studies.

Rohter, Larry. 2010. *Brazil on the Rise.* New York: Palgrave Macmillan.

Sachs, Jeffrey. 2005: *The End of Poverty: Economic Possibilities for Our Time.* New York: Penguin Books.

Skidmore, Thomas. 2010. *Five Centuries of Change.* New York: Oxford University Press.

Sweig, Julia E. November/December 2010. "A New Global Player: Brazil's Far-Flung Agenda." *Foreign Affairs.* http://www.foreignaffairs.com/articles/66868/julia-e-sweig/a-new-global-player?page=show.

Verdades Secretas. Writ. Walcyr Carrasco. Rede Globo de Televisão. June 8, 2015. Television.

World Drug Report 2014, from the United Nations. http://www.unodc.org/wdr2014/en/.

Zweig, Stefan. 1941. *Brazil: Land of the Future.* New York: The Viking Press.

Glossary

Abertura Política "Political opening." The gradual process that led to the end of the military dictatorship in Brazil. It started in 1974 and ended in 1988, when the new Constitution was promulgated.

Açaí Fruit native to the North region of Brazil. It is considered a "super fruit" because it is rich in antioxidants, vitamins, and minerals, and supposedly promotes weight loss, prevents cancer, and acts as an antiaging agent.

Acarajé Fritter similar to a falafel but larger and made with black-eyed beans, usually served stuffed with spicy dried shrimp and other concoctions typical of the Northeast region. It is common in the state of Bahia and associated with Afro-Brazilian religion *Candomblé*.

Afoxé Name given to an Afro-Brazilian musical genre and to Afro-Carnival groups with roots in Candomblé, especially in the city of Salvador.

Agregados Immediate and extended family members, domestic partners, in-laws, and adopted children who share the same household.

AI-5 ("Institutional Act number 5") Decree passed in December 1968 that severely restricted Brazilians' civil liberties. Under AI-5, the media were censored; labor unions were banned; many university faculty were dismissed; and political opponents were arrested and tortured, and "disappeared."

Antropofagia "Anthropophagy" was an artistic movement of Brazilian Modernism. It proposed a return to Brazilian indigenous and African roots, while incorporating European trends of the time and rejecting the country's colonialist past.

Arrocha Romantic musical genre, associated with dancing, that originated in the state of Bahia in the 2000s. It has received considerable media attention.

Arroz com feijão   "Rice and beans," the base of the everyday Brazilian meal: white rice and stewed brown or black beans (preferred in Rio state). See recipe on p. 292.

Atabaque   Tall wooden African hand drums associated with Candomblé and capoeira in Brazil. They can be played with hands or drumsticks.

Axé Music   Music genre from Salvador, Northeast of Brazil, associated with the city's carnival. The term (Axé) has African roots and means "peace be with you."

Baião   Musical genre original from the Northeast of Brazil that became widely known through the music of accordion player and composer Luiz Gonzaga.

Batalha dos Guararapes   Two combats between the Portuguese and the Dutch forces that led to the end of the Dutch occupation in Pernambuco. The first was in April 1648 and the second in February 1649. It is called "Batalha dos Guararapes" because the battles took place in the Guararapes Hills in Recife.

Batata-baroa   See *mandioquinha*.

Batuques   Generic term used since the 19th century to designate all sonorous percussive events in Brazil, whether they included dance or had a religious or profane nature. In general, the term is used in descriptions with a pejorative connotation.

Beiju   See *tapioca*.

Berimbau   The berimbau is a chordophone musical bow played with a drumstick, originary from the Bantu cultures in Central Africa. It is capoeira's main instrument.

Biquíni fio dental   "Dental Floss Bikini" is a kind of string bikini.

Bloco Afro   *Blocos Afros* are carnival groups that celebrate the African heritage in the city of Salvador. Ilê Aiyê and Olodum are two of the best known.

Bolsa Família   The Brazilian conditional cash transfer program.

Bossa Nova   Music genre from Brazil derived from samba and jazz. It was popularized in Brazil in the 1950s. The term loosely means "new trend."

Boteco   Bars in Brazil where beer, spirits, and appetizers are served.

Bradesco   One of Brazil's largest private banks.

Brazilian Jiu-Jitsu   A martial art practiced in Brazil and worldwide that blends distinctly Brazilian elements with the traditional Japanese art.

Brigadeiro   A mixture of sweetened condensed milk, cocoa powder, and butter cooked together and then formed into balls and rolled in chocolate sprinkles. Some link the name to the Brigadier Eduardo Gomes.

Bunda   Portuguese word of African origin (Kimbundu) for "butt."

Caatinga   It is a type of desert vegetation found in the Northeast of Brazil. It is rich in biological diversity. The term means "white forest."

Cachaça   Brazilian national alcoholic drink used in many national cocktails, such as *caipirinha*. It is distilled from sugarcane and usually 80–100 proof (40%–50% alcohol).

Caipirinha   The most traditional and well-known Brazilian cocktail, made with fresh lime muddled with extra-fine white sugarcane sugar, ice, and *cachaça*, although versions using other Brazilian fruits, such as passion fruit, are also appreciated.

Caju   Portuguese word for cashew. Both the cashew nut and the fruit/apple, used mostly in juices, cocktails and compotes, are popular in Brazil.

Calundu   Religious expression of African origins not very much discussed among external observers. It existed in Bahia in the 18th and 19th centuries and is considered one of the precursors of Candomblé.

Candomblé   Afro-Brazilian religion of oral tradition. The religion incorporates traditional Yoruba, Fon, and Bantu beliefs, as well as some elements of Catholicism.

Cangaceiros   Groups of armed men and women that traveled around the Northeast Backlands in the beginning of the 20th century. Although they were outlaws, the "Cangaço," as their movement was called, was a complex and revolutionary form of social banditry. The cangaceiros confronted the ruling power to find answers for the social problems of the region. Virgulino Ferreira da Silva, a.k.a. Lampião, was the most famous of them.

CAPES   Higher Education Personnel Improvement Coordination. It is a government agency, affiliated to the Ministry of Education, responsible for the evaluation, expansion, and improvement of higher education in Brazil.

Capitanias Hereditárias   Hereditary captaincies. Large lots donated, in the beginning of colonization, to Portuguese nobles, for them to explore and protect the land from invaders. They were passed on from father to son.

Capoeira   Popular Afro-Brazilian artistic expression that mixes martial arts, dance, and game elements. It is accompanied by music.

Carne de sol   Beef lightly cured with salt and dried outdoors, in the sun—hence the name, which literally means "meat of the sun."

Carne seca   Salt-cured air dried beef, typical of the Northeast Region, it is one of the main ingredients for the national dish *feijoada*.

Caruru   Porridge made with okra and seasoned with ground cashew and peanuts, ground salt shrimp, ginger, *dendê*, coconut milk, and malagueta hot pepper.

Catupiry   See *requeijão*.

Cavaquinho   Small four-string Portuguese musical instrument associated with samba.

CELPE-Bras   Ministry of Education's official Proficiency Certificate in Portuguese as a Foreign Language.

Cerrado   The *Cerrado* is the Brazilian savanna. It encompasses about 20 percent or more of Brazil.

Chanchadas   Burlesque popular film style that predominated in Brazil between 1930 and 1960. They were musical comedies and sometimes emulated detective or science fiction films.

Charque   Heavily salted cured beef, typical of the South Region of Brazil. See also *carne seca* and *carne de sol*.

Choro/chorinho   *Choro* means "cry" and "chorinho" means "little cry." It is an urban popular Brazilian music genre that is typically instrumental.

Churrascaria   A restaurant that sells *churrasco*, the Brazilian barbecue. See sidebar "*Churrasco* and *Rodízio*" on page 294.

Churrasco   Barbecued food, usually meat, such as *picanha* and other beef cuts, pork, chicken, and sausage, usually served with *farofa* and salad. See sidebar "*Churrasco and Rodízio*" on page 294.

Ciências Sem Fronteiras   A governmental exchange program whose aim is to provide an international studying experience for Brazilian university students majoring in the sciences. The program is called "Brazil Scientific Mobility Program" in the United States.

Cinema Novo   "New Cinema." Political and intellectual film movement in the 1960s and 1970s; marked by low-cost productions, social commitment, use of nonprofessional actors, and critique of neocolonialism.

CNPQ   National Council for Technological and Scientific Development; government agency, affiliated to the Ministry of Education, responsible for the advancement of scientific and technological research.

Cobogó   A traditional element used all over Brazil until today, is a prefabricated hollow element made of cement or pottery. It gets its name because of its inventors' surnames: Coimbra, Boeckmann, and Góis.

Comida a/por quilo   Self-service restaurants that sell "food by the kilogram." The price of the food varies according to its weight on the plate.

Concretismo   Avant-garde artistic movement in the 1950s and 1960s in literature and fine arts. Art is expressed in an abstract and geometric form.

Copacabana   The most popular beach in Rio de Janeiro.

Corcovado   The Corcovado is a mountain in the Tijuca National Park, also known for hosting the statue of Christ the Redeemer.

Cortiço   Low-income multifamily units with bathrooms and service area for community use.

CPF   "Cadastro de Pessoas Físicas." The number attributed by the Brazilian Federal Revenue to both Brazilians and resident aliens who pay taxes. The CPF number is used to calculate the income tax that is due. It is Brazil's Social Security Number.

CPLP   Founded on June 17, 1996, the *Comunidade de Países de Língua Oficial Portuguesa* (Community of Countries whose Official Language is Portuguese) is an association made of the nine countries that make up the Lusophone (Portuguese-speaking) world whose aim is to deepen friendship and cooperation among its members.

Cristãos Novos   Literally, "New Christians," it referred to Jews or Muslims who converted to Catholicism, or their descendants, particularly from the 15th to the 18th century.

Cristo Redentor   The statue of Christ the Redeemer, located on the Corcovado Mountain in Rio de Janeiro.

Cupuaçu   A "super fruit" in the same family of cocoa tree that grows in the Amazon rainforest. The silky, white pulp, fragrant and tart, is rich in antioxidants and other nutrients and can be used raw in juices or cooked to prepare many desserts. The seeds can be processed as cocoa to make a chocolate-like product called *cupulate*.

Cuscuz   In the Northeast region, a steamed corn grits cake usually eaten for breakfast, with *manteiga de garrafa* and *queijo (de) coalho*; in the Southeast region, a concoction made with a tomato-based broth, usually flavored with chicken or sardines, that is either steamed or cooked as a porridge and served molded, decorated with slices of tomato, hard boiled eggs, peas, and so on.

Dendê   Orange-colored palm oil extracted mostly from African oil palm *Elaeis guineensis*, used in many dishes of the Northeast region, especially the African-inspired ones.

Diretas Já   Popular movement between 1983 and 1984 toward the end of the military rule in Brazil. Highly attended street protests and demonstrations urged direct popular elections.

Ditadura militar   The military dictatorship refers to the period between 1964 and 1985 when every ruler of Brazil was a four-star general.

Embraer   Major producer of midsize airplanes from Brazil. Its main competitor is Bombardier, from Canada.

Empregada doméstica   The maid. In Brazil most middle- and upper-class families have a maid, who might live in the household or not. Maids play complex roles, from doing the cleaning and raising children to managing the house.

ENEM   National Secondary Education Examination. It is used to evaluate the performance of high school students and as part of the college admission process.

Espiritismo   "Spiritism." Brazilian version of the doctrine proposed by the work of French educator Allan Kardec (Hippolyte Léon Denizard Rivail), especially in his *Book of Spirits*. Believers consider it a science, philosophy, and religion. Chico Xavier and Divaldo Pereira Franco are two of the major names in Brazil.

Estado Novo   "New State." In November 1937, President Getúlio Vargas staged his own coup, dissolving the legislature and announcing a new constitution that concentrated power in his hands. The end of World War II put pressure on Vargas's dictatorship, and in 1945 the army forced him to resign.

Farinha de mandioca   Portuguese term for manioc flour, a coarse meal made from grated and toasted manioc/cassava.

Farofa   *Farinha de mandioca* toasted with seasonings and several other ingredients (e.g., onion, garlic, bacon, eggs, banana, parsley) served as accompaniment to many Brazilian dishes, including *churrasco* and *feijoada*.

Farofeiros   A derogatory name for lower classes referring to the messy custom of eating food accompanied by *farofa*.

Favela   Slum areas with no or little sanitation, public services, or urbanization that are typical of urban areas. These areas are usually marked by a strong social connection among the residents and by an intense cultural life.

Feijoada   Black beans—or brown, in Bahia—stewed with salt-cured beef and fresh, salt and smoked pork cuts and sausages, traditionally served with white rice, sautéed collard greens, *farofa* or *farinha de mandioca*, a tomato salsa called *vinagrete*, and orange slices.

FGTS Fundo de Garantia do Tempo de Serviço (Severance Indemnity Fund for Employees). Special fund for workers established in 1967 to shelter workers in case they lost their job without just cause.

Forró Musical genre and dance typical of the Brazilian Northeast that has become widely known because of the media appeal of the several nationally acclaimed bands that play the genre.

FUNAI National Amerindian Foundation. Official government organization responsible for the protection and promotion of the rights of the indigenous peoples in Brazil.

Ginga Capoeira's basic movement. Expression also used in soccer and everyday situations to express one's ability and creativity. Consider a mark of Brazilian identity.

Graviola Common fruit in certain areas of Latin America. It is called "Soursop" in English.

Guaraná Amazonian fruit (*Paullinia cupana*) of the maple family with a high caffeine content; it is used to make a soft drink with the same name, widely consumed all over Brazil.

Guerra de Canudos The tragic "Canudos War" happened between 1896 and 1897. The Brazilian government sent the army to attack the Canudos settlement in the state of Bahia. Most of the residents died.

Guerra do Paraguai "Paraguayan War" or "The War of the Triple Alliance." Military conflicts between Brazil, Argentina, and Uruguay, known as "the Triple Alliance," and Uruguay between 1864 and 1870. The number of casualties was very high, and Paraguay lost more than half of its adult male population.

Guitarra baiana The Bahian guitar is a four- or five-stringed electric musical instrument tuned like a mandolin with the scale of a *cavaquinho*. It is the oldest known solid-body electric mandolin. It is a symbol of Carnaval in Bahia.

INBEV Major brewery company from Brazil whose headquarters are in Belgium. It now owns Anheuser-Busch (AB-Inbev).

INEP Anísio Teixeira's National Institute of Pedagogical Studies. Government organization whose role is to promote research and evaluations about the Brazilian Educational System in order to develop public policies.

Ipanema A famous beach in Rio de Janeiro. It gained international fame due to the Bossa Nova hit song "The Girl from Ipanema."

Jabuticaba Purplish-black round fruit in the family of the *Myrtaceae* that grows directly from the trunk of a tree native to the states of São Paulo and Minas Gerais. The extremely perishable fruit is prized for its fragrant white, liquid center, and can be consumed raw or used to prepare jellies and drinks, such as juice, wine and *caipirinha*.

Jazes Musical groups from the 1920s to the 1940s inspired by the North American and Caribbean songs of the time that they learned through films, records, and the North American sailors who passed by Brazilian port cities.

JBS SA Brazilian corporation. The world's leading producer of meat/protein.

Jeito/Jeitinho Refers to the Brazilian philosophy that one must improvise to get by. The Brazilian style of playing soccer reflects this philosophy.

Jogo Bonito   Literally translates to "beautiful game," a term that is used to describe the Brazilian style of playing soccer.

Lanchonete   a place to eat a light meal, usually consisting of *salgadinhos*, sandwiches, juices, and/or *vitaminas*. See sidebar on p. 290.

LDBEN   The National Education Guiding Law, created in 1961 to regulate the educational activity in Brazil.

Lei Afonso Arinos   1951 law that made racism a crime.

Lei da Anistia   Amnesty law passed by the military dictatorship in 1979 that applied to all political crimes; encouraged many political exiles to return to Brazil, but also exempted soldiers and government officials who had perpetrated the crimes from prosecution.

Lei da Ficha Limpa   Brazil's "Clean Slate" Law (2010). It establishes a number of cases that make a citizen ineligible for running for public office.

Lei de Cotas   Law of Social Quotas. Affirmative action law that requires universities to destine half of their admission for poor students from public schools. The number of Afro-Brazilian students in the universities has increased significantly after that.

Lei do Audiovisual   Brazilian 1993 law that promotes investment in production and coproduction of films and audiovisual projects in general, as well as initiatives related to exhibition, distribution, and technical infrastructure.

Lei Maria da Penha   2006 law that was created to control domestic violence against women in Brazil.

Literatura de Cordel   String Literature. Booklets containing narratives about current events, and famous people, typical of the Brazilian Northeast. They are called "Cordel" (string) because they (used to) hang from a string at the place of sale.

Malandragem   Associated with the city of Rio de Janeiro, and made popular by samba songs, "malandragem" is a term to describe the lifestyle and acts of the *malandro*, an antihero character that combines charm, seduction, and laziness and illegality. The *malandro* usually cannot be trusted, as he will try to take advantage of everyone and every situation.

Mandioca   Known in English as cassava, manioc and yucca, and also called *macaxeira* and *aipim* in different regions of Brazil, it is a tuberous root containing large amounts of carbohydrates, potassium, and good-quality fibers. It also contains compounds which can combine to form hydrocyanic acid. The foliage, called *maniva*, is mostly used in the Northern region to make traditional dishes such as *maniçoba*. The same region produces *tucupi*, a fermented broth made with the juice collected from the roots when they are grated and squeezed to make *farinha de mandioca*. For more information, see sidebar on p. 287.

Mandioquinha   Same as *batata-baroa*, it is also called *mandioquinha-salsa*. Root (*Arracacia xanthorrhiza*) of Peruvian origin that looks like a parsnip, but it is much creamier in texture and has a more delicate flavor. It is known in English as white carrot and arracacha.

Mangaba   *Mangaba* (*Hancornia speciosa*) is a common fruit in the Northeast region of Brazil.

Maniçoba   A dish of the North region of Brazil, it is sometimes called the *feijoada* of Pará state. Ground *mandioca* leaves (called *maniva*) are cooked for several hours to eliminate their poisoning effect and then stewed with basically the same meats used in *feijoada*.

Manifesto Antropofágico   Literary manifest written by Oswald de Andrade, one of the founders of Brazilian Modernism, and published in 1928, whose aim was to question Brazil's cultural dependence. The text laid out the basis for the *Antropofagia* Movement.

Manteiga de garrafa   A type of clarified butter typical of the Northeast region similar to Indian ghee, usually sold in a bottle (*garrafa* in Portuguese); used in many dishes of the North and Northeast regions.

Maracanã   Soccer stadium in Rio de Janeiro built for the 1950 World Cup and remodeled for the 2014 World Cup; site of the opening and closing games for the 2016 Olympics.

Maracujá   Portuguese word for passion fruit.

Mata Atlântica   The Atlantic forest. Its vast extension ranges from the Northeast to the South along the Atlantic coast.

MEC   Brazilian Ministry of Education.

Mensalão   The *Mensalão* scandal was a corruption case involving the PT, Brazil's Workers Party, which became public in 2005, when Luiz Inácio Lula da Silva was the president of Brazil. The name was coined as a reference to a "high monthly payment."

Mercosul   Commercial agreement between the South American countries, namely Brazil, Argentina, Paraguay, Uruguay, and Venezuela.

Micareta   A carnival-like celebration that happens in different periods of the year, usually in cities or states where there is no official carnival.

Mito da Democracia Racial   Racial democracy myth; notion that there was no racial discrimination in Brazil; commonly attributed to the work of Brazilian sociologist Gilberto Freyre because of his seminal work *The Masters and the Slaves*.

Mito das Três Raças   Notion that privileges the Amerindians, Africans, and Europeans as founding elements of the Brazilian people. Often attributed to the work of anthropologist Darcy Ribeiro, the idea of myth used to be promoted in the Brazilian educational system. Today it is seen as problematic, since it excludes other important groups such as the Japanese and the Syrian-Lebanese.

Moqueca   Typical Brazilian dish of Amerindian origin, it is a stew, usually made with fish (sometimes with seafood), tomato, onion, cilantro and/or flat-leaf parsley, and, depending on the region of the country, *dendê*, coconut milk, and sliced peppers. It is usually served with *pirão* and Brazilian-style white rice (see recipe on p. 301); also called *peixada*.

Moreno/a   Usually a reference to a lighter-skinned racially mixed Brazilian.

MST—Movimento Sem Terra   Brazilian Landless Workers Movement. Their three pillars are fight for land, agrarian reform, and a more fair and brotherly society.

Mulato   Usually a reference to a darker-skinned racially mixed Brazilian.

Nheengatu   Lingua franca created at the time of the Jesuits in Brazil to promote catechism. It was formed by several indigenous languages and Portuguese. It is spoken until today in the Amazon region.

Operação Lava Jato   "Operation Car Wash." Investigation of alleged corruption and money laundering at Petrobras, the Brazilian multinational oil and energy corporation.

Orixás   African deities worshipped in Candomblé that represent ancestors and forces of nature.

Padaria   Brazilian bakery where one can find freshly baked breads and confections, as well as cold meats, cheeses, sodas, juices, milk, *salgadinhos*, cigarettes, candy, etc. Some have tables and serve quick meals, like the *lanchonetes* (see sidebar on p. 305).

Pagode   Urban musical genre derived from samba. It is popular in several regions of Brazil, especially in Rio de Janeiro, but also in São Paulo and in Bahia, where a particular dance style has arisen.

Pantanal   The world's largest tropical wetland region. It is situated in the state of Mato Grosso do Sul, with portions in Bolivia and Paraguay.

Pão de Açúcar   The Sugar Loaf Mountain is a famous tourist destination in Rio de Janeiro.

Pão de queijo   Gluten-free cheese roll typical of Minas Gerais state and the Southeast Region in general. It is made with *polvilho azedo* and *queijo (da) canastra* cheese. See recipe on p. 301.

Pardo/a   All-encompassing skin category used to describe racially mixed people in Brazil.

Peixada   See *moqueca*.

Peteca   The word for shuttlecock in the native Tupi language. It is also the name of the hand shuttlecock game played by the Tupi Indians around AD 1500.

Petrobras   The major Brazilian oil company, traded in the NYSE.

Picanha   The most popular beef cut in Brazilian *churrascarias*, it is a triangular piece of meat located close to the tail of the animal; called "rump cover," "rump cap," and "top sirloin cap" (among other names) in English.

Pinha   Common fruit in Brazil; also called "Fruta do Conde." It is called "sugar apple" in English.

Pirão   Porridge-like concoction made with a flavorful broth, either fish, chicken or meat-based, to which *farinha de mandioca* and seasonings are added.

Pirão de leite   Manioc flour cooked with milk; typical side dish for *carne de sol*.

Plano Cruzado   First large-scale economic plan launched in 1986 when José Sarney was the president of Brazil.

Plano Real   Multiphase economic plan launched in 1994 to curb inflation; created Brazil's present currency (the *real*).

Política Café com Leite   "Coffee and Milk Politics." Political bloc formed by politicians from São Paulo (dominated by coffee interests) and Minas Gerais (dominated by dairy interests) during the old republic (1890–1930), who agreed to alternate the presidency between them.

Polvilho Manioc/cassava/tapioca starch, available in two versions—*azedo* (sour) and *doce* (sweet, sold in the U.S. as tapioca starch / flour); also called *goma* in Portuguese.

Puxadinho Often illegal extension to a house to take advantage of the space. An example would be the irregular construction of an extra room in a building.

Quadrilha Traditional participative square dance groups typical of the June festivities in the Northeast of Brazil.

Queijo (da) canastra Semi-soft, sharp cheese produced in the Canastra Mountains of Minas Gerais state -- a quintessential ingredient for the authentic *pão de queijo* (see recipe on p. 301).

Queijo (de) coalho semi-soft curd cheese from the Northeast Region, it is usually grated and used as a melting cheese in dishes such as *tapioca/beiju*, or cut into logs and grilled.

Quilombo Runaway slave communities, maroon societies in Brazil.

Quilombo dos Palmares The most famous maroon community in Brazil (1605–1694), located in the state of Alagoas. Their leaders Ganga Zumba and Zumbi are mythical figures in the Afro-Brazilian imaginary.

Requeijão Creamy spreadable cheese from Minas Gerais state, also called *requeijão cremoso*; *requeijão de corte* or *crioulo* or *baiano* is made in the Northeast region and has a much firmer consistency; *catupiry*, also called *requeijão culinário*, is a type of *requeijão* named after the brand that created it—it is thermoresistant and used as an ingredient for recipes.

Retomada The term "Retomada" refers to the phase of "renewal" of the Brazilian cinema in the 1990s. *Carlota Joaquina: Princess of Brazil* is considered the first film of the period.

Revolta dos Malês The Malê Revolt happened in the city of Salvador in 1835. About 1500 Muslim slaves (called "Malê" then) rebelled unsuccessfully against slavery and the imposition of Catholicism.

Riograndenser Hunsrückisch Dialect of the German language that developed in the South of Brazil. It incorporates Portuguese words and has suffered influence from other languages. It is still spoken.

Rodízio A type of all-you-can-eat service provided in restaurant that implies eating a variety of dishes. The most common one is the *churrasco*, usually served in skewers, but pizza and sushi *rodízios* are becoming very popular as well.

Salgadinho The Brazilian word *salgadinhos*, which could be loosely rendered as "little savory/salty snacks," encompasses several kinds of preparations, baked or deep-fried, usually consisting of a wheat flour-based dough and a filling (there are many exceptions, though). Also called *salgados*, they are sold in *lanchonetes* and *padarias* (see sidebar on p. 305), and traditionally served at birthday parties and other family reunions.

Samba Samba is a musical rhythmical expression of dance and percussion considered one of the most important representations of the Afro-Brazilian tradition. It is often linked to Carnival and to the city of Rio de Janeiro, and it has become a symbol of Brazilian identity.

Samba de Roda Traditional Afro-Brazilian musical genre typical of the Recôncavo region of Bahia associated with a specific dance style.

Samba-Reggae Afro-Brazilian musical genre from the state of Bahia that mixes samba, reggae, and other musical influences. It is usually associated with Carnaval and with Bahia's Afro groups such as Olodum.

Semana de Arte Moderna "Modern Art Week." An artistic festival held in São Paulo from February 11 to 18, 1922. It introduced Brazilian Modernism to a national audience; famous for modernists' celebration of African and indigenous contributions to Brazilian culture.

Sertão Backlands region covered with *caatingas*. The people who live there are called *sertanejos*.

Sertanejo Someone who lives in the *sertão*. The word is also used to designate a country-like musical genre associated with the backlands of Brazil.

Sincretismo Religioso Common cultural phenomenon in Brazil characterized by the fusion of religious elements originated from diverse cultural contexts and different historic moments.

SNI "Serviço Nacional de Informações." National Information System, Brazil's national intelligence service that was in vogue from 1964 to 1990.

SUS "Sistema Único de Saúde." Brazil's public state-sponsored unified health system.

Talian This dialect still spoken in some regions of Brazil evolved from Venetian dialects. It was influenced by Northern Italy dialects and Brazilian Portuguese.

Tapioca In the northern regions, the name of a puffed manioc/cassava flour, also called *farinha de tapioca*; more toward the south of Brazil, the name of a dish made with moist tapioca starch (*goma*) that is baked in a thin layer, like a crepe, and served drizzled with *manteiga de garrafa* and/or coconut milk, and also stuffed with several fillings, both savory and sweet, and eaten as a light meal or breakfast item. In the northern regions, this specialty is called *beiju*.

Teatro de Revista Popular satiric theater genre marked by use of musicals, comedy, exotic clothes, and sensuality. It also provided a commentary on social reality.

Tejupares First Portuguese houses in Brazil made from straw. The name came from the Tupy-Guarani *tejy*: people and *upad*: place.

Terreiro Candomblé house of worship or sacred areas where followers gather to practice their religion.

Tropicalismo Avant-garde artistic movement of the 1960s. It merged Brazilian and foreign tendencies.

Tupi-Guarani Major indigenous language group. It was very important during the contact between Amerindians and Europeans in the colonization period of Brazil.

Tupinambá General name for the indigenous groups that inhabit the Brazilian coast.

Ufanismo Passionate expression of affirmation of nationalism.

Umbanda   Syncretic urban Afro-Brazilian religion, created over 100 years ago, that gathers elements from Kardecist Spiritism, Catholicism, Candomblé, and indigenous beliefs.

Umbu   Umbu (*Spondias tuberosa*) is a fruit that is common in the *caatinga* vegetation in the Northeast of Brazil.

Vale   Major Brazilian mining company, among the largest in the world.

Vatapá   Porridge made with bread soaked in coconut milk and seasoned with ground cashew and peanuts, ground salt shrimp, ginger, *dendê*, and malagueta hot pepper. It is usually a side dish for *Caruru* or eaten inside the *Acarajé*.

Vitamina   Fruit mixed in a blender with either water or orange juice. Some of them can be mixed with milk, and/or have more than one fruit, and sometimes instant oats added.

Facts and Figures

TABLE A1 Country Information

Location	Located in central and northeastern South America, Brazil is the continent's largest country. It is bordered to the north by Colombia, Venezuela, Guyana, Suriname, and French Guiana; to the west by Peru and Bolivia; and to the south by Paraguay, Argentina, and Uruguay. The country also has a long eastern coastline on the Atlantic Ocean.
Official Name	Federative Republic of Brazil
Local Name	Brasil
Government	Federal Republic
Capital	Brasília
Weights and Measurements	The metric system is in use.
Time Zone	Two hours ahead of U.S. Eastern Standard
Currency	Real
Head of State	President Dilma Rousseff
Head of Government	President Dilma Rousseff
Legislature	National Congress (Congresso Nacional)
Major Political Parties	The Workers' Party (PT); the Brazilian Social Democracy Party (PSDB); the Progressive Party; the Green Party; the Brazilian Democratic Movement Party (PMDB); the Democrats (DEM)

Sources: ABC-CLIO World Geography database; CIA World Factbook, http://www.cia.gov; FAO (FAOSTAT database), http://www.fao.org; World Bank, http://www.worldbank.org.

TABLE A2   Demographics

Population	204,259,812 (estimate) (2015)
World Population Rank	Fifth (2013)
Population Density	23.8 people per square kilometer (2013)
Population Distribution	87.0% urban (2011)
Age Distribution	0–14 years: 24.2%; 15–24 years: 16.7%; 25–54 years: 43.6%; 55–64 years: 8.2%; 65 years and over: 7.3% (2013)
Median Age	30.3 years (2013)
Population Growth Rate	0.8% per year (2013)
Net Migration Rate	–0.2 (2013)
Languages	Portuguese
Religious Groups	Christian (percentage not available)

Sources: ABC-CLIO World Geography database; CIA World Factbook, http://www.cia.gov; U.S. Census Bureau (International Data Base), http://www.census.gov.

TABLE A3   Geography

Area	3,286,500 square miles
Arable Land	8.5% (2011)
Arable Land per Capita	0.37 hectares per person (2011)
Coastline	4,655 miles
Land Borders	9,129 miles
Climate	In the Amazon basin's tropical rain forest, the climate is hot and wet. In the central and southern uplands' savannah grasslands, the climate is temperate, with warm summers and mild winters. Temperatures in Rio de Janeiro range from 63°F to 85°F.
Land Use	7.0% arable land; 0.8% permanent crops; 23.3% permanent meadows and pastures; 55.7% forest land; 13.1% other
Major Agricultural Products	Coffee, oranges, soybeans, sugarcane, maize, rice, cocoa, cassava, cattle, milk
Natural Resources	Bauxite, gold, iron ore, manganese, nickel, phosphates, platinum, tin, rare earth elements, uranium, petroleum, hydropower, timber

Sources: ABC-CLIO World Geography database; CIA World Factbook, http://www.cia.gov; FAO (FAOSTAT database), http://www.fao.org; World Bank http://www.worldbank.org.

TABLE A4   Economy

Gross Domestic Product (GDP) —official exchange rate	$2,340,754,000,000 (estimate) (2015)
GDP per Capita	$11,607 (estimate) (2015)
GDP—Purchasing Power Parity (PPP)	$2,466,567,000,000 (estimate) (2013)
GDP (PPP) per Capita	$12,340 (estimate) (2013)
Industry Products	Textiles, footwear, passenger and commercial vehicles, synthetic rubber, electricity, steel, machinery, cement
Unemployment	8.3% (2009)
Labor Profile	Agriculture: 15.7%, industry: 13.3%, services: 71% (estimate) (2011)
Total Government Revenues	$851,100,000,000 (estimate) (2013)
Total Government Expenditures	$815,600,000,000 (estimate) (2013)
Budget Deficit	2.7% of GDP (2012)
GDP Contribution by Sector	Agriculture: 5.2%, industry: 26.3%, services: 68.5% (2012 estimate)
External Debt	$428,300,000,000 (estimate) (2012)
Economic Aid Extended	$0 (2011)
Economic Aid Received	$870,000,000 (2011)

Sources: ABC-CLIO World Geography database; CIA World Factbook, http://www.cia.gov; ILO (LABORSTA database), http://www.ilo.org; IMF (World Economic Outlook), http://www.imf.org; World Bank, http://www.worldbank.org.

TABLE A5   Communications and Transportation

Facebook Users	74,000,000 (estimate) (2013)
Internet Users	75,982,000 (2009)
Internet Users (Percentage of Population)	45.0% (2011)
Television	35.8 sets per 100 population (2003)
Land-Based Telephones in Use	43,026,000 (2011)
Mobile Telephone Subscribers	244,358,000 (2011)
Major Daily Newspapers	532 (2004)
Average Circulation of Daily Newspapers	6,552,000 (2004)
Airports	4,105 (2012)
Paved Roads	13.5% (2010)
Roads, Unpaved	1,028,689 (2004)
Passenger Cars per 1,000 People	178 (2009)
Railroads	17,931 miles (2008)
Ports	Major: 20 (including Santos, Rio de Janeiro, Paranagua, Itajaí, Rio Grande, Salvador)

Sources: ABC-CLIO World Geography database; CIA World Factbook, http://www.cia.gov; Facebook, https://www.facebook.com/; World Bank, http://www.worldbank.org.

TABLE A6   Military

Total Active Armed Forces	327,710 (2010)
Active Armed Forces	0.2% (2010)
Annual Military Expenditures	$29,700,000,000 (2009)
Military Service	Military service is by conscription, with terms lasting 12 months.

Sources: ABC-CLIO World Geography database; Military Balance.

TABLE A7   Education

School System	Primary education begins at the age of seven in Brazil. After eight years, students continue to three years of secondary education. Students may also choose to enroll in vocational education, for programs lasting between three and five years, rather than academic secondary school.
Mandatory Education	8 years, from ages 7 to 15
Average Years Spent in School for Current Students	14 (2008)
Average Years Spent in School for Current Students, Male	14 (2008)
Average Years Spent in School for Current Students, Female	14 (2008)
Primary School–Age Children Enrolled in Primary School	6,189,500 (2010)
Primary School–Age Females Enrolled in Primary School	5,923,266 (2010)
Secondary School–Age Children Enrolled in Secondary School	19,147,579 (2010)
Secondary School–Age Males Enrolled in Secondary School	9,333,635 (2010)
Secondary School–Age Females Enrolled in Secondary School	9,813,944 (2010)
Students per Teacher, Primary School	21.0 (2011)
Students per Teacher, Secondary School	16.0 (2011)
Enrollment in Tertiary Education	6,552,707 (2010)
Enrollment in Tertiary Education, Male	2,825,509 (2010)
Enrollment in Tertiary Education, Female	3,727,198 (2010)
Literacy	90% (2009)

Source: ABC-CLIO World Geography database; Country government; UNESCO. http://www.unesco.org; World Bank, http://www.worldbank.org.

TABLE B   Population of Major Cities (July 2014)

City (Capital)	State	Population
São Paulo	São Paulo	11,895,893
Rio de Janeiro	Rio de Janeiro	6,453,682
Salvador	Bahia	2,902,927
Brasília	(Federal District)	2,852,372
Fortaleza	Ceará	2,571,896
Belo Horizonte	Minas Gerais	2,491,109
Manaus	Amazonas	2,020,301
Curitiba	Paraná	1,864,416
Recife	Pernambuco	1,608,488
Porto Alegre	Rio Grande do Sul	1,472,482
Belém	Pará	1,432,844
Goiânia	Goiás	1,412,364
São Luís	Maranhão	1,064,197
Maceio	Alagoas	1,005,319
Natal	Rio Grande do Norte	862,044
Campo Grande	Mato Grosso do Sul	843,120
Teresina	Piauí	840,600
João Pessoa	Paraíba	780,738
Aracaju	Sergipe	623,766
Cuiabá	Mato Grosso	575,480
Porto Velho	Rondônia	494,013
Florianópolis	Santa Catarina	461,524
Rio Branco	Acre	363,928
Vitória	Espirito Santo	352,104
Boa Vista	Roraima	314,900
Palmas	Tocantins	265,409
Macapá	Amapá	8,553

Source: IBGE, Diretoria de Pesquisas, Coordenação de População e Indicadores Sociais (Research Supervision and Population and Social Indicators Coordination).

TABLE C   Human Development Index Indicators

Life Expectancy at Birth (Years)	73.28 years (2014 estimate)
Adult Literacy Rate (percentage age 15 and above)	92.6% (2015 estimate)
Education Expenditures	5.8% of GDP (2010)
People Not Using an Improved Water Source (%)	2.5 (2012 estimate)
Children Underweight for Age (percentage under age five)	2.2 (2007)

Source: CIA Factbook.

TABLE D   List of Presidents of Brazil

President	Took Office	Left Office
Manoel Deodoro da Fonseca	November 15, 1889	February 25, 1891
Manoel Deodoro da Fonseca	February 25, 1891	November 23, 1891
Floriano Vieira Peixoto	November 23, 1891	November 15, 1894
Prudente José de Morais e Barros	November 15, 1894	November 15, 1898
Manoel Ferraz de Campos Salles	November 15, 1898	November 15, 1902
Francisco de Paula Rodrigues Alves	November 15, 1902	November 15, 1906
Affonso Augusto Moreira Penna	November 15, 1906	June 14, 1909
Nilo Procópio Peçanha	June 14, 1909	November 15, 1910
Hermes Rodrigues da Fonseca	November 15, 1910	November 15, 1914
Wenceslau Braz Pereira Gomes	November 15, 1914	November 15, 1918
Delfim Moreira de Costa Ribeiro	November 15, 1918	July 28, 1919
Epitácio Lindolfo da Silva Pessoa	July 28, 1919	November 15, 1922
Arthur da Silva Bernardes	November 15, 1922	November 15, 1926
Washington Luís Pereira de Sousa	November 15, 1926	October 24, 1930
Getúlio Dornelles Vargas	November 3, 1930	July 20, 1934
Getúlio Dornelles Vargas	July 20, 1934	November 10, 1937
Getúlio Dornelles Vargas	November 10, 1937	October 29, 1945
José Linhares	October 29, 1945	January 31, 1946
Eurico Gaspar Dutra	January 31, 1946	January 31, 1951
Getúlio Dornelles Vargas	January 31, 1951	August 24, 1954
João Fernandes Campos Café Filho	August 24, 1954	November 11, 1955
Carlos Coimbra da Luz	November 8, 1955	November 11, 1955
Nereu de Oliveira Ramos	November 11, 1955	January 31, 1956
Juscelino Kubitschek de Oliveira	January 31, 1956	January 31, 1961
Jânio da Silva Quadros	January 31, 1961	August 25, 1961
Paschoal Ranieri Mazzilli	August 25, 1961	September 8, 1961
João Belchior Marques Goulart	September 8, 1961	January 24, 1963
João Belchior Marques Goulart	January 24, 1963	March 31, 1964
Paschoal Ranieri Mazzilli	April 2,1964	April 15, 1964
Humberto de Alencar Castello Branco	April 15, 1964	March 15, 1967
Arthur da Costa e Silva	March 15, 1967	August 31, 1969
Emílio Garrastazu Médici	October 30, 1969	March 15, 1974
Ernesto Geisel	March 15, 1974	March 15, 1979
João Baptista de Oliveira Figueiredo	March 15, 1979	March 15, 1985
Tancredo de Almeida Neves	He did not take office	
José Sarney	March 15, 1985	March 15, 1990
Fernando Afonso Collor de Mello	March 15, 1990	October 2, 1992
Itamar Augusto Cautiero Franco	October 2, 1992	January 1, 1995
Fernando Henrique Cardoso	January 1, 1995	January 1, 1999

President	Took Office	Left Office
Fernando Henrique Cardoso	January 1, 1999	January 1, 2003
Luiz Inácio Lula da Silva	January 1, 2003	January 1, 2007
Luiz Inácio Lula da Silva	January 1, 2007	January 1, 2011
Dilma Rouseff	January 1, 2011	January 1, 2015
Dilma Rouseff	January 1, 2015	Current

Source: http://www2.planalto.gov.br/acervo/galeria-de-presidentes.

TABLE E   Basic Health and Nutrition Indicators

Maternal Mortality Rate per 100,000 (2013)	69
Infant Mortality Rate per 1,000 (2014)	19.21
Neonatal Mortality Rate per 1,000 (2012)	9
Child Mortality Rate under Five Years per 1,000 (2012)	14
Chronic Child Malnutrition under Five Years (2006)	2.2%[a]
Acute Child Malnutrition under Five Years (2006)	0.3%[a]
Antipolio Vaccination Coverage under One Year (2012)	97%
DPT Vaccination Coverage under One Year (2012)	99%
MMR Vaccination Coverage under One Year (2009)	99%

[a]http://www.indexmundi.com/brazil/child-malnutrition.html.
Source: UNICEF, http://www.unicef.org/infobycountry/brazil_statistics.html;
https://www.cia.gov/library/publications/the-world-factbook/geos/br.html.

TABLE F1   Electricity

119.1 Million kW (2011 estimate)	119.1 million kW (2011 estimate)
Electricity Production	530.4 billion kWh (2011 estimate)
Electricity Consumption	478.8 billion kWh (2011 estimate)

Source: https://www.cia.gov/library/publications/the-world-factbook/geos/br.html.

TABLE F2   Production of Electricity by Source

Type	Percentage of Total
Fossil Fuel	18.8
Hydroelectric	69.3
Nuclear Fuel	1.6
Other	10.4%

Source: https://www.cia.gov/library/publications/the-world-factbook/geos/br.html.

Major Brazilian Holidays and Festivals

JANUARY 1

New Year's Day

The first day of the New Year is an official national holiday, and government offices are either closed or operate on a reduced schedule. New Year's celebrations start on December 31, when Brazilians count down to midnight to commemorate the beginning of the Gregorian calendar year. Famous festivities of an Afro-Brazilian religious character take place in Rio de Janeiro and Salvador, but the whole country rejoices on the holiday.

FEBRUARY/MARCH (40 DAYS BEFORE GOOD FRIDAY)

Carnival (Carnaval)

This is Brazil's most famous holiday. It usually lasts from three days to a week right before Lent. As in other parts of the world, Carnival in Brazil features parades, music, and dancing. The celebration takes place throughout Brazil but is particularly famous in Rio de Janeiro, Salvador, and Recife. Carnival is not a national public holiday (only in certain states and municipalities), but most government offices are closed during part of or the whole period.

APRIL

Good Friday

This Christian holiday is not a national public holiday, but it is observed in many municipalities. It takes place on the Friday before Easter Sunday. It is a day to remember Jesus Christ's Passion, crucifixion, and death, and it is a popular occasion for beach vacations.

APRIL 19

Native Brazilian Day (Dia do Índio)

This is a holiday to pay homage to the Amerindians and celebrate their cultures and traditions. Although the holiday is very popular, it is not observed in many parts of Brazil, except generally in educational settings. Students all over the country focus their studies on the indigenous history and cultures on and near the holiday. There are more political celebrations in regions of the country where there is a larger indigenous population, such as in the states of Amazonas, Pará, Rondônia, Goiás, Mato Grosso, and Moto Grosso du Sul, as well as in the interior of the state of Bahia.

APRIL 21

Tiradentes

This is the anniversary of the death of Joaquim José da Silva Xavier, known as Tiradentes. He was the leader of the *Inconfidência Mineira*, a failed uprising movement that took place in the state of Minas Gerais. Brazilians consider Tiradentes a national hero because his group planned for Brazil to gain independence from Portugal and establish a republic.

MAY 1

Labor Day

A day to honor workers and their accomplishments. The holiday is often an opportunity for certain groups to protest against their labor conditions, to debate workers' rights, or to demand improvements such as wage increases. This is especially true for state and federal workers. It is a national public holiday, when all government offices close.

MAY

Mother's Day

This very popular holiday is not a national public holiday. It is celebrated on the second Sunday of May. It is an occasion for family gathering, when traditionally people honor their mothers with gifts, lunches, or dinners. The holiday has been criticized for stimulating consumerism, as it has become one of the country's most profitable business holidays.

MAY/JUNE

Corpus Christi

This Christian holiday is celebrated 60 days after Easter, on the second Thursday after Whitsun. Corpus Christi means literally "body of Christ," and the holiday observes the Holy Eucharist. The Catholic Church organizes religious processions throughout the country to observe the occasion. It is an optional bridge holiday, which is observed in some states and municipalities.

JUNE 12

Lover's day (Dia dos Namorados)

The Brazilian version of Valentine's Day is not a national public holiday, but it is very trendy among couples. Unlike the U.S. holiday, Brazilians do not celebrate friendship, but their romantic relationships, especially among "namorados" (boyfriends or girlfriends). Married couples also celebrate the date. There is not much of a card-giving tradition, but exchange of gifts are common and expected. Couples often go out for lunch or dinner to commemorate the occasion.

JUNE 13–29

June Festivities (Festas Juninas)

Three Catholic saints are celebrated in Brazil during the June festivities. The most popular one is the Festival of Saint John the Baptist (São João), which is celebrated on June 24, but the festivities traditionally start the day before. Saint Anthony (Santo Antonio) is celebrated on June 13, and Saint Peter (São Pedro) on June 29. They are regional holidays of religious character, but their celebrations often also include dancing and listening to country music, wearing regional clothes, and eating local foods. It is especially the case of the Saint John Festival, which is very popular in the Northeastern region of the country.

AUGUST

Father's Day

This very popular holiday is not a national public holiday. It is celebrated on the second Sunday of August. It is an occasion for family gathering, when traditionally people honor their fathers with gifts, lunches, or dinners. The holiday has also been criticized for stimulating consumerism, but it is not typically as profitable as Mother's Day.

SEPTEMBER 7

Independence Day

Brazil declared independence from Portugal in 1822 on this date. Parades and fireworks are common throughout the country. It is a national public holiday, when all government offices close.

OCTOBER 12

Nossa Senhora Aparecida, Patron Saint of Brazil

Our Lady of the Apparition, whose name is due to the fact that an image of the Catholic saint was found by two fishermen in 1717. They first caught the body of the image, whose head followed. The origin of the image is unknown, but some believe the artist was Friar Agostinho de Jesus. Her image is of a black woman. One explanation for this is the clay that she was made of. Some people believe that the saint often "appears" resembling a given oppressed people. Several miracles are attributed to her. The Catholic Church organizes religious celebrations all over Brazil on that date. Her Cathedral is located in the city of Aparecida, named after her, in the state of São Paulo.

October 12 is also "Children's Day," a popular national family holiday that was created to celebrate children's rights. However, it has become one of Brazil's most commercially profitable holidays, as it became a tradition for parents to buy gifts for their children to celebrate that day.

OCTOBER 15

Teachers' Day

This is a school holiday to pay homage to teachers and their important role in people's lives. Classes are suspended that day, and there are typically a series of educational and political activities promoted to celebrate the occasion.

NOVEMBER 2

All Souls' Day (Dia de Finados)

This is a Christian holiday to pay homage to the faithfully departed. Families pray and light candles for the departed, and friends and relatives customarily visit cemeteries and bring flowers. There are no parties, as in Mexico, for example. It is a day of introspection.

NOVEMBER 15

Proclamation of the Brazilian Republic (Proclamação da República)

This national public holiday celebrates the overthrowing of the empire of Brazil through a military coup d'état that unseated Dom Pedro II, then emperor of Brazil, in 1889.

NOVEMBER 20

Black Awareness Day (Dia da Consciência Negra)

The holiday used to be celebrated on May 13, when slavery was officially abolished in 1888. However, Afro-descendants found the day of the death of Zumbi a more suitable occasion. Zumbi was the leader of the Palmares Quilombo, a maroon society in the state of Alagoas. He is a symbol of struggle, power, and resistance. Black Awareness Day is usually marked by a series of political and educational activities related to racial identity and racial relations. It is not observed all over Brazil yet.

DECEMBER 25

Christmas Day (Natal)

Christmas is both a religious holiday and a national public holiday in Brazil. It is the celebration of the birth of Jesus Christ. It is an occasion for family gathering and gift giving, which usually takes place on the night before. It is Brazil's most commercially successful holiday.

Government offices and businesses in general are closed all over the country. The Catholic Church celebrates masses and Protestant churches congregate to commemorate the occasion. There has been some criticism about the tradition recently because of the similarities the celebration has with the festivities in the United States and European countries. Papai Noel (Father Noel), the Brazilian version of Santa Klaus, for instance, dresses very much like his Northern counterpart, despite the fact that it is summer and very hot in Brazil.

Country-Related Organizations

GOVERNMENT RESOURCES AND BRAZIL FACTS

Portal Brasil

http://www.brasil.gov.br

This is Brazil's government website, where one can find a variety of news, facts, services, and resources. The topics include, but are not at all limited to, citizenship, justice, culture, education, infrastructure, health, welfare, and tourism. There are links to information about every state of Brazil, and a form to send a message directly to the country's president.

IBGE—Instituto Brasileiro de Geografia e Estatística
(Brazilian Institute of Geography and Statistics)

http://www.ibge.gov.br

Address: Rua André Cavalcânti, 106, Centro, Rio de Janeiro, RJ, 20231-050
Phone: 55(21) 2142-4677

This institute is responsible for collecting data and producing statistics based on geographic, demographic, and economic data. The information is available to private and public organizations, and to the general public.

Itamaraty—Ministério das Relações Exteriores (Foreign Ministry)

http://www.itamaraty.gov.br

Address: Ministério das Relações Exteriores. Esplanada dos Ministérios, Bloco H, Palácio Itamaraty, Anexo II, Sala 11, Brasília, DF, 70170-900
Phone: 55(61) 2030-6713

This is Brazil's Foreign Ministry official website. It provides information on the Brazilian embassies and consulates around the world, and on other topics related to Brazil's diplomacy and international affairs.

Policia Federal (Federal Police)

http://www.dpf.gov.br/

Address: Departamento de Polícia Federal, SAS Quadra 6, lotes 09/10 – Ed. Sede/DPF, Brasília, DF, 70037.900
Phone: 55(61) 2024-8000

The Federal Police of Brazil is in charge of border control, immigration issues and customs, and issuing passports. It also acts on cases of trafficking, smuggling, corruption, and other major crimes.

HEALTH

SUS—Sistema Único de Saúde (Unified Health System)

Portal da Saúde (National Health Portal)—http://portalsaude.saude.gov.br/index.php/servicos

Address: Ministério da Saúde. Esplanada dos Ministérios, Bloco G, Brasília, DF, 70058-900
Phone: 55(61) 3315-2425

This website provides general information on Brazil's Health Ministry, and its Unified Health System (SUS), the country's public health system.

BUSINESS AND ECONOMIC RESOURCES

Banco Central do Brasil/Central Bank of Brazil

http://www.bcb.gov.br/?ENGLISH

Headquarters: Setor Bancário Sul (SBS), Quadra 3, Bloco B—Ed. Sede
Brasília—DF, 70074–900
Phone: (61) 3414–1414

The Central Bank is the main monetary institution of the country, reporting to the Ministry of Finance.

BB—Banco do Brasil (Brazil Bank)

http://www.bb.com.br

Address: SBS, Quadra 02, Bloco Q, Centro Empresarial João Carlos Saad, 11° andar, Brasília, DF, 70070-120
Phone: 55(61) 3218-6200

Partly state owned, the Brazil Bank has thousands of branches in Brazil and | in other 21 countries. Foreign currency can be exchanged in many of its branches.

BM&F BOVESPA—Bolsa de Valores, Mercadorias & Futuros de São Paulo (São Paulo's Stock Exchange)

http://www.bmfbovespa.com.br/en-us/intros/intro-about-us.aspx?idioma=en-us

Address: Rua Líbero Badaró, 471, Sé. São Paulo, SP, 01009-000
Phone: 55(11) 3272-7373

São Paulo's stock exchange is the 13th-largest stock exchange in the world. In 2008, BOVESPA merged with BM&F, the Brazilian Mercantile and Futures Exchange, becoming BM&F BOVESPA.

BNDES—Banco Nacional de Desenvolvimento (The National Bank for Economic and Social Development)

http://www.bndes.gov.br/SiteBNDES/bndes/bndes_pt/

Address: Av. República do Chile, 100, Rio de Janeiro, RJ, 20231-917
Phone: 55(21) 2172-6591

The BNDS is a federal company that reports to the Ministry of Development, Industry, and Trade. BNDS provides long-term financing and investment in all economic areas, including agricultural, industrial, commercial, infrastructure, social, regional, and environmental projects.

FIRJAN—*Federação das Indústrias do Estado do Rio de Janeiro* (*The Federation of Industries of the State of Rio de Janeiro*)

http://www.firjan.org.br

> Address: Av. Graça Aranha, 1, Centro, Rio de Janeiro, 0800 0231 231
> Phone: 55(21) 2532-3700; E-mail: faleconosco@firjan.org br

It also includes the following organizations: CIRJ—the Industrial Center of Rio de Janeiro, SESI—Industry Social Services, SENAI—National Industrial Training Service, and IEL—the Euvaldo Lodi Institute. Its Business Center Complex can be rented for congresses, seminars, industrial fairs, meetings, and other business-oriented events.

FIESP—*Federação das Indústrias do Estado de São Paulo* (*The Federation of Industries of the State of São Paulo*)

http://www.fiesp.com.br/

> Address: Av. Paulista, 1313, Cerqueira César, São Paulo, SP, 01311-923
> Phone: 55(11) 3549-4499

It also includes the following organizations: CIESP—Centro das Indústrias do Estado de São Paulo (Center of Industries of the State of São Paulo), SESI-SP, SENAI-SP, and the IRS—Roberto Simonsen Institute.

The website of the Federação das Indústrias do Estado de São Paulo brings news, videos, photos, and relevant information about FIESP, CIESP, SESI-SP, SENAI-SP, and IRS.

Receita Federal (*The Department of Federal Revenue*)

http://idg.receita.fazenda.gov.br/

> Address: SAS, Quadra 03, Bloco O, Sobreloja, Asa Sul, SP, Brasília, DF,
> 70079-900
> Phone: 55(11) 3003-0146

The Department of Federal Revenue of Brazil (English translation for Secretaria da Receita Federal do Brasil), or Receita Federal (RFB) is connected to the Ministry of Finance of Brazil.

CPF—*Cadastro de Pessoas Físicas* (*Individual Taxpayer Registration Number*)

CPF is the number attributed by the Brazilian Federal Revenue to both Brazilians and resident aliens who pay taxes. The CPF number is used to calculate income tax that is due. It is Brazil's Social Security number.

Petrobras

http://www.petrobras.com/en/home.htm

Address: Av. República do Chile, 65, Centro, Rio de Janeiro, 20031-912
Phone: 55(21) 3224-4477

Petrobras is a semipublic Brazilian multinational energy corporation. According to its website, Petrobras has "operations in the entire oil and gas productive chain, and in the production of biofuels and of other alternative energy sources."

SOCIAL RESOURCES

FUNAI—National Amerindian Foundation

http://www.funai.gov.br/

Address: SBS Quadra 02, Lote 14, Ed. Cleto Meireles, Brasília, DF, 70070-120
Phone: 55(61) 3247-6000

FUNAI is the Brazilian government's organization in charge of the development of policies related to the indigenous peoples and to safeguard their rights in the country.

Anistia Internacional Brasil (Amnesty International - Brazil)

https://anistia.org.br/home/

Address: Praça São Salvador, 5-Casa, Laranjeiras, Rio de Janeiro, 22231-170
Phone: 55(21) 3174-8601

Anistia Internacional Brasil is a major nongovernmental organization focused on human rights.

Viva Rio

http://vivario.org.br/en/

Address: Rua do Rússel, 76, Glória, Rio de Janeiro, RJ, 22210-010
Phone: 55(21) 2555-3750

According to its website, Viva Rio is an organization that "aims to foster a culture of peace and social inclusion through a commitment to research, field work, and the formulation of public policies." It works on proposing solutions to communities exposed to violence, implementing solutions for social and environmental issues, and mediating conflicts.

CULTURAL AND EDUCATIONAL RESOURCES

ANCINE—Agência Nacional de Cinema (National Cinema Agency)

http://www.ancine.gov.br/

Headquarters: SRTV Sul Conjunto E, Edifício Palácio do Rádio, Bloco I, Cobertura, Brasília, DF, 70340–901
Phone: 55(61) 3027–8521

ANCINE is Brazilian government's National Film Agency. It was created to protect, promote, and regulate the national film and video industry. The agency also fosters coproductions between the local cinema industry and other countries.

MEC—Ministério da Educação

http://www.mec.gov.br/

Address: Esplanada dos Ministérios Bloco L, Ed. Sede e Anexos, Brasília, DF, 70.047-900
Phone: 55(61) 0800 616161

Brazil's Ministry of Education.

CAPES—Coordenação de Aperfeiçoamento de Pessoal de Nível Superior (Coordination for the Improvement of Higher Education Personnel)

http://www.capes.gov.br/

Address: ERL- Norte, 4° andar, Brasília, DF, 70040-020
Phone: 55 800616161

It is a government agency, affiliated to the Ministry of Education, responsible for the evaluation, expansion, and improvement of higher education in Brazil.

CNPQ—Conselho Nacional de Desenvolvimento Científico e Tecnológico (National Council for Scientific and Technological Development)

http://www.cnpq.br/

Address: SHIS QI 1 Conjunto B, Blocos A, B, C e D, Lago Sul, Brasília, DF, 71605-001
Phone: 55 800619697

This is a government agency responsible for the advancement of scientific and technological research.

MinC—Ministério da Cultura (Ministry of Cultural Affairs)

http://www.cultura.gov.br/

Address: Ministério da Cultura – Sede, Esplanada dos Ministérios, Bloco B, Térreo, Brasília, DF, 70068-900
Phone: 55(61) 2024-2222

Brazil's Ministry of Cultural Affairs includes literature, arts in general, and folklore. The ministry is also responsible for the country's historical, archeological, artistic, and cultural assets and archives.

SPORTS

CBF—Confederação Brasileira de Futebol (Brazilian Soccer Confederation)

http://www.cbf.com.br/

Address: Avenida Luis Carlos Prestes, 130, Barra da Tijuca, Rio de Janeiro, RJ, 22775-055
Phone: 55(21) 35721900

CBF oversees the organization and promotion of soccer championships in Brazil, and the men's and women's national soccer teams.

COB—Comitê Olímpico do Brasil (Brazilian Olympic Committee)

http://www.cob.org.br/en

Address: Avenida das Américas, 899, Barra da Tijuca, Rio de Janeiro, RJ, 22631-000
Phone: 55(21) 2134-335777
COB is in charge of the development of Olympic Sports in Brazil.

Ministério do Esporte (Ministry of Sports)

http://www.esporte.gov.br/

Address: Esplanada dos Ministérios, Bloco A, Brasília, DF, 70054-906
Phone: 55(61) 3217-1800

The Ministry of Sports is responsible for projects regarding the development and expansion of sports, social inclusion through sports, free access to sports training, quality of life, and human development.

Rio 2016—Summer Olympics in Rio

http://www.rio2016.com/
The official website for the Summer Olympics in Rio de Janeiro in 2016.

TOURISM RESOURCES

DEAT—Delegacia do Turista (Tourist Precinct)

Address: Av. Afrânio de Melo Franco, 159, Leblon, Rio de Janeiro
Phone: 55(21) 2332–2924 and 55(21) 2332–2885/2889
E-mail: deat@pcerj.rj.gov br

DEAT is a special police unit for tourism support, located in the city of Rio de Janeiro. It provides multilingual assistance.

Embratur (Brazilian Institute of Tourism)

http://www.embratur.gov.br/

Address: Esplanada dos Ministérios, Bloco U, Térreo, Brasília, DF, 70065-900
Phone: 55(61) 2023-7146

The Brazilian Institute of Tourism is affiliated with the Ministry of Tourism.

Visit Brasil

http://www.visitbrasil.com
This website has travel tips and information on touristic destinations, besides information on national trade.

CNI—National Confederation of the Industry

http://www.portaldaindustria.com.br/cni

Address: SBN – Quadra 01, Bloco C, Ed. Roberto Simonsen, Brasília, DF, 70040-903
Phone: 55(61) 3317-9000

CNI is the largest trade association of Brazil. It promotes industry round-table, sectoral studies and lobbies the government and Congress for pro-industry legislations.

Ipeadata

http://ipeadata.gov.br/

This is the website of the database of the Institute of Applied Economic Research (IPEA), a public organization. It has data on Brazil's economy, as well as information about demography, geography, and social statistics and social security per state, city, and regions.

Transparência Brasil (For a Transparent Brazil)

http://www.transparencia.org.br/

Address: Rua Bela Cintra, 409, São Paulo, SP, 01415-000
Phone: 55(11) 3259-6986

Transparência Brasil is a think tank dedicated to political accountability and data transparency.

Folha de São Paulo (International Version)

http://www1.folha.uol.com.br/internacional/en/

Folha is the biggest Brazilian newspaper with coverage of politics, business, environment, and culture. It has an English version with selected stories that were also published in Portuguese.

ACCESSIBILITY

Acessibilidade Brasil

http://www.acessibilidadebrasil.org.br/

Address: Lages, Cde, R, No. 44 – SI 503, Centro, Rio de Janeiro, RJ, 20241-900
Phone: 55(21) 2232-1848

The group organizes social, educational, and fund-raising activities to promote the rights of people with disabilities.

GLBTQ

PARADASP—Associação da Parada do Orgulho LGBT (LGBT Pride Parade Association)

http://www.paradasp.org.br/

The association organizes the Pride Parade in the city of São Paulo, one of the largest in the world.

CPDS—Coordenação de Políticas Para a Diversidade Sexual (Political Coordination for Sexual Diversity)

Address: Pátio do Colégio, 148, térreo, Sé, São Paulo, SP, 01016-040
Phone: 55(11) 3291–2700
E-mail: diversidadesexual@sp.gov br

The CPDS is a state office in São Paulo in charge of creating public polices related to the rights of sexual and gender minorities. It is affiliated with the Office for Justice and Defense of Citizenship.

CEDS-Rio—Coordenadoria Especial da Diversidade Sexual (Special Coordination for Sexual Diversity)

http://www.cedsrio.com.br/home.html

Address: Rua São Clemente, 360 (Palácio da Cidade), Botafogo, Rio de Janeiro, RJ, 22260-006
Phone: 55(21) 2976-9137

The CEDS is state office in Rio de Janeiro in charge of creating public polices related to the rights of sexual and gender minorities. It also proposes to mediate the dialogue between other state organizations and the GLBTQ community.

PUBLIC RESEARCH CENTERS

Arquivo Nacional (The National Archive)

http://www.arquivonacional.gov.br/cgi/cgilua.exe/sys/start.htm?tpl=home

Address: Praça da República, 173, Rio de Janeiro, RJ, 20211-350
Phone: 55(21)2179-1227

The federal central management system of documents is in charge of the preservation of archives in Brazil. It includes information on foreign nationals who arrive in Brazil. The Arquivo Nacional also offers data, publications, tools, and services for researchers.

BN—Biblioteca Nacional (National Library)

Headquarters: Av. Rio Branco, 219, Centro, Rio de Janeiro, RJ, 20040-009
Phone: 55(21) 3095-3879

http://www.bn.br/
The BN foundation is responsible for collecting, storing, and preserving the intellectual production of Brazil. Its collection includes over 10 million items, being the world's seventh-largest national library, and the largest library in Latin America, according to UNESCO.

Selected Bibliography

Alberto, Paulina L. *Terms of Inclusion: Black Intellectuals in Twentieth-Century Brazil*. Chapel Hill: University of North Carolina Press, 2011.

Ang, Eng Tie. *Delightful Brazilian Cooking*. Seattle, WA: Ambrosia Publications, 1993, p. 176.

Anunciato, Ofélia. *A Taste of Brazil*. Cologne: Könemann. Translated by Julie Martin and First Edition Translations, 2000, p. 136.

Avelar, Idelber, and Christopher Dunn (eds.). *Brazilian Popular Music and Citizenship*. Durham, NC: Duke University Press, 2011.

Baer, Werner. *The Brazilian Economy: Growth and Development*. Boulder, CO: Lynne Rienner Publishers, 2013.

Baranov, David. *The Abolition of Slavery in Brazil. The "Liberation" of Africans through the Emancipation of Capital*. Westport, CT: Greenwood Press, 2000.

Barnitz, J. *Twentieth-Century Art of Latin America*. Austin: University of Texas Press, 2001.

Barry Ames. *The Deadlock of Democracy in Brazil: Interests, Identities, and Institutions in Comparative Politics*. Ann Arbor: University of Michigan Press, 2001.

Bartlett, Leslie. *The Word and the World: The Cultural Politics of Literacy in Brazil*. New York: Hampton Press, 2009.

Bastide, Roger. *The African Religions of Brazil*. Baltimore, MD: The John Hopkins University Press, 1978.

Bayón, D., and M. Marx. *History of South American Colonial Art and Architecture: Spanish South America and Brazil*. New York: Rizzoli, 1992.

Beattie, Peter M. *The Tribute of Blood: Army, Honor, Race, and Nation in Brazil, 1864–1945*. Durham, NC: Duke University Press, 2001.

Bethell, Leslie (ed.). *Colonial Brazil*. Cambridge: Cambridge University Press, 1987.

Burdick, John. *Looking for God in Brazil: The Progressive Catholic Church in Urban Brazil's Religious Arena*. Oakland: University of California Press, 1996.

Calirman, Claudia. *Brazilian Art under Dictatorship: Antonio Manuel, Artur Barrio, and Cildo Meireles*. Durham, NC: Duke University Press, 2003.

Candido, A. "The Brazilian Family." In T. L. Smith and A. Marchant (eds.). *Brazil: Portrait of Half a Continent*. New York: The Dryden Press, 1951.

Candido, Antonio. *On Literature and Society*. Translated by Howard Becker. Princeton, NJ: Princeton University Press, 1995.

Carvalho, Ana Maria. "Portuguese in the United States." In K. Potowsky. *Language Diversity in the USA*. Cambridge: Cambridge University Press, 2010. 223–237.

Castro, Ruy. *Bossa Nova: The Story of the Brazilian Music That Seduced the World*. Chicago, IL: Chicago Review Press, 2012.

Conrad, Robert Edgard. *Children of God's Fire: A Documentary History of Black Slavery in Brazil*. University Park: Penn State University Press, 2000.

Coutinho, Afrânio. *An Introduction to Literature in Brazil*. Trans. Gregory Rabassa. New York: Columbia University Press, 1969.

Coutinho, Eduardo F. "Postmodernism in Brazil." In Hans Bertens and Douwe Fokkema (eds.). *International Postmodernism. Theory and Practice*. Amsterdam/Philadelphia: John Benjamins Publ. Co., 1997, pp. 327–336.

Coutinho, Eduardo F. "Brazilian Modernism." In Astradur Eysteinsson and Vivian Liska (eds.). *Modernism*. 2 vols. Vol. 2. Amsterdam/Philadelphia: John Benjamins Publ. Co., 2007, pp. 759–768.

Crocitti, John J., and Monique Vallance. *Brazil Today: An Encyclopedia of Life in the Republic*. Santa Barbara, CA: ABC-CLIO, 2012.

Crook, Larry. *Brazilian Music and the Heartbeat of a Modern Nation*. Santa Barbara, CA: ABC-CLIO, 2005.

Crook, Larry, and Randal Johnson (eds.). *Black Brazil. Culture, Identity and Social Mobilization*. Los Angeles: UCLA Latin American Center Publications (UCLA Latin American Studies 86), 1999.

Da Matta, Roberto. *Carnivals, Rogues, and Heroes: An Interpretation of the Brazilian Dilemma*. Trans. John Drury. Notre Dame, IN: Notre Dame University Press, 1991.

Dean, Warren. *With Broadax and Firebrand: The Destruction of the Brazilian Atlantic Forest*. Oakland: University of California Press, 1997.

Degler, Carl N. *Neither Black nor White: Slavery and Race Relations in Brazil and the United States*. New York: Collier, Macmillan, 1971.

Draper III, Jack. *Forró and Redemptive Regionalism from the Brazilian Northeast: Popular Music in a Culture of Migration*. New York: Peter Lang, 2003.

Eakin, Marshall. *Brazil: The Once and Future Country*. New York: Palgrave Macmillan, 1998.

Fajans, Jane. *Brazilian Food: Race, Class and Identity in Regional Cuisines*. London/New York: Berg, 2012, p. 148.

Freyre, Gilberto. *The Masters and the Slaves (Casa-Grande & Senzala): A Study in the Development of Brazilian Civilization*. Trans. Samuel Putman. Oakland: University of California Press, 1987.

Goodwin, Philip L. *Brazil Builds*. New York: The Museum of Modern Art, 1943.

Green, James. *Beyond Carnival: Male Homosexuality in Twentieth-Century Brazil*. Chicago: University of Chicago Press, 2001.

Harrison, Phyllis A. *Behaving Brazilian: A Comparison of Brazilian and North American Social Behavior*. New York: Newbury House Publishers, 1983.

Hemming, John. *Red Gold: The Conquest of the Brazilian Indians*. Cambridge, MA: Harvard University Press, 1978.

Hermano, Vianna. *The Mystery of Samba. Popular Music and National Identity in Brazil*. Chapel Hill: University of North Carolina Press, 1999.

Hertzman, Marc. *Making Samba: A New History of Race and Music in Brazil*. Durham, NC: Duke University Press, 2013.

Holanda, Sérgio Buarque de. *Roots of Brazil*. Trans. G. Harvey Summ. Notre Dame, IN: University of Notre Dame Press, 2012.

Johnson, Randal, and Robert Stam. *Brazilian Cinema*. New York: Columbia University Press, 1995.

Lara, Fernando Luiz. *The Rise of Popular Modernist Architecture in Brazil*. Gainesville: University Press of Florida, 2008.

Lehnen, Leila. *Citizenship and Crisis in Contemporary Brazilian Literature*. New York: Palgrave Macmillan, 2013.

Lesser, Jeffrey. *Negotiating National Identity: Immigrants, Minorities, and the Struggle for Ethnicity in Brazil*. Durham, NC: Duke University Press, 1999.

Lever, Janet. *Soccer Madness: Brazil's Passion for the World's Most Popular Sport*. Chicago, IL: University of Chicago Press, 1983.

Levine, Robert M. *The History of Brazil*. Westport, CT: Greenwood Press, 1999.

Levine, Robert M., and John Crocitti (eds.). *The Brazil Reader: History, Culture, Politics*. Durham, NC: Duke University Press, 1999.

Lowe, Elizabeth. *The City in Brazilian Literature*. Madison, NJ: Fairleigh Dickinson University Press, 1981.

Margolis, Maxine L. *Little Brazil: An Ethnography of Brazilian Immigrants in New York City*. Princeton, NJ: Princeton University Press, 1994.

Margolis, Maxine L. *Goodbye, Brazil: Émigrés from the Land of Soccer and Samba*. Madison: University of Wisconsin Press, 2013.

Martins, Sérgio B. *Constructing an Avant-Garde: Art in Brazil, 1949–1979*. Cambridge: MIT Press, 2013.

McCann, Bryan. *The Throes of Democracy: Brazil since 1989*. Black Point: Fernwood Publishing, 2008.

McCann, Bryan. *Hard Times in the Marvelous City: From Dictatorship to Democracy in the Favelas of Rio de Janeiro.* Durham, NC: Duke University Press, 2014.

Melo, Marcus André, and Carlos Pereira. *Making Brazil Work: Checking the President in a Multiparty System.* New York: Palgrave Macmillan, 2013.

Moritz Schwartz, Lilia. *The Spectacle of the Races: Scientists, Institutions, and the Race Question in Brazil, 1870–1930.* Transl. Lilia Guyer. New York: Hill and Wang, 1999.

Murphy, John P. *Music in Brazil.* New York and Oxford: Oxford University Press, 2006 (includes CD).

Nagib, Lúcia. *The New Brazilian Cinema.* London: I.B. Tauris in Association with the Centre for Brazilian Studies, University of Oxford, 2003.

Nitti, John, and Michael Ferreira. *500 Portuguese Verbs.* New York: Barron's Educational Series, 2005.

Oliven, Ruben. "The Imaginary of Brazilian Popular Music." *Vibrant*, vol. 8, n. 1 [Dossie Music and Anthropology in Brazil. Carlos Sandroni, Hermano Vianna, and Rafael José de Menezes Bastos (eds.)], 2011, pp. 170–207.

Omari-Tunkara, Mikelle. *Manipulating the Sacred: Yorùbá Art, Ritual, and Resistance in Brazilian Candomblé.* Detroit: Wayne State University Press, 2005.

Page, Joseph. *The Brazilians.* Boston, MA: Da Capo Press, 1996.

Perini, Mário. *Modern Portuguese: A Reference Grammar.* New Haven, CT: Yale University Press, 2002.

Perrone, Charles, and Christopher Dunn (eds.). *Brazilian Popular Music and Globalization.* New York: Routledge, 2001.

Power, Timothy, and Matthew Taylor (eds.). *Corruption and Democracy in Brazil: The Struggle for Accountability.* Notre Dame: University of Notre Dame Press, 2011.

Rector, Mônica (ed.). *Brazilian Writers* (*Dictionary of Literary Biography*, v. 370). Detroit, MI, and New York: Thomson Gale, 2005.

Ribeiro, Darcy. *The Brazilian People: The Formation and Meaning of Brazil.* Gainesville: University of Florida Center for Latin American Studies, 2000.

Rocha-Coutinho, M. L. (1999). "Behind Curtains and Closed Doors: Brazilian Women in Family Relations." *Feminism & Psychology*, 9, 373–380.

Röhrig-Assunção, Mathias. *Capoeira. The History of an Afro-Brazilian Martial Art.* London and New York: Routledge, 2005.

Santiago, Silviano. *The Space In-Between: Essays on Latin American Culture.* Trans. Ana Lucia Gazolla. Durham, NC: Duke University Press, 2002.

Scheper-Hughes, Nancy. *Death without Weeping: The Violence of Everyday Life in Brazil.* Oakland: University of California Press, 1993.

Schwarz, Roberto. *Misplaced Ideas: Essays on Brazilian Culture.* Trans. John Gledson. New York: Verso, 1992.

Segawa, Hugo. *Architecture of Brazil.* São Paulo: Edusp, 2010.

Skidmore, Thomas. *Politics in Brazil 1930–1964: An Experiment in Democracy.* New York: Oxford University Press, 1988.

Skidmore, Thomas E. *Brazil: Five Centuries of Change.* New York: Oxford University Press, 2010.

Skidmore, Thomas. *Black into White. Race and Nationality in Brazilian Thought.* New York: Oxford University Press, 1974.

Sullivan, E. J., Solomon R. Guggenheim Museum, and Museo Guggenheim Bilbao. *Brazil: Body & Soul.* New York: Guggenheim Museum, 2001.

Talmon-Chvaicer, Maya. *The Hidden History of Capoeira: A Collision of Cultures in the Brazilian Battle Dance.* Austin: University of Texas Press, 2008.

Telles, Edward. *Race in Another America: The Significance of Skin Color in Brazil.* Princeton, NJ: Princeton University Press, 2004.

Tosta, Antonio Luciano de Andrade. "Modern and Postcolonial? Oswald de Andrade's *Antropofagia* and the Politics of Labeling." *Romance Notes.* 51.2 (2011): 217–26.

Twine, France W. *Racism in a Racial Democracy: The Maintenance of White Supremacy in Brazil.* New Brunswick, NJ: Rutgers University Press, 1998.

Underwood, David. *Oscar Niemeyer and Brazilian Free-Form Modernism.* New York: George Braziller, Inc., 1994.

Wasserman, Renata R. Mautner. *Exotic Nations: Literature and Cultural Identity in the United States and Brazil, 1830–1930.* Ithaca, NY: Cornell University Press, 1994.

Xavier, Ismail. *Allegories of Underdevelopment: Aesthetics and Politics in Modern Brazilian Cinema.* Minneapolis: University of Minnesota, 1997.

About the Editors and Contributors

Antonio Luciano de Andrade Tosta (PhD, Brown University) is assistant professor of Brazilian literature and culture at the University of Kansas. He is the author of *Confluence Narratives: Ethnicity, History, and Nation Making in the Americas* (2016), and has published widely on Brazilian and Brazilian American literature and film in the United States, Canada, and Brazil. Tosta also coedited the *Luso-American Anthology: Writings by Portuguese Speaking Authors in North America* (2011).

Eduardo F. Coutinho (PhD, UC-Berkeley) is professor of comparative literature at the Federal University of Rio de Janeiro, Brazil. His publications include *The Synthesis Novel in Latin America* (1991), *Em busca da terceira margem* (1993), *Literatura Comparada: textos fundadores*, ed. with T. Carvalhal (1994), *Cânones e contextos*, ed. (3 vols, 1997–1998), *Literatura Comparada na América Latina* (2003), *Beyond Binarisms*, ed. (3 vols, 2009), *Literatura Comparada: reflexões* (2013), and *Rompendo barreiras: ensaios de literatura brasileira e hispano-americana* (2014).

Simone Bohn is associate professor of political science at York University, Toronto, Canada. She researches political parties in Latin America, gender and politics, and religion and politics in Brazil.

Volnei M. Carvalho has a BA in language studies from the Federal University of Bahia in Brazil. He has a master's degree in applied linguistics from the same institution, where he is currently a PhD candidate, also in the field of applied linguistics. Mr. Carvalho is a teacher of English as a Foreign Language in Salvador. He is originally from Rio de Janeiro.

Marcos Cerdeira is a PhD student in Portuguese and Brazilian studies at Brown University. He has a master's degree in Latin American studies at the University of Illinois at Urbana-Champaign.

Rodrigo R. Coutinho holds an MS in global affairs from New York University, an MBA in international business from UCAM-RJ, and a BA in business administration from PUC-RJ. He specializes in international relations, on topics such as globalization, transnationalization, and emerging markets. As a management consultant, he advises multinational corporations in North America, Central America, and South America, and Europe.

Alice Heeren is a PhD candidate at Southern Methodist University's RASC/a: Rhetorics of Art, Space and Culture program in Dallas, Texas. She holds an MA in art history, theory, and criticism from the School of the Art Institute of Chicago, and a BA in art education and a BFA in printmaking from the Universidade Federal de Minas Gerais in Brazil. Her research focuses on Modern and Contemporary Brazilian Art and Architecture and cultural theory.

Michael Iyanaga is a Mellon Faculty Fellow in Latin American studies at the College of William and Mary and a research affiliate of the Federal University of Pernambuco. Earning his PhD in ethnomusicology from UCLA, Dr. Iyanaga is an active music researcher and Portuguese-English translator.

Joseph Abraham Levi is an academic consultant/grader at George Washington University. He holds a PhD in romance philology/linguistics from the University of Wisconsin-Madison. His publications cover Portuguese language and lusophone studies; romance philology, linguistics, and dialectology; Italian and Sephardic Jewry; missionaries in Africa and Asia (16th–18th centuries); African, Islamic, and medieval studies.

Angela Lühning has a PhD in ethnomusicology from Freie Universität in Berlin. Since 1990 she has taught at the Federal University of Bahia (UFBA), advising students on topics ranging from indigenous and Afro-Brazilian music to education and public policy. In parallel, she has published books, articles, and children's literature. In addition, she coordinates research at the Pierre Verger Foundation and organizes projects with underprivileged youth of the Pierre Verger Foundation's Cultural Space.

Erin Flynn McKenna, a doctoral candidate in the Department of Recreation, Sport and Tourism at the University of Illinois at Urbana-Champaign, is committed to expanding knowledge about Latin America through her research and teaching. In her current project she is looking at immersive tourism and globalization in Salvador, Brazil.

Doriane Andrade Meyer is a Brazilian architect and urbanist. She is currently pursuing a master's degree in architecture at the University of Kansas. Ms. Meyer has worked as a licensed architect in Brazil since she received her BA from the Federal University of Bahia. Her research interests include colonial and modern Brazilian architecture, with a focus on non-Western influences in Brazilian architecture. Her research has recently been supported by the Tinker Foundation.

Renato Lima de Oliveira is a PhD candidate in political science at MIT. He holds a master's degree in Latin American studies from the University of Illinois at Urbana-Champaign (UIUC) and a bachelor's degree in social communication from UFPE. He worked previously as a journalist in Brazil.

Maria Lúcia Rocha-Coutinho (PhD, Catholic University of Rio de Janeiro) is professor of psychology at the Federal University of Rio de Janeiro, Brazil. Her publications include *Tecendo por trás dos panos* (1994), "Behind Curtains and Closed Doors: Brazilian Women in Family Relations" (1999, *Feminism & Psychology*, 9, 373–380), "New Options, Old Dilemmas: Close Relationships and Marriage in Brazil" (in Comunian and Gielen, eds. *It's All about Relationships*, 2002), "Mulheres brasileiras em posições de liderança: novas perspectivas para antigos desafios," with Rodrigo R. Coutinho (*Global Economics and Management Review*, vol. XVI, no. 1, 2011), and "Investimento da mulher no mercado de trabalho: repercussões na família e nas relações de gênero" (in Feres-Carneiro, T., ed. *Família e casal*, 2015).

Helade Scutti Santos is currently in charge of the Portuguese program at Rice University, where she also teaches Spanish. She received her PhD in Hispanic linguistics from the University of Illinois at Urbana-Champaign in 2013. Her dissertation focused on the acquisition of Portuguese as a third language. Prior to coming to the United States, Helade taught Spanish for 14 years in a variety of instructional contexts in Brazil.

Kara D. Schultz is a PhD candidate in Latin American and Atlantic history at Vanderbilt University. She received her MA from Vanderbilt University and her BA from the University of Richmond. Her research centers on the slave trade and the African diaspora in the 17th-century South Atlantic world.

Antonio Lima da Silva has a BA in pedagogy and a master's degree in education from the Universidade Federal da Bahia, where he is currently a PhD candidate in education. He has taught at and coordinated higher education institutions in Salvador, Bahia, Brazil.

Carla Alves da Silva is a PhD candidate in Portuguese and Brazilian studies at the University of Illinois at Urbana-Champaign. She has a master's degree in literatures in English from the State University of Estado do Rio de Janeiro, and a dual BA in Portuguese and English from the Federal University of Rio de Janeiro. She currently teaches Portuguese at Loyola University Chicago. In 2015, she taught Brazilian cinema as a visiting lecturer at Northwestern University.

Simone Cavalcante da Silva is a PhD candidate in Portuguese and Brazilian studies at the University of Illinois at Urbana-Champaign. She has been the supervisor and instructor of Portuguese in the Department of Romance Languages at the University of Oregon since 2007. Her current work and interests include questions of gender and sexuality, film theory, Brazilian cinema, and second language acquisition.

Domingos Sávio Pimentel Siqueira holds a PhD in letters and linguistics from the Federal University of Bahia (UFBA) in Salvador, Brazil, where he is also assistant professor of English in the Department of Germanic Languages.

Elisa Duarte Teixeira has an AAS degree in culinary arts from El Centro College, Dallas, and a BA in linguistics from the University of São Paulo, Brazil, where she also completed specialization in translation, MA, and PhD in English. She has lived in the United States since 2008 and currently works at the University of Michigan's Brazil Platform. She coauthored a vocabulary of cooking terms and maintains the blog Authentic Brazilian Cuisine.

Index

Note: Page numbers followed by *f* indicate a figure on the corresponding page. Page numbers followed by *t* indicate a table on the corresponding page.